P9-DWY-681

Cepheids: Theory and observations

Cepheids: Theory and observations

Proceedings of the IAU colloquium no. 82

Edited by
Barry F. Madore
Associate Professor of Astronomy,
David Dunlop Observatory, University of Toronto

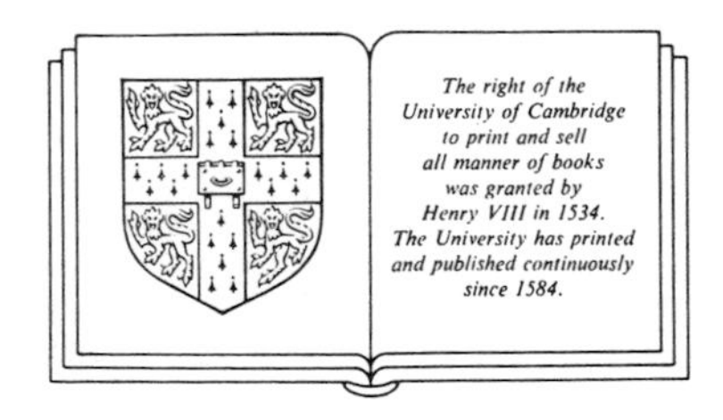

CAMBRIDGE UNIVERSITY PRESS
Cambridge
London New York New Rochelle
Melbourne Sydney

Published by the Press Syndicate of the University of Cambridge
The Pitt Building, Trumpington Street, Cambridge CB2 1RP
32 East 57th Street, New York, NY 10022, USA
10 Stamford Road Oakleigh, Melbourne 3166 Australia

First published 1985

Printed in Great Britain at the University Press, Cambridge

Library of Congress catalogue card number:

ISBN 0 521 30091 6

TABLE OF CONTENTS

INTERNATIONAL ASTRONOMICAL UNION COLLOQUIUM NO. 82: CEPHEIDS: OBSERVATIONS AND THEORY

TORONTO, CANADA MAY 28 TO JUNE 1, 1984

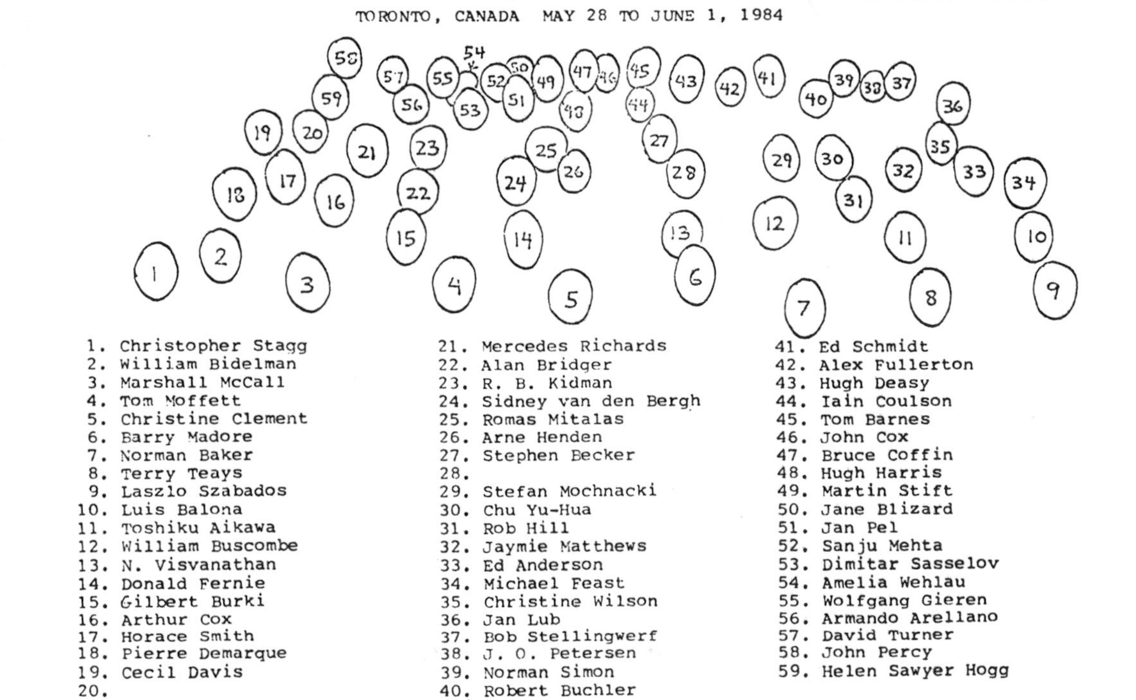

1. Christopher Stagg
2. William Bidelman
3. Marshall McCall
4. Tom Moffett
5. Christine Clement
6. Barry Madore
7. Norman Baker
8. Terry Teays
9. Laszlo Szabados
10. Luis Balona
11. Toshiku Aikawa
12. William Buscombe
13. N. Visvanathan
14. Donald Fernie
15. Gilbert Burki
16. Arthur Cox
17. Horace Smith
18. Pierre Demarque
19. Cecil Davis
20.
21. Mercedes Richards
22. Alan Bridger
23. R. B. Kidman
24. Sidney van den Bergh
25. Romas Mitalas
26. Arne Henden
27. Stephen Becker
28.
29. Stefan Mochnacki
30. Chu Yu-Hua
31. Rob Hill
32. Jaymie Matthews
33. Ed Anderson
34. Michael Feast
35. Christine Wilson
36. Jan Lub
37. Bob Stellingwerf
38. J. O. Petersen
39. Norman Simon
40. Robert Buchler
41. Ed Schmidt
42. Alex Fullerton
43. Hugh Deasy
44. Iain Coulson
45. Tom Barnes
46. John Cox
47. Bruce Coffin
48. Hugh Harris
49. Martin Stift
50. Jane Blizard
51. Jan Pel
52. Sanju Mehta
53. Dimitar Sasselov
54. Amelia Wehlau
55. Wolfgang Gieren
56. Armando Arellano
57. David Turner
58. John Percy
59. Helen Sawyer Hogg

This volume is dedicated to

Dr. John P. Cox

a pioneer in the field of variable star research,
who died of a prolonged illness
only a few short months after bravely travelling
to Toronto to deliver his review.

HISTORICAL PREFACE

J. D. Fernie
Chairman, Scientific Org. Committee
Department of Astronomy
David Dunlap Observatory
University of Toronto, Toronto, Ontario, Canada

This Colloquium marked the two hundredth anniversary of the discovery of cepheid variables. On September 10, 1784 Edward Piggot established the variability of the star we now call η Aquilae, the very same night that his friend and co-worker, John Goodricke, first found β Lyrae to vary. Only a month later, on October 20, Goodricke was "almost convinced" that δ Cephei is variable.

Goodricke is a figure much beloved of textbook writers: his affliction as a deaf-mute, the fact that he died at 21, and his work having been done in relative isolation, have led to his often being portrayed as some kind of bucolic genius. He was, in fact neither. He came of a well-connected family and one, more-over, that was surprisingly enlightened for the times. In an age when deaf-muteness was customarily equated with insanity, his parents recognized their son's affliction for what it was, and sent him for special training at an academy in Edinburgh. It is now fairly certain that by the time he was thirteen he could lip-read and perhaps even speak. He continued at a normal school, where he showed intelligence and a reasonable ability in the standard curriculum.

Edward Piggot also came of a well-connected and much-travelled family. His father, an imenent surveyor and Follow of various Royal Societies and national Academies, had a strong interest in astronomy and was the correspondent of such as Herschel and Maskelyne. He and his son referred to themselves whimsically as 'gentlement astronomers', and upon settling for a time in the English city of York established a considerable private observatory. Here it was that the great variable star discoveries would be made.

The year 1781 provided a great stimulus to observational astronomy, especially in England. This was the year in which Herschel discovered Uranus, and all who had the means were eager to observe "Mr Herschel's comet". Edward Piggot, then 28, had the means, and John Goodricke, then 17 and living nearby, had the interest. So began their cooperative efforts.

The relationship was clearly one in which the knowledgeable Piggot was introducing the inexperienced youth to new fields. Moving on from the new planet, it was Piggot who suggested variable stars as a subject ripe for investigation, and specifically Algol as a long-suspected candidate. After Goodricke had in fact established Algol's variability a year later in November, 1782, it was Piggot who verified the discovery and, through his father's connections, make it widely known. Furthermore, recent research has shown that it may even have been Piggot who first suggested the eclipsing binary theory by way of explanation. But either way, Piggot gave full and generous support to Goodricke as the discoverer, leading to Goodricke's receiving the Copley Medal of the Royal Society.

Goodricke was an arduous observer, a characteristic that probably led to his death. Following his discovery of δ Cephei's variability in late-1784, he observed it no less than a hundred times in the first ten months of 1785 -- conceivably a record in the English climate! In early 1786 he contracted pneumonia "in the consequence of exposure to night air in astronomical observations," and died on April 20. He was 21 and had been a Fellow of the Royal Society for only two weeks. His death was "an event I shall ever lament," wrote Piggot. "This worthy young man exists no more; he is not only regretted by many friends, but will prove a loss to astronomy, as the discoveries he so rapidly made evince."

Piggot gave up active observing at this time and resumed his travels abroad. He would make only two further discoveries of variable stars, notably R Coronae Borealis in 1795, and his subsequent life seem not to have been the happiest. It included three years detention in France during French/English hostilities. He was never elected on FRS, nor a Fellow of the Royal Astronomical Society, which was founded shortly before his death in 1825 at the age of 72.

I am sure that the research reported on the pages that follow would have made strange reading to Edward Piggot and John Goodricke, but it stands as a worthy salutation to them and to all who have followed in their footsteps.

FUNDAMENTAL PARAMETERS OF CEPHEIDS

J.W. Pel
Kapteyn Astronomical Institute, Groningen, The Netherlands

INTRODUCTION

In the two centuries since the discovery of δ Cephei and η Aquilae, the study of Cepheids, and of pulsating stars in general, has become a very important field in astronomy. The usefulness of Cepheids, both as distance indicators and as test objects for stellar astrophysics, has been amply demonstrated, and if one just looks at the flood of papers on Cepheids that appears each year, it is clear that the subject is not only old and respectable, but also still very much alive. This two-hundredth anniversary is therefore an appropriate occasion to evaluate what we have learnt about Cepheids, and where the remaining problems lie.

In this paper I will try to review some of our present knowledge of "fundamental parameters" of Cepheids. Which of the many parameters are "fundamental"? Since the period P_i of a given stellar pulsation mode, i, is primarily a function of mass M and radius R of the star, and since stellar evolution calculations usually give the luminosity L and effective temperature T_{eff} of a star as a function of its mass, age τ, and chemical composition (X,Y,Z), a useful list of fundamental parameters should contain at least: M, τ, (X,Y,Z), L, R, T_{eff}, P, i. Although the list has some redundancy, this is already a considerable number of parameters, and a proper discussion of these quantities would have to deal with almost every aspect of Cepheid behaviour. The scope of this paper will necessarily have to be much more limited. As a first limitation, I will discuss only the "classical" type of Cepheids, and say nothing about the W Virginis stars. Furthermore, I will concentrate on observational results, and on the constraints that can be derived from them on some of the above parameters, giving most attention to the calibration of L and T_{eff}.

REDDENING CORRECTIONS

The colour excess of a Cepheid can hardly be called a fundamental parameter, but it is usually such a difficult obstacle on the way to the intrinsic properties that I will start out with a few remarks on the reddening problem. For many years the intrinsic colours of Cepheids have been the subject of considerable debate. Uncertainties in the reddening corrections, in extreme cases as large as $0^m.2$ in E(B-V), have been mainly due to the following problems: 1) colour excesses for long-period Cepheids derived from a (spectral type)-$(B-V)_0$ relationship have been systematically larger than excesses based entirely on photometric calibrations; 2) up to recently the large majority of the available Cepheid photometry has been in the UBV system, which is not very well suited for the determination of accurate Cepheid reddenings. For Cepheids in the Magellanic Clouds the situation has been complicated further by the effects of lower line blanketing.

During the last decade a large new collection of accurate photoelectric Cepheid photometry has been obtained in photometric systems that do allow good reddening determinations for F-G supergiants. Extensive surveys of galactic and Magellanic Cloud Cepheids have been made in the Kron-Cousins BVI system (Dean et al. 1978; Martin et al. 1979; Caldwell & Coulson 1984a), in the Walraven VBLUW system (Pel 1978), and in the Strömgren uvbyβ system (Feltz & McNamara 1980; Eggen 1983 a,b). A more limited set of data in the DDO system has been published by Dean (1981). The reddening results of this new generation of photometry show a much improved mutual agreement. It seems clear now that in the earlier studies Cepheid reddenings were generally overestimated, particularly in methods based on a spectral type - colour calibration, and at the longer periods. As for the Magellanic Cloud

Fig.1. Comparison of colour excesses for galactic Cepheids in five photometric systems. Since the largest common overlap is in the VBLUW data, all colour excess differences are given with respect to the VBLUW reddenings. $E(V-B)_{VBLUW}$ has been transformed to the other scales with the following "theoretical" transformations (computed from extinction law, passbands, and energy distributions): $E(b-y) = 1.75E(V-B)$,and

$$E(B-V)/E(V-B) = 2.41 -0.11(\log P -1) -0.18E(V-B).$$

Open symbols indicate stars with known or suspected companions or with other peculiarities.

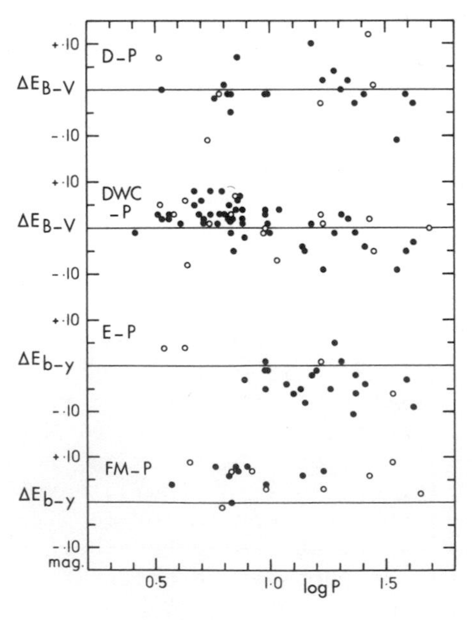

Abbreviations used:

D: Dean (1981); DDO photometry, published excesses given in E(B-V).
DWC: Dean et al. (1978); Kron-Cousins BVI system.
E: Eggen (1983 a,b); Strömgren uvbyβ + RI.
FM: Feltz & McNamara (1980); Strömgren uvbyβ + khg.
P: Pel (1978); Walraven VBLUW system.

Cepheids, better estimates for the abundance effects can now be made, and most new evidence indicates very small reddenings, at least in the outer regions of the Clouds (Caldwell & Coulson 1984 b; see also the review by Feast 1984).

This convergence in the newer reddening results is of course very good news, but let me point out that it is too early to call the reddening problem "solved". In Fig.1 I have compared reddenings for galactic Cepheids from BVI, DDO, Strömgren, and Walraven photometry. Evidently there still exist systematic zeropoint differences and period-dependent trends. These effects are much smaller than they used to be, but they are still disappointingly large compared to the photometric accuracy of well below $0^m.01$ of which all these systems are capable. The residuals can only partly be ascribed to shortcomings of the transformations that were used, and we can only conclude that systematic uncertainties of up to $0^m.05$ in E(B–V) are probably still present. Are errors at that level really disturbing? This can be judged from Table 1: the effects on radii and masses are not very serious, but those on T_{eff} and L are too large for a proper comparison of the empirical and theoretical positions of the Cepheid strip in the HR-diagram.

Table 1. Effect of $\Delta E(B-V) = +0^m.05$ for an average 10-day Cepheid.

parameter	relation used	resulting change	
temperature	$(B-V)_o - T_{eff}$	$\Delta T_{eff} = +130$ K	1)
gravity	typical photometric Balmerjump index	$\Delta \log g = +0.2$	2)
bol.corr.	$BC(T_{eff}, \log g)$	$\Delta BC = +0^m.01$	3)
luminosity	$A_V/E(B-V)$, 3), known distance	$\Delta L = +14$ %	4)
luminosity	$P-L-T_{eff}$ (theoretical), 1)	$\Delta L = +12$ %	5)
distance	P-L, with 4) for apparent L	$\Delta d = -7$ %	6)
distance	$P-L-T_{eff}$, 4) for app. L, 5) for true L	$\Delta d = +1$ %	7)
radius	$R \propto L^{\frac{1}{2}} . T_{eff}^{-2}$, 1), 4)	$\Delta R = +2$ %	8)
radius	Baade-Wesselink method	ΔR small, $\ll 1$ % ?	9)
mass	evolutionary M-L, 4)	$\Delta M_{evol} = +3.5$ %	10)
mass	pulsation relation $P-L-M-T_{eff}$, 1), 4)	$\Delta M_{puls} = +4.5$ %	11)

Fig.2. Temperature-colour relations for Cepheids and supergiants. $\Theta_{eff} \equiv 5040/T_{eff}$.

Supergiants:

Johnson I	: Johnson (1966).
Böhm-Vitense I	: Average for I^a and I^b from Böhm-Vitense (1972).
Schmidt I	: Schmidt (1972).
Van Paradijs I^b	: Van Paradijs (1973); 7 individual I^b supergiants.
Blackwell & Shallis I	: Individual supergiants from Blackwell & Shallis (1977). E(B-V) for δ CMa (at $(B-V)_o = 0^m.62$) derived from VBLUW data. The other stars are within 200 pc, and small reddening corrections were made according to $E(B-V)=0^m.3\ kpc^{-1}$.
Flower I	: Flower (1977).
Bell & Gustafsson	: The hotter supergiant models from Bell & Gustafsson (1978), for Doppler broadening velocity 3.5 km s^{-1}.

Cepheids:

Oke-Kraft-Parsons	: Oke (1961); Kraft (1961); Parsons (1971, 1974). The relation by Rodgers (1970) is not shown, as it lies very close to the Oke-Kraft-Parsons line.
Schmidt Ceph.	: Schmidt (1972).
Pel	: Average relation derived from Pel (1978).

For the sake of clarity the Böhm-Vitense (1981) "best fit" relation is not shown; it lies halfway the Flower I and Böhm-Vitense I lines. Also not shown is the Bell & Parsons (1974) relation, which runs parallel to the Oke-Kraft-Parsons line, but $0^m.05$ bluer. For the "Wesselink temperatures" of U SGR and S NOR see text.

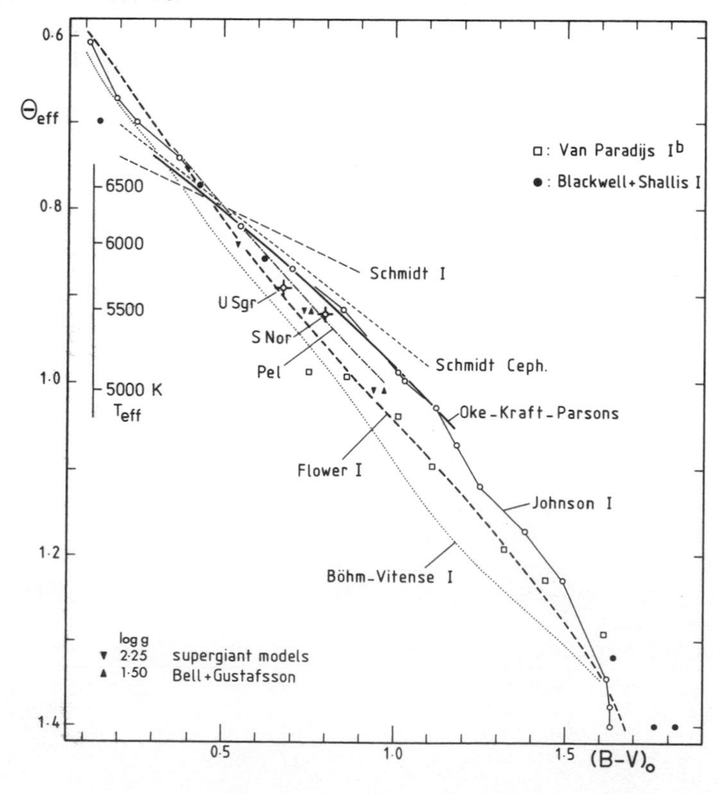

That reddenings can be determined with an accuracy much better than $0^m.05$ is demonstrated by the following examples. From Strömgren photometry of NGC 6087, the cluster containing S NOR, Schmidt (1980 b) finds for the cluster members $E(b-y) = 0^m.123$, and Eggen (1980) $E(b-y) = 0^m.142$, whereas my own VBLUW photometry (Pel 1984) gives $E(V-B) = 0.076$, or $E(b-y) = 0^m.134$. The standard deviation for a single star is $0^m.01$ in each case. In M 25, with U SGR, the reddening is much larger, and inhomogeneous. For 18 cluster members within 6 arcmin from the Cepheid, Schmidt (1982) finds $E(b-y) = 0^m.358$, $\sigma = 0^m.031$. The VBLUW reddenings for 27 members within the same area correspond to $E(b-y) = 0^m.338$, $\sigma = 0^m.025$, showing that even in this difficult field a satisfactory agreement can be reached with different techniques.

EFFECTIVE TEMPERATURES

As can be expected from the Stefan-Boltzmann law, the effective temperature of a Cepheid is a very important quantity, and the estimates for some other, not directly observable parameters can be very sensitive to the accuracy of T_{eff}. An extreme example is the "pulsation mass", which depends very strongly on T_{eff} because of the pulsation relationship from which it is derived: $P \propto L^{0.83}.M^{-0.66}.T_{eff}^{-3.45}$ (Iben & Tuggle 1972). In this respect some numbers in Table 1 are misleading, because they suggest that T_{eff} has only minor effects on R and M. Without the compensating effect of the luminosity increase by A_V, however, $\Delta T_{eff} = +130$ K results in $\Delta R = -4.5$ % and $\Delta M_{puls} = -11.5$ % (Table 1 lines 8 and 11). Also in the HR-diagram, an error as small as 130 K is by no means negligible, as it corresponds to about one fifth of the width of the Cepheid strip at constant L.

By far the most common method of determining Cepheid temperatures is the use of a temperature-colour relation, where the colour is traditionally $(B-V)_o$. In order to estimate how well Cepheid temperatures are known, I have therefore collected in Fig.2 most of the $T_{eff}-(B-V)_o$ calibrations that exist in the literature for Cepheids and cool supergiants. Although my own photometric temperatures for Cepheids are actually based on a two-dimensional calibration, I have also represented those temperatures by a mean $T_{eff}-(B-V)_o$ relation, to allow a comparison.

Fig.2 should be interpreted with caution for several reasons. Firstly, not all of these relations are completely independent. The Flower calibration, for example, is based partly on the Van Paradijs data. Secondly, the problems with reddening also enter this diagram in some cases, via the reddening corrections for calibrating stars. Thirdly, it is well known that $(B-V)_o$ is not an ideal temperature index, because it is also sensitive to gravity and chemical composition. On the other hand, the uncertainties related to reddening are small compared to the very large differences in the temperature relations, and as long as we restrict ourselves to nearby classical Cepheids, differential line-blanketing can safely be neglected. Gravity effects within the Cepheid region do cause scatter in $(B-V)_o$, as can be seen from the Bell & Gustafsson colours in Fig.2 (see also Pel 1980), but also this scatter is small compared to the wide range in temperature scales in Fig.2. With some limitations, the comparison in this figure is therefore still instructive. But can it tell us what the most reliable temperature scale for Cepheids is?

The most refined synthetic spectra presently available for intermediate-type supergiants are those by Gustafsson & Bell (1979) and Kurucz (1979). The latter models are represented in Fig.2 by my own temperature relation for Cepheids, which is based on the Kurucz data. The overlap between these two sets of models is limited, but the agreement is reasonable, with differences that stay within 100 K over most of the common range. On the other hand, in her review of the effective temperature scale, Böhm-Vitense (1981) gives a best-fit relation for A-M supergiants which is mainly based on the calibrations by Böhm-Vitense (1972), Van Paradijs (1973), Bell & Parsons (1974), Luck (1977 a,b), and on a few stars with direct angular diameter measurements. The temperatures by Luck are in turn based on the Bell _et al_. (1976) model atmospheres. In the region of the Cepheids, Böhm-Vitense's "most probable" relation lies halfway between the Böhm-Vitense 1972 and the Flower relation, about 200-300 K cooler than my Cepheid temperatures.

How can we resolve this discrepancy? It is well known that the agreement between real and synthetic stellar spectra is still far from perfect. In particular, when synthetic (blue-visual) colours are normalized around A0 V, they systematically produce too low temperatures for F-G stars (cf. Relyea & Kurucz 1978; Gehren 1981). This problem is probably due to insufficient near-UV opacity in the models. I have tried to estimate the size of this effect in the Kurucz VBLUW colours, taking also into account improved photometric calibrations and results for new (unpublished) Kurucz models with improved treatment of convection. These estimated corrections, which are still very uncertain, indicate that my Cepheid temperatures may have to be decreased by about 100 K at the blue end of the colour-T_{eff} relation, and increased by a smaller amount at the red end. The gap in temperature between the Böhm-Vitense best-fit relation and my "best guess" would then become nearly constant over the whole Cepheid range, but it would still be about 250 K wide.

Unfortunately there are only two F-G supergiants for which interferometric diameter data allow a direct determination of T_{eff}: α CAR and δ CMA. Code _et al_. (1976) derive T_{eff}= 7460 K for α CAR and 6110 K for δ CMA, but with large uncertainties (σ = 450 K). Blackwell & Shallis (1977) find 7206 ± 173 K and 5877 ± 390 K for the same stars (see Fig.2), but also these numbers are too uncertain to improve the situation.

Since we are dealing with Cepheids, there is one other "direct" way available to estimate T_{eff}: via the Baade-Wesselink (BW) radius. The BW method in its original form, which assumes a unique correspondence between T_{eff} and colour, is entirely independent of temperature. This is not quite true for some modern versions of the BW method, but in all cases the BW radius depends very little on the adopted temperature scale. Moreover, temperatures derived from $L \propto R^2.T_{eff}^4$ are relatively insensitive to errors in R and L, and the errors in R_{BW} and L should be uncorrelated when L is derived from cluster distances.

In Fig.2 I have indicated "Baade-Wesselink" temperatures for the two cluster-Cepheids with the best cluster data, S NOR and U SGR. The BW radii of these stars were taken from the recent discussion of Cepheid radii by Fernie (1984); they are based on many BW determinations (14 for U SGR, 9 for S NOR). The luminosities were derived from the cluster photometry (cf. next section). The resulting temperatures are 5470 K for S NOR and 5655 K for U SGR. I am of course pleased to see that these temperatures fit well with my own temperature calibration, but unfortunately this agreement may be entirely accidental. The errors in the BW temperatures are still large (σ = 180 K for S NOR, 210 K for U SGR), mainly due to the uncertainties in L (10-15 %). BW radii may also be systematically too small (Fernie 1984). An upward correction of the BW radii by 10 % would lower the BW temperatures by 275 K. We are therefore left with the conclusion that the slope of the $T_{eff}-(B-V)_o$ relation is probably fairly well established, but that the zeropoint remains uncertain within a range of about 250 K.

The preceding discussion may leave a rather disappointing impression about the status of temperature determinations for Cepheids. With 250 K uncertainty in the temperature scale, and reddening errors adding another 130 K, Cepheid temperatures could be wrong by 380 K. This would correspond to more than half of the width of the Cepheid strip, and it would make mass estimates or comparisons with theoretical blue edges almost impossible. Some optimism is also justified, however. It has been shown that a very good fit to theoretical blue edges can be obtained with Cepheid temperatures derived from multicolour photometry and synthetic spectra (e.g. Pel & Lub 1978). This can not be used to support a particular calibration of reddenings and temperatures, of course, but it indicates at least that a systematic error in T_{eff} as large as 380 K is unlikely. It should also be remembered that relative temperature determinations are much more accurate than the absolute scale. Within a given photometric calibration, an accuracy of $0\overset{m}{.}03$ in E(B-V) is well attainable (see Fig.1), which means that the relative positions of Cepheids inside the strip can be determined to within ± 12 % of the strip width.

LUMINOSITIES

It is very difficult to discuss the luminosities of Cepheids without entering into a discussion of the Cepheid distance scale. Since the role of Cepheids as distance indicators will be the subject of a separate review at this colloquium, I will concentrate here on only a few aspects of the luminosity calibration. This I will do mainly from the physical viewpoint, i.e. the calibration of bolometric luminosities (L) rather than of absolute magnitudes (M_V).

The transformation of M_V into L requires only one step, the bolometric correction (BC), but for most stars this is by no means an easy one. Fortunately Cepheids have nearly solar temperatures, and consequently their bolometric corrections are on average small. On the other hand, the temperature variations of large-amplitude Cepheids are so large - 1000 K or more - that the differences between visual and bolometric lightcurves can become quite significant. Since the "equilibrium luminosity" L of a Cepheid is the time average of the bolometric lightcurve, it is good to check whether bolometric corrections are indeed only a minor source of uncertainty in the calibration of L.

In order to estimate how well the BC scale for Cepheids is determined, I have compared in Fig.3 the empirical BC calibration of Flower (1977) with theoretical BC values based on the Kurucz (1979) model atmospheres. Flower's BC relations in this temperature range are mainly based on the data of Johnson (1966) and Code et al. (1976). The BC values for the Kurucz models were taken from Lub & Pel (1977); the supergiant relations correspond to a microturbulence of 4 km s^{-1}, and the BC scale was normalized by adopting $BC_{\odot} = -0^m\!.07$. The agreement between both BC calibrations is quite satisfactory. The Flower supergiants relation and the "mean locus" occupied by Cepheids in the Kurucz calibration differ by about $0^m\!.02$, but the important thing is that this difference is nearly constant with T_{eff}. We can therefore conclude that the shape of the BC curve for Cepheids is quite accurately known, and that the uncertainty in the zero-point is probably $\lesssim 0^m\!.02$, corresponding to an error in L of about 2 %. This is not entirely negligible, but it is only a small factor in the total uncertainty of L, as we will see.

For the calibration of Cepheid parameters, and particularly for the luminosity scale, the Cepheids in open clusters and associations remain of crucial importance. In recent years there has been a strong renewed effort to increase the number of these calibrating Cepheids, and to improve the quality of the cluster data. I refer especially to the photometric programs by Turner (1977 1978 1980 1983) and Schmidt (1983 and op. cit.) The results of this new work are in many respects promising. Although for very long it seemed impossible to increase the number of 13 "classical" calibrators of Sandage & Tammann (1969), Fernie & McGonegal (1983) now list 27 Cepheids for which cluster/association membership is likely. These stars cover the whole range of Cepheid periods, and they define a P-L relationship with the remarkably small r.m.s. scatter of $0^m\!.16$ in M_V. It is not certain, however, that this small scatter really reflects the present level of accuracy in the Cepheid luminosity scale.

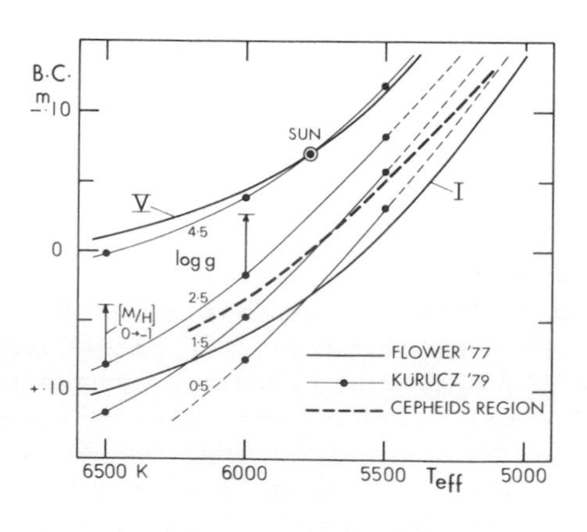

Fig. 3
Empirical (Flower 1977) and theoretical (Kurucz 1979) BC calibrations for Cepheids. The arrows [M/H] 0 → -1 indicate the effect of a factor 10 decrease in metallicity.

Two persistent problems cause considerable uncertainty in the distances of galactic open clusters. Firstly, there are the well-known complications related to the Hyades zeropoint. The recent increase of the Hyades distance appears well established now (Hanson 1980), but it remains uncertain how the distances to other clusters should be corrected for the effect of a high metallicity of the Hyades, and even the composition of the Hyades is still a matter of dispute (Cayrel de Strobel 1980; Flower 1980). Secondly, there exists a discrepancy between the cluster distances derived from UBV photometry and those from the new uvbyβ data of Schmidt, in the sense that Schmidt's distances are systematically smaller by 20-40 % (Schmidt 1980a; Caldwell 1983). This discrepancy is alleviated partly by the recent recalibration of the Hβ index by Balona & Shobbrook (1984), but even these last corrections leave systematic differences corresponding to about 20 % in distance. It is obvious that an error of this size would ruin the Cepheid distance scale, and that it would also be very serious for comparisons with Cepheid theory: a 20 % distance error means 40 % in L, 20 % in R (at given T_{eff}), 10 % in M_{evol}, and 50 % in M_{puls}.

In an attempt to shed some new light on this problem, I will now discuss recent results from VBLUW photometry of NGC 6087 and M 25, the parent clusters of S NOR and U SGR. These clusters are probably the two most reliable points in the calibration of Cepheid luminosities. The ages of NGC 6087 and M 25 are so close to that of the Pleiades ($\tau \sim 7.5 \times 10^7$yr), and the main-sequences of the three clusters are so similar, that it should be possible to derive very accurate differential distance moduli with respect to the Pleiades. Since the Pleiades distance can be determined from a direct fit to nearby parallax stars (Van Leeuwen 1983), this offers the very attractive opportunity of distance determinations for U SGR and S NOR that are entirely independent of the Hyades. The uncertainty due to possible composition differences between the clusters cannot be bypassed in this way, of course. On the other hand, with respect to metallicity the Pleiades are probably much more typical for young open clusters than the Hyades (cf. Nissen 1980).

Let us see now what comes out of such a Hyades-independent calibration. The available VBLUW data consists of the extensive Pleiades photometry by Van Leeuwen (1983), and my own VBLUW photometry for about 140 stars in each of the clusters NGC 6087 and M 25 (Pel 1984). It is not possible to discuss any details of the photometric analysis here, but a few important points should be mentioned. The question of cluster membership causes no problems for the brighter Pleiades, but this is not so for the other two clusters. Similarly to the uvbyβ photometry, the VBLUW system provides 4 criteria to decide about cluster membership: position in the HR-diagram, colour excess, and spectral classifications based on two reddening-independent two-colour diagrams. The latter diagrams are calibrated in terms of T_{eff} and log g, and they allow two independent 2-dimensional classifications for each star. Accurate individual reddening corrections can then be made, and a de-reddened magnitude-T_{eff} diagram can be constructed.

Fig.4 shows the HR-diagrams of the three clusters, for the upper main-sequences down to about A0. The T_{eff} scale used here is based on the synthetic spectra by Kurucz (Kurucz 1979; Lub & Pel 1977), but I want to stress that this is unimportant for the distance determination. A diagram with the reddening-independent temperature index [B–U] along the abscissa looks extremely similar to Fig.4, and its leads to the same results. A comparison of Fig.4 with theoretical isochrones (e.g. Hejlesen 1980) confirms that the three clusters differ very little in age, $\Delta\tau \lesssim 25$ %. This comparison shows also that differential evolution effects in the late B and A stars are extremely small. In the main-sequence fits I have therefore given the highest weight to the lower parts of the HR-diagrams in Fig.4. Small corrections for the age differences were applied, based on the theoretical isochrones. The final results for the distances of M 25 and NGC 6087 are given in Table 2, together with results from other studies.

Can we learn anything from Table 2 with respect to the discrepancy between UBV and uvbyβ distances? After adjusting Schmidt's distances with the corrections by Balona & Shobbrook, there appears to be excellent agreement between the VBLUW and uvbyβ results for M 25, but not for NGC 6087. I have the strong suspicion that the disagreement for NGC 6087 is related to the fact that the uvbyβ distance is based on only 11 of the brightest cluster stars, all near the main-sequence turnoff, and clearly evolved. It could very well be that the Hβ calibration for such stars is less reliable. The uvbyβ distance for M 25 should be much more accurate, since it is based on 27 cluster stars which cover the main-sequence nearly down to $\log T_{eff} = 4.0$.

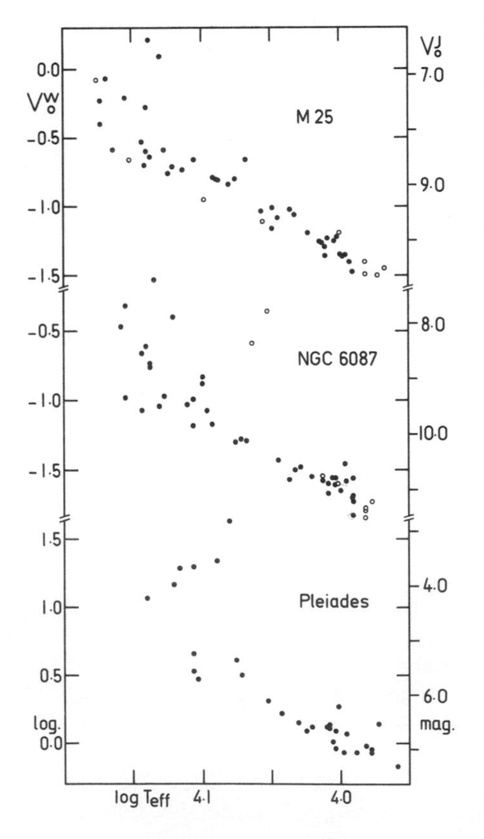

Fig.4.
Reddening-free HR-diagrams for M 25, NGC 6087, and Pleiades, as determined from VBLUW photometry. The Walraven V_o scale is in log units, the (transformed) UBV V_o scale in magnitudes. Temperatures have been derived from reddening-independent colour-indices. Open circles denote stars with less accurate reddening values, or cases where there is some doubt about cluster membership. Binaries have not been removed from the Pleiades diagram.

In this respect it is probably significant that Schmidt's preliminary result for M 25 (Schmidt 1980a), which was still based on only 12 cluster members, gave a modulus of only $8^m_.44$. It is likely that this first set of 12 stars covered mainly the bright, evolved cluster stars.

After all the work on the Hyades modulus and the cluster distance scale over the last fifteen years, it is ironic that the new VBLUW distances for S NOR and U SGR are so close to the 1969 Sandage & Tammann values. This is no argument against the new Hyades distance, but it indicates strongly that the increase in the Hyades distance is indeed to a large fraction offset by composition effects (cf. Martin et al. 1979; Caldwell 1983). Since more reliable theoretical data on composition effects in cluster main-sequences are now available (VandenBerg & Bridges 1984), we can show that the Pleiades distance used in the VBLUW method does not conflict with the revised Hyades modulus. Adopting a Hyades modulus of $3^m_.30 \pm 0^m_.06$ from Hanson (1980), and a modulus difference Pleiades-Hyades $\Delta(m-M)_0 = 2^m_.52 \pm 0^m_.05$ (mean value from Turner 1979 and Jones 1981), the uncorrected Pleiades modulus becomes $5^m_.82$. If the Hyades are more metal-rich by $\Delta[Fe/H] = +0.15$, the Pleiades modulus should be corrected by $-0^m_.22$ ($\pm 0^m_.03$?) according to the VandenBerg & Bridges data. This gives a final Pleiades modulus of $5^m_.60 \pm 0^m_.08$. Van Leeuwen's fit to parallax stars gave $5^m_.57 \pm 0^m_.08$.

Returning now to the aim of this whole exercise – the calibration of Cepheid luminosities – I will modestly adopt the VBLUW moduli as probably the most reliable distances for S NOR and U SGR. From the VBLUW photometry of the Cepheids (Pel 1978) we finally get $\log(L/L_\odot) = 3.536$ for S NOR, and 3.436 for U SGR (see Table 3). Taking into account the uncertainties in BC and reddening of the Cepheids, the r.m.s. error in both luminosities is $\sigma(\log L) = 0.044$, i.e. 11 % in L. This represents the best accuracy that is

Table 2. Distance moduli for M 25 and NGC 6087.

	M 25	σ	NGC 6087	σ
VBLUW photometry, fit to Pleiades				
$\Delta(m-M)_0$ relative to Pleiades	$3^m_.41$	$0^m_.06$	$4^m_.29$	$0^m_.06$
differential evolution corrections	-0.03	0.01	-0.02	0.01
$(m-M)_0$ Pleiades (Van Leeuwen 1983)	5.57	0.08	5.57	0.08
total distance modulus $(m-M)_0$	8.95	0.10	9.84	0.10
Distances from other methods				
Sandage & Tammann (1969); UBV, old Hyades distance $(m-M)_0 = 3^m_.03$	8.98		9.76	
Caldwell (1983); UBV, Hyades at $(m-M)_0 = 3^m_.28$, correction for high metallicity Hyades	9.14		9.81	
Fernie & McGonegal (1983); UBV, Hyades at $3^m_.29$, no correction for metallicity	9.21		9.95	
Schmidt (1980 b, 1982); uvbyβ, no explicit fit to Hyades	8.76		9.60	
Schmidt, as corrected by Balona & Shobbrook (1984)	8.91		9.64	

attainable for a calibrating Cepheid. It should be kept in mind, however, that this 11 % standard error does not yet account for the intrinsic scatter in main-sequences due to variations in composition, rotation, binary frequency. The contributions from these extra error sources are very uncertain, but they could very well increase σ_L to about 15 %.

MASSES AND RADII

The excellent review of Cepheid masses given by A.N.Cox a few years ago (Cox 1980) is still fully relevant to the present situation, so I will not attempt to repeat that analysis here. In his review Cox came to the important conclusion that the "primary" Cepheid mass problem - the fact that $M_{evol} > M_{puls}$ for all Cepheids with known L - had been solved by the increased Hyades distance and by lower Cepheid temperatures due to improved reddening and temperature calibrations. The factor that contributed most in removing the discrepancy was the larger Hyades modulus. We can expect therefore that the reduction in cluster distances that I have just discussed will cause mass problems again.

Table 3 lists the masses and radii for U SGR and S NOR that follow from the temperatures and luminosities that I adopted as "best values". The different mass estimates (following Cox's nomenclature) were computed for (Y=0.28,Z=0.02) from the same relations used by Cox: the evolutionary M-L relation for Cepheids from Becker *et al*. (1977), and the Faulkner (1977) fitting formulae for the P-M-L-T_{eff} relation. The results in Table 3 are not very surprising. The "theoretical" masses M_{th} are hardly affected by the new calibration, but the reduction in L moves M_{evol} and M_{puls} apart. This brings back a mass discrepancy that is not very large, but nevertheless significant.

Table 3. Fundamental parameters of U SGR and S NOR. Numbers in parentheses are corresponding values from Cox (1980). Masses and radii are in solar units.

parameter	U SGR		S NOR	
period	$6^{d}.745$		$9^{d}.755$	
$(m-M)_o$	$8^{m}.95$		$9^{m}.84$	
T_{eff} K	5754	(5734)	5425	(5502)
$\log(L/L_\odot)$	3.436	(3.59)	3.536	(3.65)
$R(L,T_{eff})$	52.5	(63.8)	66.4	(74.2)
M_{evol}	6.44	(7.09)	6.85	(7.36)
M_{th}	6.74	(6.72)	7.22	(7.37)
M_{puls}	5.15	(8.79)	5.45	(7.31)
M_{evol}/M_{puls}	1.25	(0.81)	1.26	(1.01)
M_{th}/M_{puls}	1.31	(0.76)	1.32	(1.01)
R_{BW} Baade-Wesselink radius (Fernie 1984)	54.3	(55.3)	65.2	(53.4)
$M_{BW} = M_{puls}$ from R_{BW}	5.61	(6.63)	5.25	(3.59)

If we assume that the discrepancy is caused by the observational data, and not by the theoretical relations, the problem can be cured with the well-known remedy: a decrease in T_{eff}, an increase in L, or a combination of both (Iben & Tuggle 1975; Cox 1980). In view of the uncertainty limits in the T_{eff} and L scales, and of the blue edge position in the HR-diagram, the least unattractive compromise would be ΔT_{eff}= -155 K and ΔL= +10 %. This would bring perfect agreement (within 1 %) between M_{evol}, M_{puls}, and M_{th} for both Cepheids (note that the new mass ratios in Table 3 are nearly identical for both stars), and it would not spoil the good fit of observed and theoretical blue edges too much. We should not accept this solution too easily, however. It clearly would not solve all problems, as R(L,T_{eff}) would increase by 10 %, and M_{BW} would remain too low. We should also keep in mind that the data in Table 3 can hardly be called representative for the whole Cepheid strip, as they are based on only two Cepheids with not too different periods.

The determinations of M and R for Cepheids are so closely related, that the problems with Cepheid masses could also be described as radius-discrepancies. In order to see what the situation is at the long- and short-period ends of the Cepheid strip, I will now switch from M to R, and discuss some results from a recent excellent review of Cepheid radii by Fernie (1984). Fernie discusses the following four types of radii:
1) Baade-Wesselink radii, R_{BW}, for which average values are taken from many sources.
2) R(L,T_{eff}) for cluster/association Cepheids, R_{CL}, based on the data by Fernie & McGonegal (1983).
3) "beat/bump radii", R_{BB}, based on R values from mixed-mode period ratios and phases of bumps in velocity curves (Cogan 1978).
3) "theoretical radii", R_{TH}, determined by solving the Becker et al.(1977) M-L relation, and the Iben & Tuggle (1975) P-M-L-T_{eff} relation at the blue and red edges of the Cepheid strip.
For these different radii he finds the following mean P-R relations:

$$\log R_{BW} = 1.244\ (\pm 0.023)\ +\ 0.587\ (\pm 0.022)\ \log P$$
$$\log R_{CL} = 1.042\ (\pm 0.015)\ +\ 0.824\ (\pm 0.010)\ \log P$$
$$\log R_{BB} = 0.833\ (\pm 0.018)\ +\ 0.956\ (\pm 0.022)\ \log P$$
$$\log R_{TH} = 1.179\ (\pm 0.006)\ +\ 0.692\ (\pm 0.006)\ \log P$$

Fernie describes this as "a sorry situation", even if one omits the R_{BB} radii on the grounds that the correct interpretation of the beat and bump phenomena is still unclear. For periods between 5 and 10 days the R_{BW}, R_{CL} and R_{TH} radii agree reasonably well, but outside this range there are serious discrepancies, of up to 50 % at the longest periods.

These results show that the good agreement between R(L,T_{eff}) and R_{BW} in Table 3 is misleading. It is also clear now that a simple zeropoint shift in L and T_{eff} can not solve all mass and radius problems over the whole period range. Adopting the distance scale of Table 3, instead of the uncorrected Hyades modulus of $3^{m}.29$ used by Fernie, would make all R_{CL} 12 % smaller. This would bring no real improvement; it would merely shift the point of best agreement between R_{TH}, R_{CL} and R_{BW} towards longer periods.

The R values that stand out most strongly are R_{BB} at the short periods, and R_{BW} at the long periods. The solution for the very small beat and bump radii (and masses) almost certainly has to come from the theory, but it is still difficult to decide whether surface He-enrichment (cf. Cox 1980) is the real answer to this problem. As for the BW radii, I should point out that Fernie's mean R_{BW} relation is an average for several rather different versions of the BW method, and that in general the more refined versions give steeper relations. It is particularly noteworthy that the BW method of Sollazzo et al. (1981) gives radii that agree very well with R_{TH}. Their method does not depend at all on the assumption of a unique colour-temperature correspondence, but it uses the detailed surface brightness variation derived from multicolour photometry and model atmospheres. Also the "Barnes-Evans" version of the BW method (Barnes et al. 1977), which is based on a (V-R) vs. surface brightness calibration, gives a much steeper R-P relation; in a recent paper Gieren et al. (1984) find $\log R \propto 0.79 \log P$ from this method. It seems therefore that the R_{BW} radii can be improved to agree much better with R_{CL} and R_{TH}. The $R_{TH} - R_{CL}$ discrepancy is probably more difficult to solve, however, as is also shown in Fernie's discussion. Changes in the theoretical relations are not sufficient (opacities, chemical composition), or go the wrong way (mass loss), whereas the empirical calibrations of L and T_{eff} allow some shift in the zeropoint of the R_{CL} relation, but hardly in its slope.

It should be noted that in a recent paper by Burki (1984) it is also concluded that good agreement can be reached for R_{BW} and R_{TH}, but in a very different way. Burki shows that R_{BW} and R_{TH} agree satisfactorily for short-period Cepheids if one corrects R_{BW} for the effects of duplicity. The long-period R_{BW} values remain smaller than R_{TH}, but Burki argues that these small R_{BW} are in fact the correct radii, and that R_{TH} should be reduced by taking into account mass loss. Although this reconciles R_{BW} and R_{TH}, I am afraid that it leaves a large discrepancy of both R_{BW} and R_{TH} with respect to the R_{CL} values (not considered by Burki) at long periods.

The conclusion from all this can only be that full agreement between empirical and theoretical data for Cepheids has not yet been reached, and that the solution of all mass and radius problems will require improvements on both the observational and the theoretical side. The above discussion also emphasizes once again how important the Cepheids in clusters and associations are in this respect. The basic parameters of these stars should be determined with the highest accuracy possible. In the list of 27 cluster/association Cepheids of Fernie & McGonegal (op.cit.) there are still several cases for which more reliable data are needed, particularly among the long-period Cepheids in associations. The cluster distance scale can hopefully be improved significantly within a few years with the Hipparcos satellite, which should give the distances to Hyades, Pleiades, and a few other nearby clusters to better than 1 %. For a small number of nearby Cepheids direct parallax measurements of useful accuracy should even be possible. Of the various discrepancies mentioned, those related to the double-mode Cepheids are probably the most disturbing. This is primarily a challenge to the theoreticians, but observers could make an important contribution if double-mode Cepheids in the Magellanic Clouds could be found.

At the end of this review, I realize that I have been talking mainly about uncertainties and discrepancies. This may have give the wrong impression that in the field of Cepheids we are just discovering problem after problem. If you would conclude that there has been no real progress, it is good to remember that after the discovery of the first Cepheids it took 130 years before it became clear that Cepheids pulsate, and 170 years before it was known why. It also took 170 years before Cepheid luminosities were known to better than a factor four. It is only fair that it takes a bit longer to solve the more subtle problems of these fascinating stars.

This review is partly based on papers that are still in press; I am very grateful to drs. L.A. Balona, G. Burki, J.A.R. Caldwell, J.D. Fernie, and W. Gieren for sending me preprints of their work. I thank Dr.H.R. Butcher for stimulating discussions and helpful comments on the manuscript.

REFERENCES

Balona, L.A. & Shobbrook, R.R. (1984). Mon. Not. R. astr. Soc. In press.
Barnes, T.G., Dominy, J.F., Evans, D.S., Kelton, P.W., Parsons, S.B. & Stover, R.J. (1977). Mon. Not. R. astr. Soc. 178, 661.
Becker, S.A., Iben, I. & Tuggle, R.S. (1977). Astrophys. J. 218, 633.
Bell, R.A. & Parsons, S.B. (1974). Mon. Not. R. astr. Soc. 169, 71.
Bell, R.A., Eriksson, K., Gustafsson, B. & Nordlund, A. (1976). Astron. Astrophys. Suppl. 23, 27.
Bell, R.A. & Gustafsson, B. (1978). Astron. Astrophys. Suppl. 34, 229.
Blackwell, D.E. & Shallis, M.J. (1977). Mon. Not. R. astr. Soc. 180, 177.
Böhm-Vitense, E. (1972). Astron. Astrophys. 17, 335.
Böhm-Vitense, E. (1981). Ann. Rev. Astron. Astrophys. 19, 295.
Burki, G. (1984). Astron. Astrophys. 133, 185.
Caldwell, J.A.R. (1983). Observatory 103, 244.
Caldwell, J.A.R. & Coulson, I.M. (1984a). SAAO Circ. In press.
Caldwell, J.A.R. & Coulson, I.M. (1984b). Mon. Not. R. astr. Soc. In press.
Cayrel de Strobel, G. (1980). In Star Clusters, IAU Symp. 85, ed. J.E. Hesser, p. 91, Reidel, Dordrecht.
Code, A.D., Davis, J., Bless, R.C. & Hanbury Brown, R. (1976). Astrophys. J. 203, 417.
Cogan, B.C. (1978). Astrophys. J. 221, 635.
Cox, A.N. (1980). Ann. Rev. Astron. Astrophys. 18, 15.
Dean, J.F., Warren, P.R. & Cousins, A.W.J. (1978). Mon. Not. R. astr. Soc. 183, 569.
Dean, J.F. (1981). Mon. Not. R. astr. Soc. 197, 779.
Eggen, O.J. (1980). Astrophys. J. 238, 919.
Eggen, O.J. (1983a). Astron. J. 88, 361.
Eggen, O.J. (1983b). Astron. J. 88,1187.
Faulkner, D.J. (1977). Astrophys. J. 218, 209.
Feast, M.W. (1984). In Structure and Evolution of the Magellanic Clouds, IAU Symp. 108, eds. S. van den Bergh & K.S. de Boer, p. 157, Reidel, Dordrecht.
Feltz, K.A. & McNamara, D.H. (1980). Publ. astr. Soc. Pacif. 92, 609.
Fernie, J.D. & McGonegal, R. (1983). Astrophys. J. 275, 732.
Fernie, J.D. (1984). Astrophys. J. In press.
Flower, P.J. (1977). Astron. Astrophys. 54, 31.

Flower, P.J. (1980). In Star Clusters, IAU Symp. 85, ed. J.E. Hesser, p. 119, Reidel, Dordrecht.
Gehren, T. (1981). Astron. Astrophys. 100, 97.
Gieren, W., Coulson, I.M. & Caldwell, J.A.R. (1984). Preprint; cf. also this conference.
Gustafsson, B. & Bell, R.A. (1979). Astron. Astrophys. 74, 313.
Hanson, R.B. (1980). In Star Clusters, IAU Symp. 85, ed. J.E. Hesser, p.71, Reidel, Dordrecht.
Hejlesen, P.M. (1980). Astron. Astrophys. Suppl. 39, 347.
Iben, I. & Tuggle, R.S. (1972). Astrophys. J. 178, 441.
Iben, I. & Tuggle, R.S. (1975). Astrophys. J. 197, 39.
Johnson, H. (1966). Ann. Rev. Astron. Astrophys. 4, 193.
Jones, B.F. (1981). Astron. J. 86, 290.
Kraft, R.P. (1961). Astrophys. J. 134, 616.
Kurucz. R.L. (1979). Astrophys. J. Suppl. 40, 1.
Lub, J. & Pel, J.W. (1977). Astron. Astrophys. 54, 137.
Luck, R.E. (1977a). Astrophys. J. 212, 743.
Luck, R.E. (1977b). Astrophys. J. 218, 752.
Martin, W.L., Warren, P.R. & Feast, M.W. (1979). Mon. Not. R. astr. Soc. 188, 139.
Nissen, P.E. (1980). In Star Clusters, IAU Symp. 85, ed. J.E. Hesser, p.51, Reidel, Dordrecht.
Oke, J.B. (1961). Astrophys. J. 134, 214.
Parsons, S.B. (1971). Mon. Not. R. astr. Soc. 152, 121.
Parsons, S.B. (1974). In Stellar Instability and Evolution, IAU Symp. 59, eds. P. Ledoux, A. Noels, A.W. Rodgers, p.56, Reidel,Dordrecht.
Pel, J.W. & Lub, J. (1978). In The HR-Diagram, IAU Symp. 80, eds. A.G. Davis Philip & D.S. Hayes, p. 229, Reidel, Dordrecht.
Pel, J.W. (1978). Astron. Astrophys. 62, 75.
Pel, J.W. (1980). In Current Problems in Stellar Pulsation Instabilities, eds. D.Fischel, J.R.Lesh & W.M.Sparks, p.1, NASA Tech.Mem.80625.
Pel, J.W. (1984). In preparation.
Relyea, L.J. & Kurucz, R.L. (1978). Astrophys. J. Suppl. 37, 45.
Rodgers, A.W. (1970). Mon. Not. R. astr. Soc. 151, 133.
Sandage, A. & Tammann, G.A. (1969). Astrophys. J. 157, 683.
Schmidt, E.G. (1972). Astrophys. J. 174, 605.
Schmidt, E.G. (1980a). Space. Sci. Rev. 27, 449.
Schmidt, E.G. (1980b). Astron. J. 85, 158.
Schmidt, E.G. (1982). Astron. J. 87, 650.
Schmidt, E.G. (1983). Astron. J. 88, 104.
Sollazzo, C., Russo, G., Onnembo, A. & Caccin, B. (1981). Astron. Astrophys. 99, 66.
Turner, D.G. (1977). Astron. J. 82, 163.
Turner, D.G. (1978). J. R. astr. Soc. Canada, 72, 248.
Turner, D.G. (1979). Publ. astr. Soc. Pacif. 91, 642.
Turner, D.G. (1980). Astrophys. J. 235, 146.
Turner, D.G. (1983). J. R. astr. Soc. Canada. 77, 31.
VandenBerg, D.A. & Bridges, T.J. (1984). Astrophys. J. 278, 679.
Van Leeuwen, F. (1983). Thesis, Leiden University.
Van Paradijs, J.A. (1973). Astron. Astrophys. 23, 369.

THE DOUBLE-MODE CEPHEIDS

L.A. Balona
South African Astronomical Observatory

Abstract. Recent observations of double-mode Cepheids and proposed candidates are reviewed. It appears that the change of modal content reported in some stars may not be real. Observations show that the double-mode Cepheids are indistinguishable from normal Cepheids of similar period. This poses a severe problem for pulsation theory which predicts very low masses from the ratio of first overtone to fundamental period. Attempts to resolve this problem and the problem of modal selection are discussed.

INTRODUCTION

The phenomenon of double-mode pulsation is common among the lower luminosity stars in the Cepheid instability strip. Most δ Scuti stars are in fact multimode pulsators; those with larger amplitudes usually pulsate in two modes and are sometimes known as AI Vel stars or dwarf Cepheids. Among the Cepheids, the double-mode pulsators comprise nearly half the number of stars in the 2 - 4 day period range.

The double-mode Cepheids pose two serious problems for stellar pulsation theory. Firstly, the masses inferred from the ratio of their periods (presumed to be the fundamental and first overtone radial modes) are about half the value deduced from the theory of stellar evolution. This is by far the most serious of the various mass discrepancy problems encountered among the Cepheids. Secondly, the very existence of these stars is a problem because stable double-mode behaviour has not yet been produced in any Cepheid model.

Because of these problems, the double-mode Cepheids have received much observational and theoretical attention in recent years.

RECENT OBSERVATIONAL STUDIES

In Table 1 we summarize some data on the eleven known double-mode Cepheids. The data are mainly from Stobie & Balona (1979) and from references cited below in the discussion of individual stars. Despite intensive efforts to discover further candidates (Pike & Andrews 1979; Henden 1979, 1980; Barrell 1982b), none have been found. Three stars recently proposed as possible members of this class or closely related to them are also discussed below.

TU Cas. Faulkner (1977) analysed photoelectric observations of TU Cas and suggested the presence of a third periodicity which was interpreted as the second overtone at P_2 = 1.25246 days. Subsequently, Hodson, Stellingwerf & Cox (1979) repeated the analysis and concluded that P_2 was just an artifact of the data. More recently, Faulkner (1979) has disputed this conclusion.

The presence of a third mode would have very important consequences as it further constrains the theory. We have re-analysed all photoelectric

Table 1. Periods, period ratios and semiamplitudes of the light and radial velocity variations of double-mode Cepheids. Standard errors of these quantities are given on the second line whenever available.

Name	P_0	P_1	P_1/P_0	$A(V_0)$	$A(V_1)$	$A(RV_0)$	$A(RV_1)$
TU Cas	2.13931	1.518285	0.709708	0.292	0.103	13.5	6.6
	2	12	9	5	5	.9	.9
U TrA	2.568425	1.824876	0.7105037	0.297	0.125	14.7	7.5
	3	3	14	13	12	.4	.4
VX Pup	3.0109	2.1390	0.7104	0.166	0.144	6.3	9.0
	10	5	3	4	4	.7	.7
AP Vel	3.12776	2.19984	0.70333	0.275	0.138	13.2	9.8
	10	5	3	7	7	.8	.8
BK Cen	3.17387	2.22297	0.70040	0.245	0.106	13.2	10.0
	10	5	3	13	13	1.2	1.4
UZ Cen	3.33435	2.35529	0.70637	0.308	0.070	14.7	6.2
	11	6	3	10	9	.9	.8
Y Car	3.63981	2.55954	0.70321	0.266	0.120	11.1	6.8
	13	7	3	7	7	.9	1.0
AX Vel	3.673170	2.592924	0.705909	0.116	0.148	3.6	7.5
	14	5	3	5	5	.5	.5
GZ Car	4.15885	2.93372	0.70542	0.150	0.086	5.3	7.1
	17	9	4	5	4	1.0	1.0
BQ Ser	4.27073	3.01205	0.7053				
	?	?	?				
V367 Sct	6.29307	4.38466	0.696744	0.143	0.137		
	4	2	5	?	?		

observations including an early set by Bahner & Mavridis (1971) which was overlooked by the previous workers. Our conclusion is that the third periodicity definitely does not exist.

Hodson *et al.* also found that the light amplitude of the first overtone has decayed with respect to that of the fundamental by about forty per cent over the last sixty years. Niva (1979) found a similar, though smaller, trend in the radial velocities. The observational evidence in both cases is weak. The conclusion of the former workers rests entirely on visual observations conducted during the early decades of this century. Photoelectric observations obtained during the last twenty years do not show any trend. The amplitude changes in the radial velocities found by Niva are not statistically significant, differing by no more than one standard deviation.

U Tra. Faulkner & Shobbrook (1979) found an increase in amplitude of the first overtone with respect to the fundamental in the photoelectric data prior to 1977. More recently, the same authors (Faulkner & Shobbrook 1983) repeated the analysis in a different manner and concluded that the change in modal content reported earlier was not significant, though their most recent data do show some increase in first overtone amplitude. They concluded that the apparent changes in modal amplitudes may not be real, but are probably due to seemingly random fluctuations in the light curve. They also find that the first overtone is subject to a greater degree of incoherence in phase and variability in amplitude than is the fundamental.

AX Vel. This is the only double-mode Cepheid in which the first overtone light amplitude is larger than that of the fundamental. As Table 1 shows, there are other double-mode Cepheids in which the first overtone predominates in radial velocity amplitude. Shobbrook & Faulkner (1982) obtained new observations of AX Vel. They conclude that, like U TrA, the light variations of the fundamental mode are stable and coherent, but that the first overtone suffers small phase fluctuations which can be as large as 0.07 periods over a five year interval.

Y Car. This is the only double-mode Cepheid known to be a spectroscopic binary (Stobie & Balona 1979). Balona (1983) obtained an orbital period of 993 days, which is typical for binary Cepheids. The mass function leads to a lower limit of 1.2 solar masses for the companion if the evolutionary mass is used for the primary.

PROPOSED CANDIDATES

CO Aur. This star was first observed by Smak (1964) to check its classification as an RV Tau star. He could not detect any periodicity in his photoelectric observations. The same result was obtained by DuPuy & Brooks (1974) who re-observed the star. Recently, Mantegazza (1983) has concluded that CO Aur is a double-mode Cepheid by re-analysing Smak's data. He finds P_0 = 1.784, P_1 = 1.4255 days. If this star is indeed a double-mode Cepheid, then it has the shortest period of the group, but more important it has the unusual ratio P_1/P_0 = 0.80.

This ratio is incompatible with current ideas on double-mode Cepheids and could lead to a reappraisal of the subject. Antonello & Mantegazza (1983) obtained further photometry which is claimed to confirm CO Aur as a double-mode Cepheid.

We have analysed all available photometry for this star including the data by Antonello & Mantegazza which they kindly put at our disposal prior to publication. We confirm that the P_0 period is indeed present in all the data sets, though the alias at P = 2.273 days is almost as strong. However, when the data is prewhitened by P_0, there are no peaks in the power spectrum at or near P_1 or in fact any peak much above noise level. This not only applies to the set of combined data, but also to the individual data sets. The strongest indication for P_1 is in Smak's data, but this peak is only some 10 or 20 per cent stronger than the noise level and can hardly be regarded as significant. We were nevertheless puzzled by a diagram in Mantegazza (1983) and Antonello & Mantegazza (1983) which purports to show the variation of the fundamental prewhitened by the first overtone and vice versa. We could not reproduce this diagram by the normal prewhitening procedure. Instead, if we assumed the presence of the second period, found the best fitting coefficients by a double-mode least squares solution, and removed the coefficients pertaining to P_0 or to P_1, we could reproduce their diagram by prewhitening with the remaining coefficients. This procedure is incorrect, since the inclusion of an arbitrary periodicity distorts the least squares solution in such a way as to reflect the periodicity of the missing coefficients in the prewhitened data.

In conclusion, CO Aur cannot be regared as a double-mode Cepheid at this stage. The 1.78 day periodicity is however certainly real, but further observations are required to elucidate the nature of this star.

HD161796. This is an F3Ib star located at high galactic latitude. Burki, Mayor & Rufener (1980) found a semi-regular variation in both photometry and radial velocities with a characteristic period of about 54 days. Percy & Welch (1981) found a period of about 60 days for the light variations during 1979, but in 1980 the period seemed to have shortened to about 40 days.

Fernie (1983) obtained further photometric data during 1980 and confirmed a period of 43 days for this season. He concludes that the 60 day to 43 day ratio for the two seasons is what one expects for a star in the process of switching from fundamental to first overtone radial pulsation. Takeuti (1983) has calculated that a Cepheid of 30 solar masses would have the observed periods.

In a semi-regular variable, one could expect to find by chance the correct period ratio in two or even more consecutive cycles. To conclude that this star is a Cepheid undergoing mode transition one would have to show that for a long time the star was pulsating in predominantly one period and after an interval in predominantly the other period. We analysed the available photometric and radial velocity data, but failed to confirm this behaviour. Fernie has tried to explain the 54 day periodicity in the data by Burki et al. as an aliasing effect, but we cannot support this conclusion. Their radial velocity data show a clear unaliased peak at 53 ± 1 days. The combined photometric data shows a peak at 56.7 days; in both cases a period near 60 days can be excluded.

HR7308. This star is unique among the Cepheids. Not only does it have the shortest period (1.49 days) but the amplitude itself varies with a period of about 1200 days (Breger 1981). This amplitude modulation can also be interpreted as a beating of two very nearly equal periods. In this case HR7308 could be classified as a double-mode Cepheid, though with P_1/P_0 = 0.999, the pulsation modes cannot be the same as in normal double-mode Cepheids.

We have analysed the radial velocity observations of Burki, Mayor & Benz (1982) which constitute a large, accurate and homogeneous body of data with excellent phase coverage over nearly one complete beat cycle. Table 2 shows the results of a periodogram analysis with successive prewhitening. At each stage the frequency range between 0 and 1 cycle day^{-1} was searched for peaks. Except for those in the table, no others were found.

It is apparent that the data cannot be represented adequately by just two periods. The results of Table 2 show that the main pulsation at 0.6709 cycles day^{-1} is flanked symmetrically on each side by a pair of equally spaced frequencies, the spacing being equal to the beat frequency. When all five oscillations are used, the standard deviation of the least squares fit is 0.7 km s^{-1} per observation.

There are two interpretations of this frequency spectrum. If all five frequencies are real oscillations, then the most obvious interpretation is that they are caused by rotationally split quadrupole or higher order nonradial oscillations. On the other hand one could intepret this spectrum as asymmetric amplitude modulation of a single radial pulsation. It is easy to show that in this case the spectrum should consist of the main pulsation flanked by equally spaced components with symmetric decaying amplitudes. The results of Table 1 seem to confirm these conditions within the errors. In this interpretation there should also be a relationship between the phases, but this could not be checked as the phases are strongly affected by small uncertainties in the frequencies. An analysis of the available photometry confirms the pattern shown by the radial velocities, though the smaller amount of data and the uneven phase coverage pose some problems.

Table 2. Frequencies of radial velocity variation in HR7308.

f (day^{-1})	Semiamplitude (km s^{-1})	Indentification
0.6709	4.98 ± 0.07	f_0
0.6717	2.07 ± 0.07	$f_0 + \Delta f$
0.6699	1.82 ± 0.07	$f_0 - \Delta f$
0.6727	1.00 ± 0.07	$f_0 + 2\Delta f$
0.6689	0.61 ± 0.07	$f_0 - 2\Delta f$

The evidence from the power spectrum of the radial velocities and the fact that Burki et al. (1982) were able to perform a Baade-Wesselink analysis assuming radial pulsation strongly suggests that HR7308 is pulsating in a single amplitude modulated radial mode. It is certainly not possible to interpret the observations in terms of a double-mode pulsation.

PHYSICAL AND PULSATIONAL PROPERTIES

The evidence that double-mode Cepheids are Population I objects of high mass is very compelling. Perhaps the best proof of this is V367 Sct which is a member of the young open cluster NGC6649. Barrell (1980) measured the mean radial velocity of the Cepheid and showed that it was the same as that of the cluster stars. The three brightest main sequence stars were found to be emission-line objects, removing an objection to cluster membership (Flower 1978).

Barrell (1982a) established that the iron abundance in ten double-mode Cepheids is consistent with that of the sun. Balona & Stobie (1979a) obtained Baade-Wesselink radii for eight double-mode Cepheids and showed that they obey the same period - radius relationship as normal Cepheids. Niva & Schmidt (1979) found the same result from the radius of TU Cas.

Barrell (1981) estimated the effective temperatures of ten double-mode Cepheids by measuring the Hα line profiles and comparing them with profiles calculated from model atmospheres. This method is independent of interstellar reddening and insensitive to differences in surface gravity. Within the uncertainties, Barrell found all double-mode Cepheids to have the same mean effective temperature. Balona & Stobie (1979a) used BVRI photometry to estimate the reddening and determine the intrinsic colours of eight double-mode Cepheids. Comparison with single-mode Cepheids (Fig. 1) shows that there is no preferred location for the double-mode Cepheids in the instability strip. The double-mode RR Lyrae variables in M15 also show no preference for a particular location in the instability strip. Unlike the double-mode Cepheids, they lie in a narrow period range in which no single mode RR Lyraes exist (Cox, Hodson & Clancy 1983).

The only observational evidence which distinguishes the double-mode Cepheids from their single mode counterparts is the remarkable discovery by Barrell (1978) of strong Hα emission occuring at seemingly random phases. There is a possibility that this emission may be an instrumental effect since the observations were made during the early days of operation of a new instrument. Barrell subsequently observed these stars for several more nights without finding any emission (Feast, priv. comm. via Barrell). Nevertheless, there is no reason to suspect any fault in the instrument and confirmation of this effect would be most important. Henden, Cornett & Schmidt (1982) could not find any abnormalities in spectra of TU Cas which included the Hα line.

The pulsational properties of double-mode Cepheids are not significantly different from those of normal Cepheids. Both have the same phase lag between the radius and light variation (Balona & Stobie 1979a). The ratio of temperature variation to radius variation, which in normal Cepheids depends on the mean temperature or period (Balona & Stobie 1979b), follows the same relation for the double-mode Cepheids. The ratio of radial velocity to light amplitude, which is different for the fundamental mode and for the first overtone mode, is easily explained as a result of radial pulsation (Stobie & Balona 1979). Balona & Stobie (1979a) find that the fractional radius variation amplitude of the first overtone dominates that of the fundamental for the hotter double-mode Cepheids. The double-mode RR Lyraes seem to follow this pattern: being hotter than any of the double-mode Cepheids the first overtone always has a larger light amplitude than the fundamental mode (Cox et al. 1983).

Fig. 1. The colour - magnitude diagram for normal Cepheids (closed circles and double-mode Cepheids (open circles). The transformed fundamental (F) and first overtone (1H) blue edges calculated by King et al. (1973) are shown.

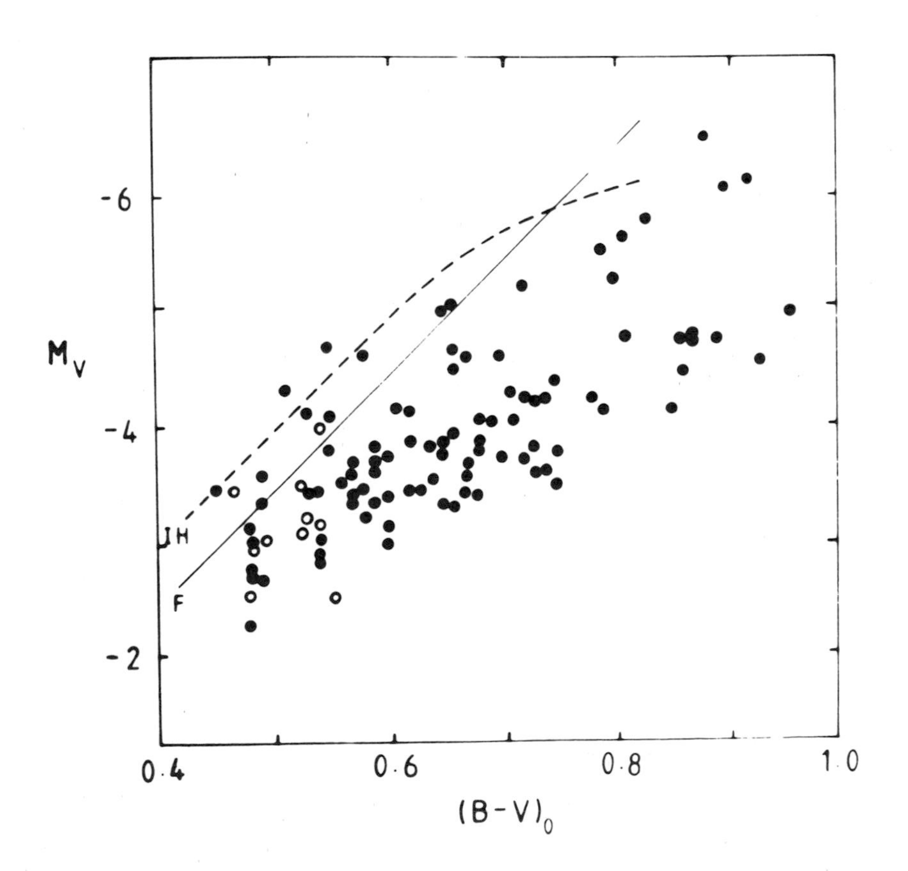

THE MASS DISCREPANCY PROBLEM

The well known mass discrepancy problem for double-mode Cepheids is most easily understood from the diagram of Fig. 2, adapted from Petersen (1979). This shows the variation of period ratio with fundamental period. Also shown are Petersen's (1979) calculations for standard Cepheid models of various masses. It is clear that the predicted masses of double-mode Cepheids lie in the range 1 to 3 solar masses, while standard evolution theory predicts masses in the range 5 to 7 solar masses for models with the same fundamental period. The models with evolutionary masses have a considerably larger period ratio than is actually observed. Inclusion of convection (Cogan 1977; Deupree 1977; Saio et al. 1977) or rotation (Cox et al. 1977; Deupree 1978) has only a marginal effect on the period ratio.

Simon (1982) has proposed an idea which may solve the mass discrepancy problem without abandoning the assumptions made in standard evolution and pulsation models. He suggests that the present metal opacities may be underestimated. Increasing these opacities by a factor of 2 to 3 in the region 10^5 °K to $2X10^6$ °K would reduce the period ratio of models with evolutionary masses to the observed values. It will also leave intact the good agreement obtained for the double-mode RR Lyrae period ratios (Cox et al. 1983) since these stars have a small metal abundance. Furthermore, it may help to solve the long standing problem of β Cep variability.

Fig. 2. The period ratio vs. the logarithm of fundamental period showing the loci of homogeneous composition models (labeled in solar masses) by Petersen (1979).

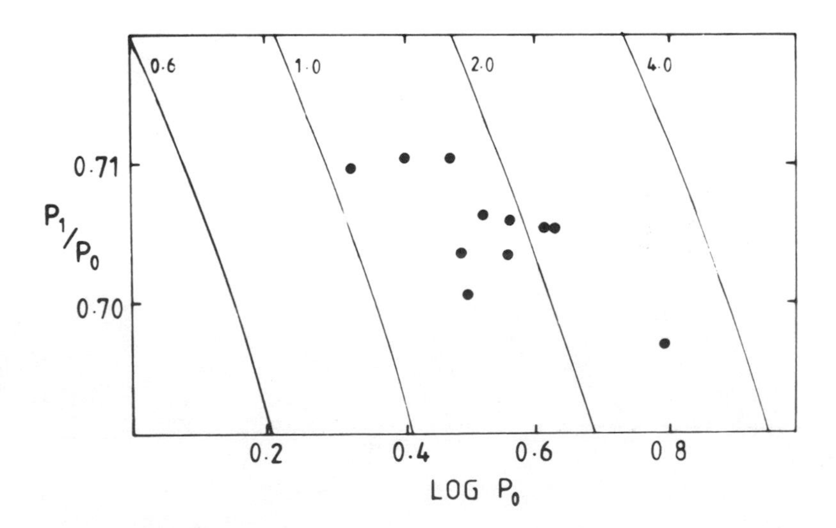

There are reasons to suspect that the present metal opacities are too low, but whether they could be increased to the required level has not been demonstrated.

If one accepts the present metal opacities, then one can invert the problem and look for unusual physical properties in the outer layers where the effect on the period ratio will be at its greatest.

Stothers (1979) has proposed that a magnetic field of several hundred gauss at the surface could reconcile the masses. The magnetic fields observed in Cepheids are much lower than this, but it is expected that the magnetic field will be tangled by convection so that the observed field will indeed be smaller. Recently, Stothers (1982) found that the presence of magnetic fields will also lead to better agreement in the predicted and observed phases of secondary bumps in many Cepheids. However, the calculated models have larger amplitudes than actually observed. The required field is also much lower than that needed to explain the period ratio in double-mode Cepheids.

The modification which has received the most attention is that proposed by Cox et al. (1977). They found that by increasing the helium abundance in the outer layers of the star the density gradient is reduced and the period ratio can be lowered to the observed values. The abundance required to match the observed ratios is Y = 0.65 up to a temperature of 10^5 °K. They suggest that preferential depletion of hydrogen by a stellar wind could produce the required helium enrichment (Cox, Michaud & Hodson 1978). A surface helium enrichment of Y = 0.75 is required to match the phases in bump Cepheids. A serious problem in this model is the instability of the enriched helium zone due to the inverted molecular weight gradient which should lead to rapid mixing. Also it is not certain that the rate of hydrogen depletion is high enough to compensate for this effect. Henden et al. (1982) looked for, but could not find, any spectroscopic evidence for a stellar wind in TU Cas. Luck & Lambert (1981) deduced an appreciable helium overabundance in some Cepheids, particularly TU Cas, from the observed depletion of oxygen. This finding does not help the mass discrepancy problem since the helium enrichment is not confined to the outer layers and the density gradient is not appreciably affected by homogeneous changes in composition. Takeuti (1980) has suggested that enhanced helium abundance in the outer layers would result in increased convectional energy. Spontaneous Hα emission as reported by Barrell (1978) could then occur.

THE MODAL SELECTION PROBLEM

The second major problem, one which has received much attention recently, is the cause of double-mode pulsation. One possibility which must be rejected is that these stars are in a state of mode transition. The expected lifetime of double-mode pulsation is much too short to explain the observed numbers of double-mode Cepheids (Stellingwerf 1975).

One of the problems which any theory of modal selection must attempt to answer is the very narrow range of observed period ratios. A range in magnetic field stengths or helium abundances in the outer layers will lead to a much larger variation in this ratio. The most promising answer to this problem and one which probably underlies the modal selection problem is Simon's (1979) resonance hypothesis. He suggests that the double-mode phenomenon involves a resonant interaction between the fundamental and first overtone modes when the sum of their frequencies is close to or equal to the frequency of the third overtone. This hypothesis might also underlie the Hertzprung progression of bumps in the light curves of normal Cepheids. In this case the resonance is between the fundamental and second overtone modes, the latter having nearly twice the frequency of the former. Resonance may also be one of the factors determining the limiting amplitude in Cepheids.

In the resonance theory, the double-mode Cepheids should occupy a region in the instability strip where a close resonance with the third overtone is possible. Simon, Cox & Hodson (1978) examined a series of nonlinear pulsation models of double-mode Cepheids near this resonance, but could not find sustained double-mode behaviour. Petersen (1979, 1980) found that with normal composition the resonance condition did confine the period ratio to a narrow range, but only models with unrealisticly high helium enrichment in the outer layers could have period ratios in the observed range if evolutionary masses are assumed. Takeuti & Aikawa (1980) and Uji-Iye (1980) have found that for a reasonable helium enrichment only models with masses much smaller than evolutionary masses could simultaneously satisfy the period ratio and resonance conditions.

Aikawa (1984) investigated the two and three-mode resonances analytically. The approach to resonance is accompanied by mutual period changes in the models involved, but the change in period ratio is too small to be of significance. Dziembowski (1982) has developed a second-order theory of nonlinear mode coupling in oscillating stars. He obtains an explicit formula for the coupling coefficient and finds an equilibrium solution for two and three interacting modes and develops stability criteria.

Regev & Buchler (1981) and Buchler & Regev (1981) have developed the two-time formalism. This is a useful analytical and computational tool for finding stable double-mode pulsation and for studying the evolution towards it. In addition, this formalism has the advantage of making more apparent the physical significance of various factors which is normally hidden in numerical simulations. Regev & Buchler (1981) suggest that persistent double-mode pulsation may not necessarily require the presence of internal resonances, but these resonances might play an important role in the evolution towards such pulsations. Regev, Buchler & Barranco (1982) include the treatment of resonances in the two-time formalism. Buchler (1983) finds that near resonance a steady state solution where only one amplitude is non-zero does not exist. He suggests that Simon *et al*. (1980) may have made an unfortunate choice for their stellar model.

CONCLUSIONS

Our understanding of double-mode pulsation in Cepheids is still very limited, though some progress has been made. The mass discrepancy problem has still not been solved satisfactorily. The theories which have been proposed suffer from several weaknesses, not the least being that they involve too many free parameters and are practically impossible to disprove. Simon's (1982) idea of enhanced metal opacities seems to be the most promising line of investigation at present.

Until numerical models have been constructed which undergo persistent double-mode behaviour, the cause of double-mode pulsation will remain a problem. Resonances seem to play a very important role in this process, but their effect is still not fully understood. It is particularly important to study the behaviour of the pulsations during the evolution towards and away from the resonance condition.

It is of great importance to confirm the presence of Hα emission in double-mode Cepheids as reported by Barrell (1978), as there is some suspicion that this could have been an instrumental effect. It would also be very useful to extend the method of temperature determination by measuring the Hα profiles to normal Cepheids. This could allow more accurate definition of the location of the double-mode Cepheids within the instability strip.

Further photometric and radial velocity observations of these stars is highly desirable. It is of great importance to verify the finding of Faulkner & Shobbrook (1983) that the first overtone is subject to greater instability and phase jitter than the fundamental. This has cast doubt on the reality of changes in modal content which have been reported in some double-mode Cepheids. It also appears to offer a clue to the resonance hypothesis.

Finally, it would be of great interest to search for double-mode Cepheids in the Magellanic Clouds. This could lead to new insights into these enigmatic objects.

REFERENCES

Aikawa, T., 1984. Mon. Not. R. astr. Soc., 206, 833.
Antonello, E. & Mantegazza, L., 1983. Inf. Bull. Var. Stars, 2411.
Bahner, K. & Mavridis, L.N., 1971. Annals of the Faculty of Technology, Univ. Thessaloniki, 5, 65 (Contr. Dept. Geodetic Astr. Univ. Thessaloniki No. 3).
Balona, L.A., 1983. Observatory, 103, 163.
Balona, L.A. & Stobie, R.S., 1979a. Mon. Not. R. astr. Soc., 189, 659.
Balona, L.A. & Stobie, R.S., 1979b. Mon. Not. R. astr. Soc., 189, 649.
Barrell, S.L., 1978. Astrophys. J., 226, L141.
Barrell, S.L., 1980. Astrophys. J., 240, 145.
Barrell, S.L., 1981. Mon. Not. R. astr. Soc., 196, 357.

Barrell, S.L., 1982a. Mon. Not. R. astr. Soc., 200, 127.
Barrell, S.L., 1982b. Mon. Not. R. astr. Soc., 200, 139.
Breger, M., 1981. Astrophys. J., 249, 666.
Buchler, J.R., 1982. Astr. Astrophys., 118, 163.
Buchler, J.R. & Regev, O., 1981. Astrophys. J., 250, 776.
Burki, G., Mayor, M. & Rufener, F., 1980. Astr. Astrophys. Suppl., 42, 383.
Burki, G., Mayor, M. & Benz, W., 1982. Astr. Astrophys., 109, 258.
Cogan, B.C., 1977. Astrophys. J., 211, 890.
Cox, A.N., Deupree, R.G., King, D.S. & Hodson, S.W., 1977. Astrophys. J., 214, L127.
Cox, A.N., Michaud, G. & Hodson, S.W., 1978. Astrophys. J., 222, 621.
Cox, A.N., Hodson, S.W. & Clancy, S.P., 1983. Astrophys. J., 266, 94.
Deupree, R.G., 1977. Astrophys. J., 215, 232.
Deupree, R.G., 1978. Astrophys. J., 223, 982.
DuPuy, D.L. & Brooks, R.C., 1974. Observatory, 94, 71.
Dziembowski, W., 1982. Acta Astronomica, 32, 147.
Faulkner, D.J., 1977. Astrophys. J., 218, 209.
Faulkner, D.J., 1979. IAU colloq. 46, "Changing trends in variable star research" (ed. F.M. Bateson, J. Smak & I.H. Urch), Univ. Waikato, Hamilton, New Zealand, 322.
Faulkner, D.J. & Shobbrook, R.R., 1979. Astrophys. J., 232, 197.
Faulkner, D.J. & Shobbrook, R.R., 1983. Proc. astr. Soc. Australia, 5, 217.
Fernie, J.D., 1983. Astrophys. J., 265, 999.
Flower, P.J., Astrophys. J., 224, 948.
Henden, A.A., 1979. Mon. Not. R. astr. Soc., 189, 149.
Henden, A.A., 1980. Mon. Not. R. astr. Soc., 192, 621.
Henden, A.A., Cornett, H. & Schmidt, E.G., 1982. "Pulsations in Classical and Cataclysmic Variable Stars" (ed. J.P. Cox & C.J. Hansen), JILA, Univ. Colorado, 193.
Hodson, S.W., Stellingwerf, R.F. & Cox, A.N., 1979. Astrophys. J., 229, 642.
King, D.S., Cox, J.P., Eilers, D.D. & Davey, W.R., 1973. Astrophys. J., 182, 859.
Luck, R.E. & Lambert, D.L., 1981. Astrophys. J., 245, 1018.
Mantegazza, L., 1983. Astr. Astrophys., 118, 321.
Niva, G.D., 1979. Astrophys. J., 232, L43.
Niva, G.D. & Schmidt, E.G., 1979. Astrophys. J., 234, 245.
Percy, J.R. & Welch, D.L., 1981. Publ. astr. Soc. Pacific, 93, 367.
Petersen, J.O., 1979. Astr. Astrophys., 80, 53.
Petersen, J.O., 1980. Astr. Astrophys., 84, 356.
Pike, C.D. & Andrews, P.J., 1979. Mon. Not. R. astr. Soc., 187, 261.
Regev, O. & Buchler, J.R., 1981. Astrophys. J., 250, 769.
Regev, O., Buchler, J.R. & Barranco, M., 1982. Astrophys. J., 257, 715.
Saio, H., Kobayashi, E. & Takeuti, M., 1977. Sci. Rep. Tohoku Univ., Ser. I, 60, 144 (Science Reports Sendai No. 180).
Shobbrook, R.R. & Faulkner, D.J., 1982. Proc. astr. Soc. Australia, 4, 400.
Simon, N.R., 1979. Astr. Astrophys., 75, 140.
Simon, N.R., 1982. Astrophys. J., 260, L87.
Simon, N.R., Cox, A.N. & Hodson, S.W., 1980. Astrophys. J., 237, 550.

Smak, J., 1964. Publ. astr. Soc. Pacific, 76, 40.
Stellingwerf, R.F., 1975. Astrophys. J., 199, 705.
Stobie, R.S. & Balona, L.A., 1979. Mon. Not. R. astr. Soc., 189, 627.
Stothers, R., 1979. Astrophys. J., 234, 257.
Stothers, R., 1982. Astrophys. J., 255, 227.
Takeuti, M., 1980. Sci. Rep. Tohoku Univ. Ser. I, 62, 115 (Science Reports Sendai No. 213).
Takeuti, M., 1983. Observatory, 103, 292.
Takeuti, M. & Aikawa, T., 1981. Sci. Rep. Tohoku Univ., Ser. 8, 2, 106 (Science Reports Sendai No. 239).
Uji-Iye, K., 1980. Sci. Rep. Tohoku Univ., Ser. 8, 1, 155 (Science Reports Sendai No. 222).

CEPHEID TEMPERATURES DERIVED FROM ENERGY DISTRIBUTIONS

T. J. Teays
Department of Physics and Astronomy, University of Nebraska
Lincoln, Nebraska 68588-0111 USA

E. G. Schmidt
Department of Physics and Astronomy, University of Nebraska
Lincoln, Nebraska 68588-0111 USA

A number of previous studies of the relation between observed colors and temperatures of Cepheids have been done (Kraft 1961; Johnson 1966; Parsons 1971; Bohm-Vitense 1972; Schmidt 1972; Pel 1978). It was the discrepancies between these various temperature scales, especially at the cooler end, that led us to undertake the present recalibration. We felt some improvement on the previous work would result from our access to better scan data, reddening information, and model atmospheres. The results presented here are preliminary, as they represent only a sample of the data we have obtained.

The energy distributions were obtained from spectrum scans, using the Intensified Reticon Scanner (IRS) of Kitt Peak National Observatory. The high sensitivity of the IRS, as well as the fact that all wavelengths are measured simultaneously, generates very high quality data. Scans were made between approximately 3600 and 8000Å. The scans were reduced with standard Kitt Peak IRS reduction software.

The stars chosen for this study are Cepheids located in galactic clusters which have been well observed, and so their other properties are relatively well known. In this discussion we will present some results, at a few phases, for CF Cas, which is a member of NGC 7790. Schmidt (1981) has obtained four-color photometry of the early type stars in NGC 7790 and determined the color excess for CF Cas. To correct the scans this color excess was used in conjunction with the reddening curve for Cassiopeia given by Nandy (1968).

In this report we have compared the scans to the model atmospheres of Kurucz (1970). The color index of the star, at the corresponding phase, was determined from a B-V color curve of the star and compared to the temperature derived from the scan. The results of a few of these comparisons are shown in Figure 1, along with the results of earlier authors.

These data were obtained while the authors were guest observers at Kitt Peak National Observatory, National Optical Astronomy Observatories, which is operated by the Association of Universities for Research in Astronomy, Inc., under contract with the National Science Foundation.

References

Bohm-Vitense, E. (1972). Astr. Ap., 17, 335
Johnson, H. L. (1966). Ann. Rev. Astr. Ap., 4, 193.
Kraft, R. P. (1961). Ap. J., 134, 616.
Kurucz, R. L. (1979). Ap. J. Suppl., 40, 1.
Nandy, K. (1968). Publ. Roy. Obs. Edinburgh, 7, 177.
Parsons, S. B. (1971). M. N. R. A. S., 152, 121.
Pel, J. W. (1978). Astr. Ap., 62, 75.
Schmidt, E. G. (1972). Ap. J., 174, 605.
Schmidt, E. G. (1981). A. J., 86, 242.

Fig. 1 - Log of effective temperature vs. B-V. The temperature scales of various authors are indicated by: a solid line, Kraft (1961) and Parsons (1971); a broken line, Pel (1978); triangles, Johnson (1966); crosses, Bohm-Vitense (1972); dots, Schmidt (1972). The boxes represent the new data for CF Cas.

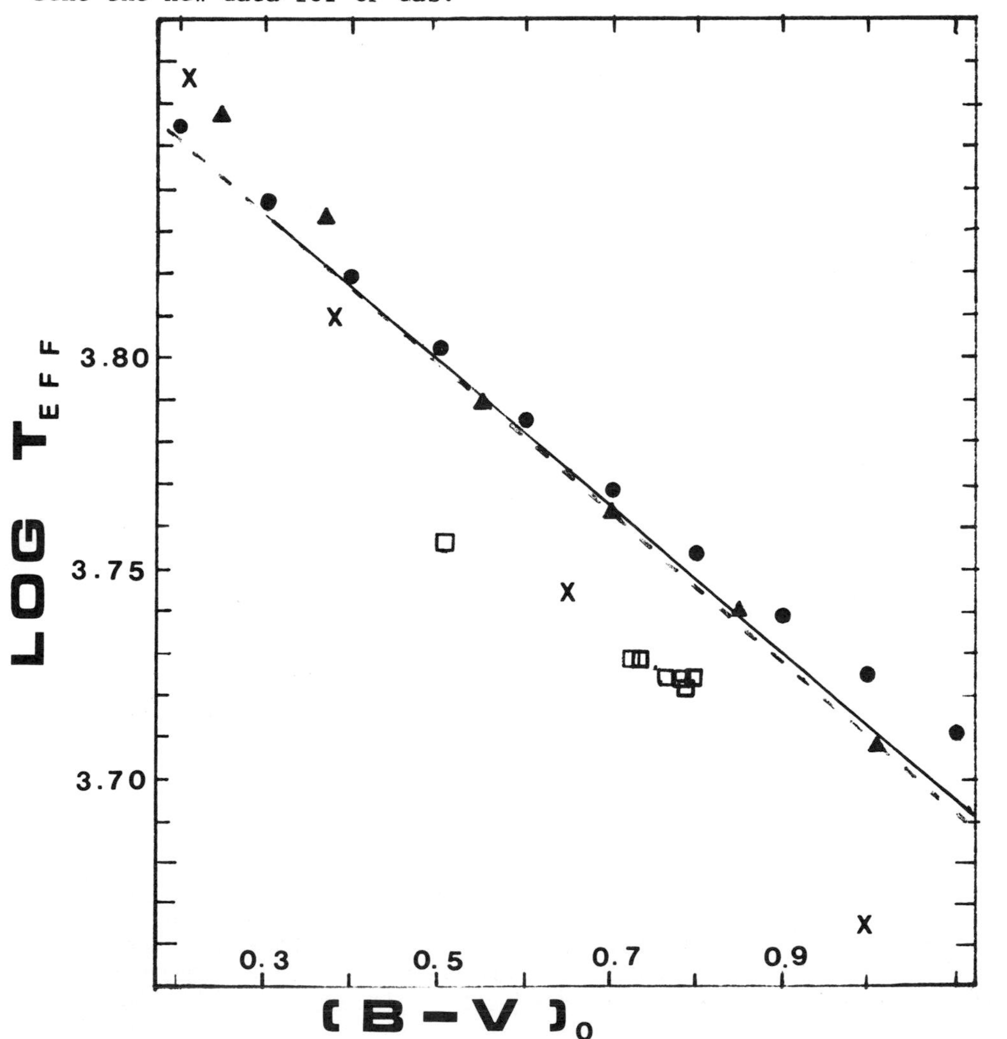

RADIAL VELOCITIES OF CLASSICAL CEPHEIDS

T.G. Barnes III
The University of Texas at Austin, Austin, Texas 78712 USA

T.J. Moffett
Purdue University, West Lafayette, Indiana 47907 USA

Approximately 2000 radial velocities of 88 classical Cepheids have been observed using a photoelectric radial velocity meter. During the same time interval, these same Cepheids were intensively observed in the BVRI bandpasses, as reported elsewhere in these proceedings. This provides a homogeneous set of phase-locked radial velocity and photometric data which are useful in several contexts. We present here a sample of these results which will be published in their entirety elsewhere.

Introduction

One of the most important uses of Cepheid radial velocity data is in application to the determination of linear radii and distances. These studies require excellent phase-locking between the radial velocity data and the photometric data (Fernie & Hube 1967). We have attempted to achieve this for 88 of the Cepheids contained in the recent BVRI photometric study of Moffett & Barnes (1984).

Observations

From the 112 variables in the Moffett & Barnes (1984) study, we selected 88 which were brighter than 10.7 mag. at minimum for the radial velocity study. The number of observations for each was determined by the quality of existing radial velocity data. In many cases only a few new velocities were needed to bring an excellent, but old, velocity curve into phase with our photometry.

The observations were acquired with a photoelectric radial velocity meter on the coudé spectrograph of the 2.1m telescope at McDonald Observatory (Slovak *et al.* 1979). All observations were made during the same six year period over which the photometric data were taken.

Results

The velocities for the brightest 22 Cepheids in our program have now been merged with existing radial velocity data. The older data were shifted in radial velocity and in phase until they matched our observations. A mean curve was then drawn by eye through the combined data. Figure 1 shows a sample result of this process. (Following the citation to the reference, we show the corrections in radial velocity and in phase that were applied.)

The uncertainties in our results can be determined from the scatter about the mean curve. For SZ Tau this is $\pm$ 2.6 km S^{-1}, which is typical.

Figure 1. Radial Velocities for SZ Tau

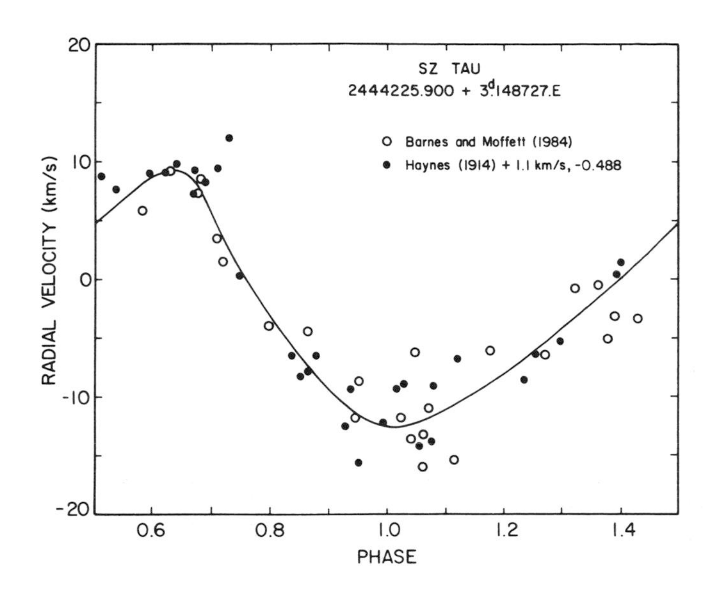

References

Barnes, T.G. & Moffett, T.J. (1984). In preparation.
Fernie, J.D. & Hube, J.O. (1967). Publ. Astr. Soc. Pacific, 79, 95-101.
Haynes, E.S. (1914). Lick Obs. Bull., 8, 85-86.
Moffett, T.J. & Barnes, T.G. (1984). Ap. J. Supp., 55, July 1.
Slovak, M.H., Van Citters, G.W. & Barnes, T.G. (1979). Publ. Astr. Soc. Pacific, 91, 840-847.

RADIUS DETERMINATION FOR NINE SHORT PERIOD CEPHEIDS

G. Burki
Geneva Observatory
CH-1290 Sauverny
Switzerland

From November, 1981 to March, 1982, nine Pop I and Pop II cepheids of periods between 1.5 and 4.2 days have been monitored, in Geneva photometry from La Silla Observatory (Chile) and in radial velocities from the Haute-Provence Observatory (France). These cepheids are listed in Table 1. Figure 1 shows an example of the light, colour and velocity curves for EU Tau, a small amplitude cepheid with a period of 2.10 days.

Table 1 Period in days, mean radius in $R_\odot$ and cepheid type according to Szabados (1977) for the nine cepheids studied.

	P	R⊙	Rem.	Type
SW Tau	1.583	10.5		II
EU Tau	2.103	46		Is
BB Gem	2.307	26:		II
BE Mon	2.705	30		I
V465 Mon	2.7132	((42))	Bin	Is
DX Gem	3.135	(40:)	Bin?	Is
SZ Tau	3.150	36		Is
ST Tau	4.0347	39		I
V508 Mon	4.134	42		Is

The mean radii have been determined with the Wesselink's method described by Burki and Benz (1982). These mean radii are given in Table 1. The radius variation of EU Tau is reproduced in Figure 1d: the solid line is taken from the integration of the velocity curve and the points are calculated from the magnitude and colour values. The best value of the mean radius is chosen by minimizing the dispersion of the points around the curve.

The radial velocity analysis reveals that V465 Mon and possibly also DX Gem are spectroscopic binaries. Thus, the Wesselink radius of these two stars is affected by the presence of a companion (Burki, 1984).

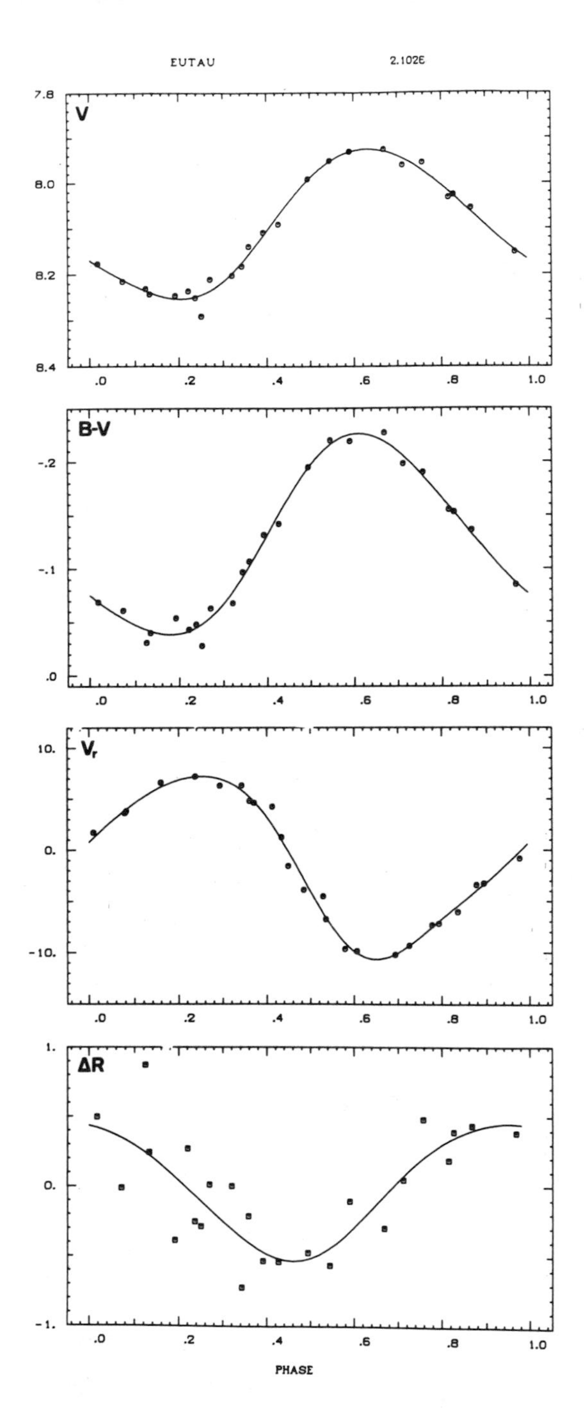

Figure 1: Variation in magnitude V, Geneva index [B-V] , radial velocity V_r in km/sec and radius in solar unit in the case of EU Tau.

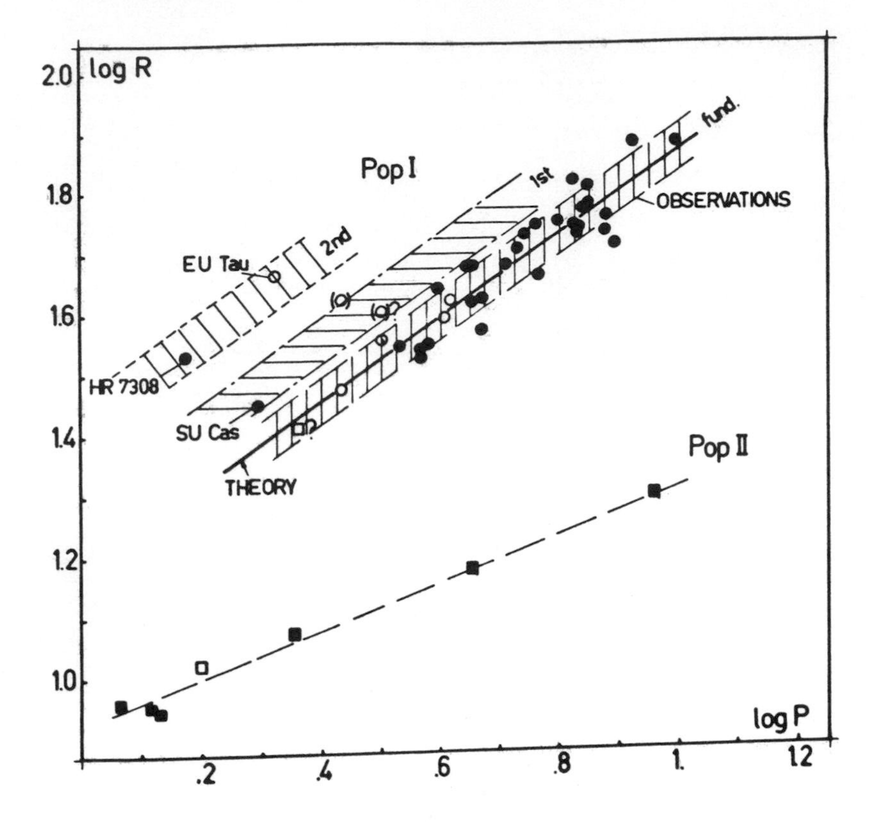

Figure 2: The period-radius relation for Pop I and Pop II cepheids with period shorter than 10 d. The radii stem from Fernie (1984), Cox (1979), Gieren (1982), Imbert (1981, 1983, 1984), Burki and Benz (1982), Burki et al. (1982), Wolley and Carter (1973). The cepheids analysed in this paper are noted by open symbols. Binary cepheids have been excluded, except for the stars analysed in this paper, V465 Mon and DX Gem, marked within parentheses.

Figure 2 shows the period-radius relations for the cepheids with P < 10d. One can see that:

1) Four cepheids, BE Mon, SZ Tau, ST Tau and V508 Mon are well placed in the relation for Pop I cepheids.
2) The two binaries V465 Mon and DX Gem are clearly above this relation.
3) SW Tau is in agreement with the relation for Pop II cepheids.
4) BB Gem, classified Pop II by Szabados (1977) agrees with the relation for Pop I cepheids. For this cepheid, new measurements are required because the radius determination is very uncertain.

The linear regression gives for Pop I cepheids in the fundamental mode:

$$\log R = 0.701 \log P + 1.175$$

with a residual dispersion of 0.039 in log R. This relation is indicated by vertical hachures in Figure 2. We would like to point out the perfect agreement between this observational relation and the theoretical relations of Cogan (1978):

$$\log R = 0.70 \log P + 1.17$$

and Fernie (1984):

$$\log R = 0.694 \log P + 1.177$$

The observational relations for the first and second overtones have been estimated by displacing the relation for the fundamental mode by respectively 0.14 and 0.35 in log P. We see that:

1) SU Cas seems to be indeed a 1st overtone pulsator, as first claimed by Gieren (1976).

2) The 2nd overtone pulsation of HR 7308 (V473 Lyr) is confirmed (see Burki et al., 1982). Recall that this star is the classical cepheid with the shortest period known (1.5 d) and that its amplitude varies by a factor of 15 in a period of 1200 d.

3) EU Tau is another probable cepheid pulsating in the 2nd overtone.

References

Burki, G.: 1984, Astron. Astrophys. 133, 185
Burki, G., Benz, W.: 1982, Astron. Astrophys. 115, 30
Burki, G., Mayor, M., Benz, W.: 1982, Astron. Astrophys. 109, 258
Cogan, B.C.: 1978, Astrophys. J. 221, 635
Cox, A.N.: 1979, Astrophys. J. 229, 212
Fernie, J.D.: 1984, preprint
Gieren, W.: 1976, Astron. Astrophys. 47, 211
Gieren, W.: 1982, Astrophys. J. 260, 208
Imbert, M.: 1981, Astron. Astrophys. suppl. 44, 319
Imbert, M.: 1983, Astron. Astrophys. suppl. 53, 85
Imbert, M.: 1984, preprint
Szabados, L.: 1977, Mitt. Sternw. Ungar. Akad. Wissen 70
Wolley, R., Carter, B.: 1973, Monthly Not. R. Astron. Soc. 162, 379

SURFACE BRIGHTNESS RADII AND DISTANCES OF CEPHEIDS AND THE PERIOD-RADIUS RELATIONSHIP

W.P. Gieren
Observatorio Astronómico, Universidad Nacional, and Physics Department, Universidad de los Andes, Bogotá, Colombia

INTRODUCTION

Recently Gieren (1984) has derived accurate radii and distances of a sample of short-period classical Cepheids using the surface brightness (SB) method introduced by Barnes & Evans (1976). The results indicated that the period-radius (P-R) relationship obtained from SB radii might possess a slope close to the value of 0.82 defined by the Cepheids in clusters and associations (Fernie, 1983) and in conflict with the values obtained from Baade-Wesselink radii and from theoretical models (see Fernie, 1983). Since this finding would lend considerable support to the presently accepted absolute magnitudes of the cluster Cepheids, it was decided to obtain SB radii and distances of well-observed long-period Cepheids in order to strengthen the P-R relationship obtained from the SB technique.

NEW SB RADII AND DISTANCES OF LONG-PERIOD CEPHEIDS

Coulson et al. (1984) have published extensive contemporaneous $UBV(RI)_c$ and photoelectric radial velocity data of the 6 Cepheids given in Table 1 which have been used for the present analysis. These stars span a period range from 10 to 17 days. Truncated Fourier series of appropriate order were fitted to the velocity curves which were then integrated to obtain the displacement curves, using a projection factor of 1.31 (Parsons 1972).

Table 1 Radii and Distances of Long-Period Cepheids

Cepheid	P(days)	$R(R_\odot)$	d(pc)	$\langle V-R\rangle_{OJ}$	s
AQ Car	9.77	64.1±4.0	2971±186	0.589	-0.378±0.008
XX Cen	10.95	55.4±3.9	1388± 96	0.589	-0.383±0.012
XY Car	12.44	80.0±5.9	2830±211	0.643	-0.387±0.012
TT Aql	13.75	91.3±5.2	1101± 63	0.685	-0.391±0.012
XX Car	15.71	92.0±3.4	3868±145	0.616	-0.371±0.012
XZ Car	16.65	108.7±6.3	2714±156	0.683	-0.406±0.015

The Johnson (V-R) color curves of the stars were obtained using the transformation equation

$$(V-R)_J = 0.587\ (V-R)_C + 0.413\ (V-I)_C + 0.03 \quad (1)$$

given by Cousins (1981) whose validity has been checked by Gieren (1984). The reddening-free $(V-R)_o$ colors of the Cepheids on the Johnson system were obtained adopting $E(V-R) = 0.9\ E(B-V)$ and using the E(B-V) values derived by Coulson et al. (1984). The unreddened apparent V magnitudes were calculated with the A_V values listed by the same authors. Surface brightness radii and distances of the stars were then calculated using the set of equations

$$F_V = 4.2207 - 0.1\ V_o - 0.5 \log \phi \quad (2)$$

$$F_V = 3.956 - 0.363\ (V-R)_o \quad (3)$$

$$D_o + \Delta D = 10^{-3}\ r\ \phi \quad (4)$$

The symbols have their usual meaning (see Gieren 1984). Equation (3) is the most recent calibration of the surface brightness parameter in terms of $(V-R)_o$ given by Barnes (1980).

Table 1 contains the radii and distances and their standard deviations which were obtained from the new data.

THE PERIOD-RADIUS RELATIONSHIP

The new radii of Table 1 for the long-period Cepheids were combined with the SB radii of 13 stars given by Gieren (1984) to construct a period-radius relationship. The radius value of V496 Aql (Gieren, 1984) was omitted for reasons discussed in that paper. At the short-period end, the star EU Tau ($P = 2^d.1$) was added whose SB radius is 18.4 $R_\odot$ (Gieren; in preparation). This leaves a total of 20 Cepheids in the period range of 2 to 17 days. A least squares fit to the data yields the period-radius relationship

$$\log R = \underset{\pm .047}{0.786} \log P + \underset{\pm .039}{1.040} \quad \text{(s.d.)} \quad (5)$$

which is shown in Figure 1 (solid line). This relationship is not compatible with the ones obtained from Baade-Wesselink radii, theoretical models or mixed-mode pulsation, but agrees within the errors of the coefficients with the P-R relation obtained from the cluster and association Cepheids which is, according to Fernie (1983)

$$\log R = \underset{\pm .020}{0.824} \log P + \underset{\pm .020}{1.042} \quad (6)$$

Turning to the distances of the present long-period Cepheids, it is found that they are almost exactly on the scale of the distances given in the catalog of Fernie & Hube (1968); the mean distance ratio is 1.01±0.04 (s.d.) . Adding the other 13 Cepheids discussed by Gieren (1984) (omitting again V496 Aql), the mean distance ratio (in the sense d(SB)/d(FH)) is 0.983±0.023 (s.d.). In Figure 2, the distance ratio of the Cepheids is plotted against the pulsation period, and

Figure 1. P-R relationship obtained from surface brightness radii (solid line). Broken line is the relationship obtained from cluster Cepheids.

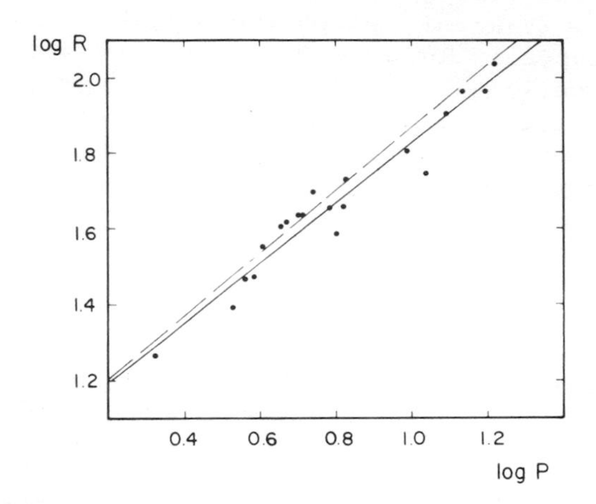

Figure 2. Distance ratio against period, for 19 Cepheids with accurate surface brightness distances.

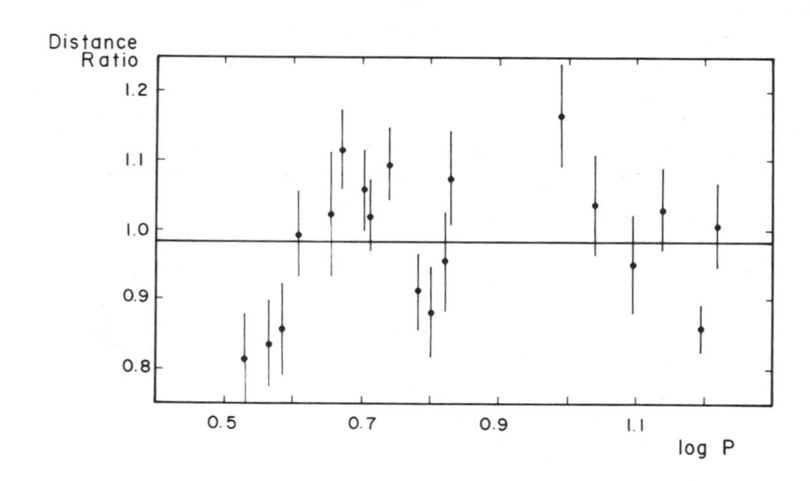

clearly no period dependence is visible in these data. This means that for pulsation periods up to 17 days, the distances of Cepheids derived with the SB technique, and therefore their absolute magnitudes, are in almost perfect agreement with the scale of the period-luminosity relation for classical Cepheids. This finding increases confidence, on the other hand, in the correctness of the radii derived from the same method.

THE SLOPE OF THE SURFACE BRIGHTNESS - $(V-R)_J$ RELATION

Barnes (1980) has suggested that the slope of -0.363 of the F_V - $(V-R)_o$ relation (3) may not be applicable when dealing with the redder, longer-period Cepheids whose intrinsic $(V-R)_J$ colors exceed ~ 0.6. Since several of the present stars have intrinsic $(V-R)_J$ colors close to 0.7 (see Table 1), it was desirable to derive their slopes in order to check if significant deviations from the value used in (3) do occur. For this purpose, the method devised by Thompson (1975) was used which permits to calculate the variations of the visual surface brightness S_V during the pulsation cycle from the known displacements r and a known mean radius R according to

$$S_V = V + 5 \log (1 + r/R) + const \quad (7)$$

For each of the present long-period Cepheids, S_V was determined in this way as a function of phase and a least squares fit yielded the slope of the S_V - $(V-R)_J$ relation. An excellent linear relationship between S_V and $(V-R)_J$ was found to hold for each of the stars. From this, the slope s of the F_V- $(V-R)_J$ is obtained by multiplication with -1/10. The values obtained for s and their standard deviations are given in Table 1. The values range from -0.371 (XX Car) to -0.406 (XZ Car) and the average slope defined by the present 6 Cepheids is -0.386±0.005. This value may be compatible with the value of -0.363±0.011 quoted by Barnes (1980) for the shorter period Cepheids, but the present data suggest that for the redder Cepheids the slope of the F_V - $(V-R)_J$ relation might become increasingly more negative. However, more long-period Cepheids need to be studied to confirm this trend.

CONCLUSIONS

The present period-radius relation from SB radii of classical Cepheids agrees within the errors with the one derived from 27 cluster Cepheids (Fernie 1983), but its slightly smaller slope could mean an alleviation of the discrepancy with theoretical radii. The present results imply that the radii derived from the cluster Cepheids, as well as from the SB technique are basically correct, implying that Baade-Wesselink radii are definitively too small for periods larger than 10^d. This is supported by the finding that SB distances, independently of period, agree very well with the scale of the P-L-C relation.

The present work confirms that the SB method is able to yield standard deviations of radii and distances in the order of 5%. Figure 3 shows that the σ decrease slightly with increasing amplitude of the Cepheid's (V-R) color curve. This means that for the large-amplitude, long-period, distant Cepheids the most accurate distances can be obtained, a welcome result for galactic structure investigations.

Figure 3. Standard deviation of surface brightness radii against the (V-R) amplitude of Cepheids.

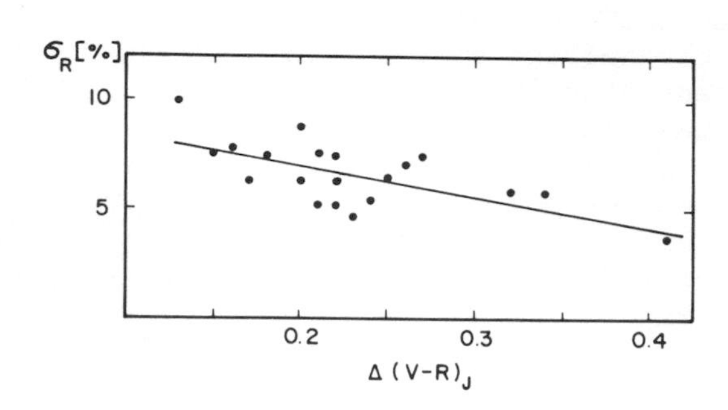

ACKNOWLEDGMENTS

My thanks are due to I.M. Coulson and J.A.R. Caldwell who obtained a large fraction of the observations used in this paper, to the directors of the South African Astronomical Observatory and of the European Southern Observatory for observing time at these institutions, and to Colciencias for research grants Nos. 20004-1-35-81 and 20004-1-39-82.

REFERENCES

Barnes, T.G. (1980). Highlights Astr., 5, 479.
Barnes, T.G., and Evans, D.S. (1976). M.N.R.A.S., 174, 489.
Coulson, I.M., Caldwell, J.A.R., and Gieren, W.P. (1984). Ap.J. Suppl., submitted.
Cousins, A.W.J. (1981). M.N.A.S. South Africa, 40, 37.
Fernie, J.D. (1983). Preprint.
Fernie, J.D., and Hube, J.O. (1968). A.J., 73, 492.
Gieren, W.P. (1984). Ap.J., July 15.
Parsons, S.B. (1972). Ap. J., 174, 57.
Thompson, R.J. (1975). M.N.R.A.S., 172, 455.

A GENERALIZATION OF THE CORS METHOD TO DETERMINE CEPHEID RADII: THEORY AND APPLICATION

B. Caccin, B. Buonaura, A. Onnembo
Istituto di Fisica Sperimentale dell'Universita'
Pad. 20 Mostra d'Oltremare, 80125 Napoli, Italy

G. Russo
Osservatorio Astronomico di Capodimonte
via Moiariello 16, 80131 Napoli, Italy

A.M. Sambuco
Consiglio Nazionale delle Ricerche, Roma, Italy

C. Sollazzo
Physics Department, University of Victoria, Canada
and
Osservatorio Astrofisico, Catania, Italy

Abstract The CORS method for the empirical determination of the radii of pulsating variables (Caccin et al., 1981; Sollazzo et al., 1981) is discussed in the framework of the quasistatic approximation to the variations of the atmospheric parameters (Unno, 1965) and reformulated in a way that does not make direct use of theoretical calibrations of the photometric system in terms of model atmospheres. The radii calculated with this approach are in good agreement with those previously obtained by means of Pel's calibrations of the VBLUW system (Sollazzo et al., 1981) which lead to a period-radius relation coinciding, within the errors, with the theoretical one (Cogan, 1978) and, consequently, mass determinations consistent with pulsational masses (Cox, 1979). The method can be immediately applied to any other multicolour system, and very promising preliminary results are obtained by using recent UBV data by Gieren (1982).

INTRODUCTION

The introduction of a two-colour dependence for the physically significant atmospheric parameters in multicolour photometry of pulsating variables, has its observational basis in the existence of closed loops in the colour-colour planes, the area of which shows a definite trend with the pulsation period. In two previous papers (Caccin et al., 1981; Sollazzo et al., 1981, hereafter paper I and II, respectively) we have shown that in the case of the surface brightness, this two-colour dependence allows a more precise determination of cepheid radii by means of the CORS method. We will show here that this assumption has its theoretical basis in the quasistatic approximation (QSA) by Unno (1965), and that colour-colour loops may be used to evaluate the area of the loop surface brightness vs. colour (necessary in the CORS method) without resorting to theoretical calibrations.

THE CORS METHOD

For the sake of clarity, we will recall in the following the

CORS method described in paper I and II. The definition of the surface brightness, s_i, is:

$$m_i - s_i - 2\underline{a} \log_{10} R = \text{const.} \qquad (1)$$

where m_i is the apparent magnitude in the bank i, a is a constant (equal to -2.5 or 1.0 according to the scale of the photometric system used) and R is the radius. Manipulating this relation, we finally obtain the following equation, hereafter defined as the CORS equation:

$$B_{i,jk} - \Delta B_{i,jk} + [R]_{jk} = 0 \qquad (2)$$

were

$$B_{i,jk} = \oint m_i \, d\, c_{jk} \qquad (3)$$

$$\Delta B_{i,jk} = \oint s_i \, d\, c_{jk} \qquad (4)$$

$$[R]_{jk} = 2\, \underline{a} \oint \log(R_\circ - K\, P \int_{\phi_\circ}^{\phi} V_r(x)\, dx\,)\, d\, c_{jk} \qquad (5)$$

where o is the phase, $c_{jk} = (m_j - m_k)$ is the colour index, K is the ratio of pulsational to radial velocity, P is the period and $R_\circ$ is the radius at an arbitrarily chosen phase $\phi_\circ$. Eqn. (2) can be solved to obtain the radius, $R_\circ$, if one can evaluate the term B, the area of the loop surface brightness vs. colour. In paper II, this has been accomplished, for a set of data in the Walraven system, by using the theoretical calibrations by Pel (1978) of T_{eff} and g_{eff} in terms of $(V-B)_\circ$.

We show in the next section that, with the assumption of QSA, the term ΔB is proportional to the area of the colour-colour loop, a fact which, with suitable hypotheses, may allow the computation of R in any photometric system.

QUASI STATIC APPROXIMATION

The groundwork of the QSA for the atmospheric layers of pulsating variables has been set up by Ledoux and Whitney (1961) and by Unno (1965). Within this approximation, the photosphere of these stars is described at any time by a classical hydrostatic, plane-parallel model in radiative equilibrium and LTE. Each such model is identified by only two parameters (within the assumption of constant chemical composition): i) the effective temperature, T_e corresponding to the instantaneous values of luminosities and radius, and ii) the effective gravity g_e defined by

$$g_{eff} = \frac{G\, M}{R^2} + \frac{d^2\, R}{d\, t^2} \qquad (6)$$

As a consequence, the whole sequence of states can be suitably represented by a point describing a closed loop in a two dimensional space. This loop is described by two equations in functionaly dependant on the parameter ϕ:

$$T_{eff} = T_{eff}\ (\phi) \qquad (7)$$

$$g_{eff} = g_{eff}\ (\phi) \qquad (8)$$

with the condition of periodicity, $T_{eff}\ (0) = T_{eff}(1)$, and $g_{eff}(0) = g_{eff}(1)$.

Furthermore, since the emergent flux is uniquely determined by T_{eff} and g_{eff}, then any photometric quantity derived from the spectra can be expressed in terms of these two same parameters, in particular:

$$s_i = s_i(T_{eff}, g_{eff}) \qquad (9)$$

$$c_{jk} = c_{jk}(T_{eff}, g_{eff}) \qquad (10)$$

$$c_{hl} = c_{hl}(T_{eff}, g_{eff}) \qquad (11)$$

Supposing the invertibility of these equations, we may write:

$$S_i = s_i\ (c_{jk}, c_{hl}) \qquad (12)$$

and therefore the area of the loop surface brightness vs. colour (see Onnembo ét al., 1984) is:

$$\Delta\ B_{i,jk} = \alpha\ .\ \oint c_{hl}\ \ d\ c_{jk} \qquad (13)$$

being $\alpha = (\ s_i/\ \ c_{hl})_\circ$.

The hypothesis of invertibility of Eqns. (9), (10), (11) is satisfied if the two colour indices are independent quantities. This is true, e.g., for the pairs (V-B,B-U) in the Walraven system,(B-V,U-B) in Johnson photometry and $(b-y, c_1)$ in Strömgren's system.

APPLICABILITY AND RESULTS

The important of this method arises from the fact that it is essentially the first successful attempt to keep full account of the well established theoretical and observational result that the surface brightness if a function of two colours. The effect of not considering this point is an underestimation of the radius, which is most sensitive at longer periods. Most of the Baade-Wesselink type methods suffer from this shortcoming. The Δ B term in Eqn. (2) expresses this improvement. It can be evaluated, at the moment, in either of the following two ways:

i) using theoretical calibrations to compute s_i from (9) and ΔB from (4)- as we did in paper II. This is the most direct and safest method. Such calibrations, which express T_{eff} and g_{eff} as a function of two suitable colours, are now available for most photometric systems. Dereddened colours are needed.

ii) computing the area of the colour-colour loop, and then estimating the proportionality constant as described in

Onnembo et al., (1984) (paper III). Dereddened colours are not crucial to this procedure.

It is worth noting that, as a first approximation, even if ΔB is set to be zero, the radius is intrinsically better determined because of the global treatment of the data.

In the following table we present some preliminary results using procedure ii).

Star		Period	R_{BW}	R_{CORS}	Phot. System	Reference
V419	Cen	$5^{d}.51$	39.2	45.3	Walraven	Pel, 1976
β	Dor	9.84	64.6	76.4	Walraven	Pel, 1976
AD	Pup	13.59	84.3	91.6	Walraven	Pel, 1976
SV	Mon	15.23	84.1	98.6	Walraven	Pel, 1976
SZ	Aql	17.14	98.8	117.0	Walraven	Pel, 1976
RU	Sct	19.70	94.7	109.7	Walraven	Pel, 1976
U	Car	38.77	137.1	163.7	Walraven	Pel, 1976
S	TrA	6.32	44.5	55.5	Johnson	Gieren, 1982
U	Sgr	6.74	60.1	59.2	Johnson	Gieren, 1982
V496	Aql	6.81	48.0	65.2	Johnson	Gieren, 1982

APPENDIX: PRACTICAL SUGGESTIONS

Feeling the need for a clearer exposition of the practical application of the method presented here, we will outline below a step-by-step procedure for the actual computation of radii.

For a given star, a radial velocity curve, V_r, (baricentric velocity subtracted) and photometry in three passbands, i, j, k, (e.g. V, B, U) are needed. The first step is to perform (possibly interactively) some kind of fitting (Fourier, spline fit etc.) of the observed data, to obtain $V(\phi)$, $(B-V)(\phi)$, $(U-B)(\phi)$, $V_r(\phi)$, as smooth equispaced curves. These smoothed quantities are used to generate, with some reliable numerical algorithm, the integral function of the radial velocity and the derivative functions of magnitude and colours with respect to the phase. Then, the term B defined by (3) can be calculated in a trivial way; for (5), the reference phase can be profitably chosen to be the minimum of V_r. As far as ΔB (in Eqn. (4)) is concerned, it can be computed in either of the two ways discussed in the previous section. The left-hand side of Eqn. (2) is now defined as a function of R_o and can be solved to obtain the radius of the Cepheid at ϕ_o. Finally, the function $R(\phi)$ can be obtained from the integral of the radial velocity curve, and the mean radius can be unambiguously calculated.

The FORTRAN code for the complete application of the CORS method is available upon request from G.R. or C.S. (tape in VAX format).

References

Caccin B., Onnembo A., Russo G., Sollazzo C. (1981), Astron. Astrophys. 97, 104
Cogan B.C. (1978), Astrophys. J. 221, 635
Cox A.N. (1979), Astrophys. J. 229, 212
Gieren W. (1982). Astrophys. J. Suppl. 49, 1
Ledoux P., Whitney, C. (1961), IAU Symp. No. 12, ed R.N. Thomas (Balogna) P. 131
Onnembo A., Buonaura B., Caccin B., Russo G., Sambuco A.M., Sollazzo C. (1984), submitted
Pel J.W. (1976), Astron. Astrophys. Suppl. 24, 413
Pel J.W. (1978), Astron. Astrophys. 62, 75
Sollazzo C., Russo G., Onnembo A., Caccin B. (1981), Astron. Astrophys. 99, 66
Unno W. (1965), Pub. Astron. Soc. Japan 17, 205

BAADE-WESSELINK RADII OF LONG PERIOD CEPHEIDS NEW OBSERVATIONAL RESULTS

Iain M. Coulson, John A. R. Caldwell
South African Astronomical Observatory
P. O. Box 9 Observatory 7935

Wolfgang Gieren
Observatorio Astronomico Universidad Nacional
and Physics Department, Universidad de Los Andes,
Begota, Colombia

Abstract The radii of Galactic Cepheids as determined from a version of the Baade-Wesselink technique are shown to depend upon the colour index used to define the temperature scale. The (V-I) radii are systematically larger than the commonly used (B-V) radii and are probably better estimators of the true radii. There still remains a problem, however, in reconciling the period-radius relation with the period-luminosity-colour relation.

1. Observations

A programme of observations of galactic Cepheids with P>10 days has recently been completed at SAAO and results will appear in the literature in due course (Coulson & Caldwell. in preparation). Preliminary results for 9 of these stars are discussed here together with those for 6 other stars, with 9<P<17 days (Coulson, Caldwell & Gieren 1984, in press). These are best compared with similar data for shorter period Cepheids obtained by Gieren (1982, et op cit) and reanalysed here.

Broadband UBVRI photometric measures have been obtained at Sutherland (SAAO) and La Silla (ESO). The measures are on the Cousins' standard system (Menzies, Banfield & Laing 1980). They have been combined with older photometry from the literature to enable redeterminations of the periods and with the best of this older data to allow full specification of the photometric properties around the pulsation cycle.

The radial velocities were all obtained with the radial velocity spectrometer at the coudé focus of the SAAO 1.9m telescope at Sutherland. They are essentially measures of the FeI lines (Coulson 1983) and are on the Wilson (1953) standard system. Older radial velocites in the literature proved to be of considerably poorer quality and have not been combined with the new data.

Phase coverage for the stars discussed here is practically complete and since the photometry and radial velocities have been obtained contemporaneously they allow radius determinations free from the uncertainties of phase mismatching.

2. Radii

Radii were obtained using Balona's (1977) maximum-likelihood version of the Baade-Wesselink method. We have modified his original formulation to account for large radial amplitudes ($\delta R/R > 0.1$) and have used not only V and B-V as magnitude and colour, but also B,R,I and V-R, V-I and R-I. The derived radii are found to be essentially independent of the choice of magnitude, but show systematic differences with colour. Fig. 1 shows the radius difference

$$\Delta = \log R_{V-I} - \log R_{B-V}$$

plotted against log P. Clearly Δ is significantly larger than zero and is probably period dependent.

$$\Delta = (0.023 \pm 0.016) + (0.029 \pm 0.016) \log P \qquad (1)$$

(se/pt = 0.023).

If the (V-I) results are less affected by gravity than the (B-V) results then they may be considered better estimates of the sizes of the stars and we may correct (B-V) radii by equation (1). This we have done for the 29 radii derived above plus those for the 32 stars studied by Balona (1977) for which there are reddening estimates (Dean, Warren & Cousins 1978). These radii yield a mean relation:

$$\log R = (0.637 \pm 0.021) \log P + (1.215 \pm 0.028) \qquad (2)$$

(s.e./pt = 0.049).

3. The PLC Zero Point

With these corrected mean radii we may also determine the mean absolute magnitude

$$\langle M_V \rangle = A \langle B-V \rangle_o - 5 \log R + C \qquad (3)$$

(see Balona 1977, equation 1.3).

In this equation C is a term usually given as

$$C = M_{bol\odot} + 10 \log T_{eff\odot} - 10a_o - b_o \qquad (4)$$

were a_o, b_o are assumed constants of the relations

$$\log T_{eff} = a_o + a_1 \langle B-V \rangle_o \qquad (5)$$

$$M_{bol} - M_V = b_o + b_1 \langle B-V \rangle_o \qquad (6)$$

From M_V we may then derive estimates of the zero-point of the P-L-C relation, γ :

$$\langle M_V \rangle = \alpha \log P + \beta \langle B-V \rangle_o + \gamma \qquad (7)$$

Equations (3) & (7) yield C-γ in the terms of several observables:

$$C-\gamma = \alpha \log P + (\beta - A) \langle B-V \rangle_o + 5 \log R \qquad (8)$$

From the 61 corrected (B-V) - radii derived above we have calculated C-γ using nominal values of α =-3.80, β = 2.70, A =2.15 and plot the results against log P in Figure 2. Cleary C-γ is not a constant for all Cepheids. This implies that at least one of the above assumptions is incorrect. It seems most likely that the temperature and/or bolometric correction calibrations may depend upon surface gravity and hence give rise to a period dependence of the type seen here.

Until this dependence is fully quantified the determination of γ from Cepheid radii will remain impracticable.

References

Balona, L.A., 1977. M.N.R.A.S., 178, 231.
Coulson, I.M., 1983. M.N.R.A.S., 200, 925.
Dean, J.F., Warren, P.R. & Cousins, A.W.J., 1978. M.N.R.A.S., 183, 569.
Gieren, W.P., 1982. Ap.J., 260, 208.
Menzies, J.W., Banfield, R.M. & Laing, J.D., 1980. SAAO Circ. No. 5, p 149.
Wilson, R.E., 1953. General Catalogue of Stellar Radial Velocities, Carnegie Inst. of Washington, D.C.

Figure 1.
Radius Differences
For 29 Galactic Cepheids

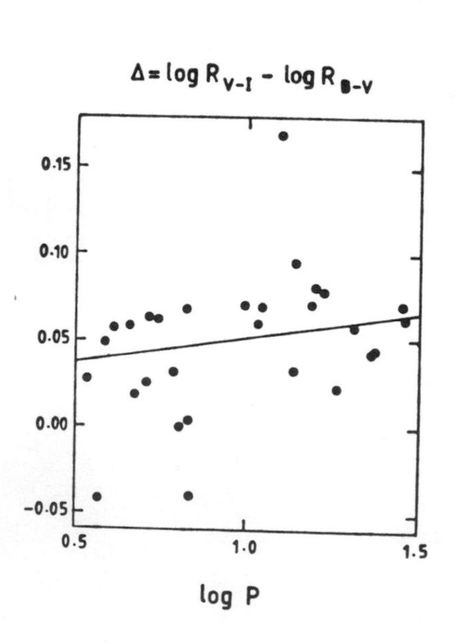

Figure 2.
C-γ for 61 Galactic Cepheids using corrected (B-V)-Radii
$C-\gamma=-3.8 \log P+0.55 \langle B-V \rangle_o + 5 \log R$.

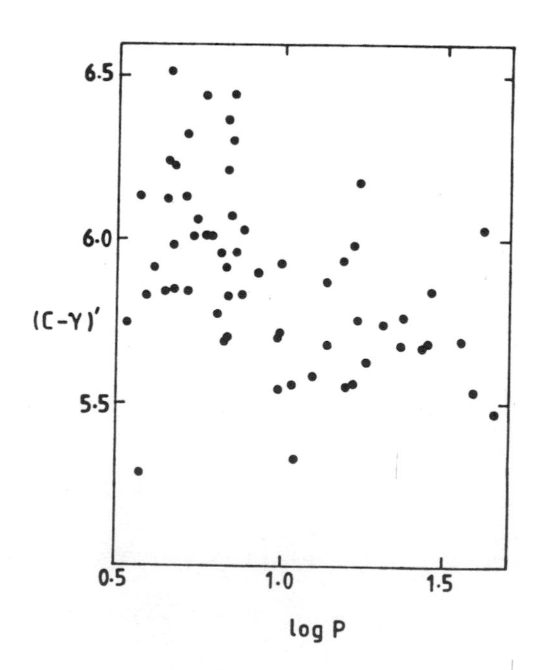

CEPHEID RADII FROM INFRARED PHOTOMETRY

D.L. Welch
Dept. of Astronomy, University of Toronto

Nancy Remage Evans
Dept. of Astronomy, University of Toronto

G. Drukier
Dept. of Astronomy, University of Toronto

Abstract We have determined radii for Cepheids using near-infrared K(2.2 μm) photometry and Balona's approach to the Baade-Wesselink method of radius determination. The lower sensitivity of the K flux to temperature variations results in improved sensitivity to radius changes. We present the results of maximum-likelihood radius solutions for Cepheids with modern radial velocity curves. We point out limitations of the Balona technique as it is currently applied in the optical and suggest improvements.

Introduction

The goal of Baade-Wesselink techniques of radius determination is to remove the effects of temperature from the total flux. The remaining flux variation can then be interpreted in terms of radius changes. It makes sense to choose a photometric bandpass where temperature variations affect the flux as little as possible. The flux at V is strongly influenced by temperature as witnessed by the strong resemblance of the V and B-V curves for Cepheids. We suggest that near-infrared bandpasses, such as K, are to be preferred for radius determination.

Procedure

We adopt here the technique of Balona (1977). This method results in a single equation of condition for light, color, and radius displacement. The equations are solved using the method of maximum likelihood as all quantities contain uncertainties. The Balona method assumes that both log T_e and the bolometric correction are linearly dependent on $(B-V)_0$. The method is discussed in more detail in a paper presently in preparation.

Long recognized as a major complication of radius determination is the problem of phase-mismatch between the light/color curves and the radial velocity curve. We used radial velocity curves for which the uncertainty in the phase match is 0.01P or less. Photometry is from Welch *et al.* (1984); radial velocity curves have been taken from a number of sources; color curves have been taken from Gieren (1981) and Moffett and Barnes (1980). We avoided Cepheids with bright blue companions. Radii determined from K photometry and optical (B-V) photometry are shown in Table 1, together with surface brightness coefficients, S_K. One sigma errors are given.

Table 1
K Surface Brightness Coefficients and Radii for Cepheids

	log P	S_K		$R/R_\odot$	
SS Sct	0.565	0.26	±0.06	37.3	±4.6
T Vul	0.647	0.42	0.03	39.1	1.6
BB Sgr	0.822	0.56	0.03	42.1	1.8
U Sgr	0.829	0.40	0.06	51.6	4.3
V496 Aql	0.833	0.73	0.08	41.5	3.7
S Sge	0.923	0.39	0.02	67.1	2.9
X Cyg	1.215	0.65	0.02	85.7	2.5
Y Oph	1.234	0.40	0.04	105.9	7.1

Discussion

A problem with the interpretation of the slope of the surface brightness-color relation (A in Balona's formulation) arises because of the wavelength dependence of reddening. If E(B-V) is taken to be constant, the ratio of total-to-selective absorption (R) is a function of $(B-V)_0$. Olson (1975) and Grieve (1983) find values for $dR/d(B-V)_0$ of 0.25 and 0.50, respectively. Therefore, Balona's A will be the sum of the true slope of the surface brightness-color relation and $[dR/d(B-V)_0]E(B-V)$. This means that the true slope A is typically smaller than Balona's average (2.15) by 0.05 to 0.15. This results in a small but systematic shift in the absolute magnitude calibration of Martin, Warren, and Feast (1979) in the sense that the derived magnitudes are brighter. It is possible to remove the wavelength dependence of the reddening by assuming a functional form for $R((B-V)_0)$, choosing E(B-V) and then iterating to $(B-V)_0$. Balona has previously interpreted high values of A as evidence for a companion. The above discussion suggests that reddening may often be the cause.

References

Balona, L.A. 1977, M.N.R.A.S., 178, 231.
Gieren, W. 1981, Ap. J. Suppl., 47, 315.
Grieve, G.R. 1983, Ph.D. Thesis, Univ. of Toronto.
Martin, W.L., Warren, P.R., and Feast, M.W. 1979, M.N.R.A.S., 188, 139.
Moffett, T.J. and Barnes, T.G. 1980, Ap. J. Suppl., 44, 427.
Olson, B.I. 1975, P.A.S.P., 87, 349.
Welch, D.L., Wieland, F., McAlary, C.W., McGonegal, R., Madore, B.F., McLaren, R.A., and Neugebauer, G. 1984 Ap. J. Suppl., 54, 547.

DISTANCES AND RADII OF CLASSICAL CEPHEIDS

T.G. Barnes III
The University of Texas at Austin, Austin, Texas 78712 USA

T.J. Moffett
Purdue University, West Lafayette, Indiana 47907 USA

We have used new BVRI photometry and radial velocities of a selection of bright classical Cepheids to determine their distances and radii through the surface brightness method. The improved photometry permitted, through the visual surface brightness relation, high-quality angular-diameter values for each Cepheid throughout its pulsation. The simultaneous radial velocities permitted the linear displacement curve to be phase-locked to the angular diameter variation. The results are individual distances and radii with considerably smaller uncertainty than could be obtained previously.

Introduction

Pulsating variables, especially Cepheids, play a major role in distance determinations, both within and external to our Galaxy. At present the Cepheid distance scale is based principally upon the cluster Cepheids, although statistical parallax analyses add a valuable check. The cluster distance scale is based, of necessity, on a small number of stars and the method relies heavily on the adopted ZAMS, which in turn depends strongly on the adopted Hyades distance.

Barnes *et al.* (1977) introduced an alternate method for determining *individual* distances to Cepheids which is based upon the visual surface *brightness*-color index relation. The distances so determined are completely independent of all other astrophysical distance scales. They also showed that the method is nearly geometric in the sense that even large errors in the interstellar extinction have a negligible effect upon the distances.

The results given in the above paper, which were based upon photometry and radial velocities from the literature, agreed with the Cepheid distance scale given by Fernie & Hube (1968), although the scatter was large. Contributing to the scatter were uncertainty in the Cepheid surface brightnesses, the uncertain phase-matching between the radial velocities and the photometry, and the errors of the photometry.

In the present work, we have addressed each of these sources of uncertainty. The method developed by Barnes (1980) is used to calibrate a visual surface brightness-color index relation from the Cepheids themselves. The BVRI photometry of Moffett & Barnes (1984) for 112 Cepheids provides a high-quality photometric data set. Radial velocities obtained simultaneously with the photometry enable secure phase-matching for 88 of these Cepheids. We summarize here our results for the first 20 Cepheids studied.

Cepheid Surfaces Brightnesses

We have followed exactly the method given by Barnes (1980) to establish a relation between the visual surface brightness parameter, F_V, and the Johnson (V-R) color index. Our data sets were the BVRI photometry of Moffett & Barnes (1984) and our unpublished radial velocities (Barnes & Moffett 1984).

In this method the photometry and radial velocities are combined to compute F_V throughout the pulsation cycle to within an unknown additive constant. The variation of F_V with (V-R) is thus determined. In every case F_V is found to vary linearly with (V-R). The mean slope for the 20 Cepheids so far analyzed is -0.362 ± 0.005 (s.e.m.) in superb agreement with Barnes' (1980) value of -0.363 ± 0.011 (s.e.m.)

To obtain the zero point of the F_V - (V-R) relation we make use of the angular diameters published by Parsons (1970) and Parsons & Bouw (1971). Parsons computed these by fitting Cepheid model atmosphere fluxes to observed fluxes in the blackbody six-color system. Knowledge of the mean angular diameter and mean V magnitude then specifies the mean F_V. A plot of F_V against mean (V-R) shows a linear distribution which agrees in slope with the independently determined value of -0.362. Adopting this value, the model atmosphere results were used only to determine the zero point of the F_V - (V-R) relation, 3.956 ± 0.003 (s.e.m.) Not surprisingly this is the same result found in the previous work, although the uncertainty is reduced by half.

Distances and Radii for Cepheids

Having established the visual surface brightness as a function of (V-R), it is a simple procedure to determine the angular diameter of the Cepheid throughout the pulsation cycle. Figure 1 shows a representative result. The smooth curve is the integrated radial velocity curve, corrected by a factor of 1.31 to pulsational velocity. With distance and mean radius as free parameters, it has been fit to the photometrically determined angular diameters.

We have carried out this analysis on 20 bright Cepheids with periods in the range 2 days to 45 days. The radii so obtained are in good agreement with Baade-Wesselink results of Balona (1977). Our results average 0.05 ± 0.02 larger in log R/R. than Balona's. The distances we obtain yield a distance scale 0.26 mag $\pm$ 0.17 mag larger in the distance modulus than the Fernie & Hube (1968) and Sandage & Tammann (1969) scales.

Figure 1. Angular Diameter Variation of T Vul

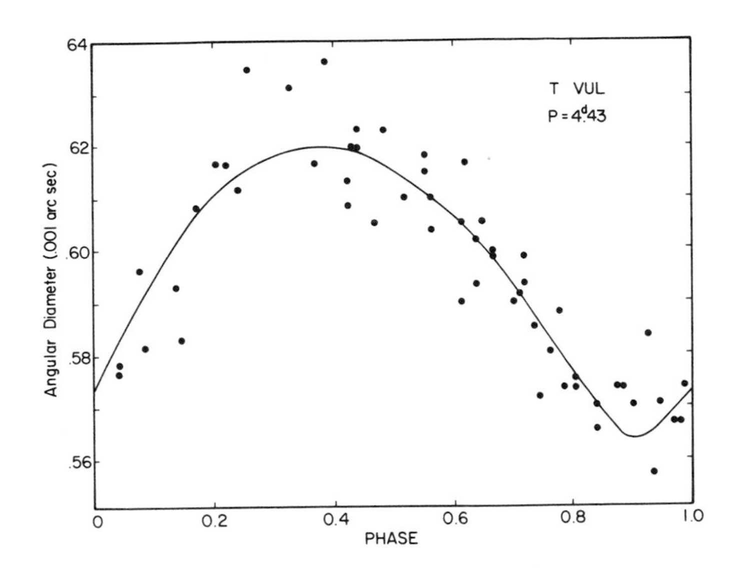

References

Balona, L.A. (1977). Mon. Not. R. Astr. Soc., 178, 231-243.
Barnes, T.G. (1980). Highlights of Astronomy, 5, 479-482.
Barnes, T.G., Dominy, J.F., Evans, D.S., Kelton, P.W., Parsons, S.B. and Stover, R.J. (1977). Mon. Not. R. Astr. Soc., 178, 661-674.
Barnes, T.G. & Moffett, T.J. (1984). In preparation.
Fernie, J.D. & Hube, J.O. (1968). Astr. J., 73, 492-499.
Moffett, T.J. & Barnes, T.G. (1984). Ap. J. Supp., 95, July 1.
Parsons, S.B. (1970). Warner and Swasey Obs., Preprint No. 3.
Parsons, S.B. & Bouw, G.D. (1971). Mon. Not. R. Astr. Soc., 152, 133-144.
Sandage, A.R. & Tammann, G.A. (1969). Ap. J., 157, 683-708.

LIGHT-CURVE PARAMETERS OF NORTHERN GALACTIC CEPHEIDS

Thomas J. Moffett
Department of Physics, Purdue University
West Lafayette, Indiana 47907 USA

Thomas G. Barnes III
McDonald Observatory, The University of Texas
Austin, Texas 78712 USA

Abstract. The BVRI light-curve parameters of 112 galactic Cepheids are determined by Fourier analysis using over 4,000 differential photoelectric observations. This catalog is similar to Schaltenbrand and Tammann's except that it is based on a homogeneous data set and the Fourier coefficients are given.

Moffett and Barnes (1984) have obtained over 4,000 differentially determined photoelectric BVRI observations of 112 Cepheids accessible from northern hemisphere observatories. In order to make these data convenient for Cepheid research, we have produced a catalog giving the light-curve parameters for these stars. The format and techniques used to generate this catalog are similar to those employed by Schaltenbrand and Tamann (1971).

The observed light curves were fitted, by least squares, to a Fourier series of the form

$$A_o + A_j \cos [j\omega t + F_j]$$

where, $\omega = 2\pi/P$, t = (HJD of observation - epoch), and j is the order of the series. The fitting program is identical to the one used by Simon and Lee (1981).

The procedure was to try various orders, 2nd to 8th, until a resonably good fit to the observed points was found. The order found for the V-mag light curve was adopted for all other magnitude and intensity fits. Separate orders were determined for each of the color indices. In a few cases, even an 8th order series did not fully represent the observed light curve. The adopted epoch was determined from the Fourier series of the V-mag light curve. Phase zero was defined as the maximum of the V-mag series. Table 1 shows a sample entry in the catalog which will be published in the Astrophysical Journal Supplement Series.

We are indebted to Dr. N. R. Simon for kindly providing us with the Fourier decomposition program.

Table 1. Light curve parameters for DT Cyg

Period	2.499035	$\overline{B-V}$	0.544
log P	0.397772	$\bar{B}-\bar{V}$	0.544
Epoch	4049.400	(B-V)max	0.475
Number	56	(B-V)min	0.616
Orders	5/3/2/2	Δ(B-V)	0.141
$\bar{V}$	5.778	ϕ(max)	0.972
V(max)	5.626	ϕ(min)	0.517
V(min)	5.914	σ	0.008
ΔV	0.288	<B-V>	0.543
ϕ(min)	0.549	<B> - <V>	0.539
σ	0.008	$\overline{V-R}$	0.470
<V>	5.774	$\bar{V}-\bar{R}$	0.470
$\bar{B}$	6.322	(V-R)max	0.421
B(max)	6.099	(V-R)min	0.514
B(min)	6.525	Δ(V-R)	0.093
ΔB	0.427	ϕ(max)	0.996
ϕ(max)	0.989	ϕ(min)	0.516
ϕ(min)	0.546	σ	0.010
σ	0.010	<V-R>	0.469
<B>	6.312	<V> - <R>	0.468
$\bar{R}$	5.308	$\overline{R-I}$	0.296
R(max)	5.203	$\bar{R}-\bar{I}$	0.295
R(min)	5.403	(R-I)max	0.260
ΔR	0.200	(R-I)min	0.320
ϕ(max)	0.028	Δ(R-I)	0.061
ϕ(min)	0.556	ϕ(max)	0.984
σ	0.010	ϕ(min)	0.572
<R>	5.306	σ	0.013
$\bar{I}$	5.013	<R-I>	0.295
I(max)	4.944	<R> - <I>	0.294
I(min)	5.082		
ΔI	0.138		
ϕ(max)	0.004		
ϕ(min)	0.575		
σ	0.011		
<I>	5.012		

REFERENCES

Moffett, T.J., and Barnes, T.G. 1984, Ap. J. Suppl., (in press).

Schaltenbrand, R., and Tammann, G.A. 1971, Astr. Ap. Suppl., 4, 265.

Simon, N.R., and Lee, A.S., 1981, Ap. J., 248, 291.

INFRARED OBSERVATIONS OF GALACTIC CEPHEIDS

J. A. Fernley, R.F. Jameson, M.R. Sherrington
University of Leicester, Leicester, United Kingdom

Abstract. Infrared photometry enables radii, and hence absolute magnitudes, to be derived more accurately than is possible using purely optical photometry. In this paper we present BVJHK observations of the galactic cepheids T Vul, U Vul and T Mon. Using this data we find reasonable agreement with current versions of the Period-Luminosity relations, in both the optical and infrared.

From these relations and existing optical and infrared data for LMC Cepheids we find for the LMC (i) $A_V = 0.35$ and (ii) a distance modulus of 18.50. In addition we show that the P-L-C relation has the form

$$\langle M_V \rangle = -3.10 \log P + 1.70 (\langle B \rangle - \langle V \rangle) - 2.37$$

1. Introduction

One of the methods of calibrating the Cepheid period-luminosity relation is to use Baade-Wesselink techniques to obtain radii, and hence absolute magnitudes, for individual stars. Recently there have been several modifications to the original Baade-Wesselink technique (e.g., Balona 1977). In addition it has been shown by Fernley et al. (1984) that the accuracy with which radii can be derived using these techniques can be improved by using near infrared observations. We have obtained BVJHK photometry of the galactic Cepheids T Vul and U Vul and using these data and Wisniewski and Johnson's (1968) BVJK photometry of the longer-period Cepheid T Mon, we derive radii (section 2) and absolute magnitudes (section 3). By plotting these results in Period-Luminosity diagrams, in both the optical and infrared, we derive the reddening and distance modulus to the LMC (section 4). Finally we use these results to derive the coefficients of the P-L-C relation (section 5).

2. Radii.

To obtain radii we use the relation derived by Balona (1977)

$$\Delta M = a\Delta C + b\Delta R + c \qquad (1)$$

where a, b and c are constants, ΔM the change in magnitude, ΔC the change in a colour and ΔR the radius variation defined by

$$\Delta R = KP \int_{\phi_1}^{\phi_2} (V_r - \gamma)\, d\phi \qquad (2)$$

where P is the period, V_r the radial velocity deduced from the Doppler shifts of the spectral lines, γ the centre of mass velocity of the star, ϕ the phase of the pulsation cycle and K a conversion factor to account for geometrical projection and limb darkening. Following Karp (1975) we assign a value K = 1.31. The radial velocity data for the three stars was taken from Lust-Kulka (1954), Sanford (1951,6) and Wallerstein (1972).

Using a maximum-likelihood technique we fit the observables ΔM, ΔC and ΔR in equation (1) to obtain values for the constants a, b and c. The constant a is the slope of the colour-surface brightness relation and the constant b is related to the mean radius, $\bar{R}$, of the star

$$b = (5\log_{10} e)/\bar{R} \qquad (3)$$

Fernley et al.(1984) showed that superior results could be obtained using infrared data. There are two reasons for this. Firstly infrared magnitudes are less temperature sensitive than optical magnitudes and are thus more sensitive to the radius variation. Secondly, the optical-infrared colour indices (V-J, V-K) have a larger amplitude than the purely optical indices (B-V) and are thus more "stable" temperature indicators. There are illustrated in Figure 1.

3. Absolute Magnitudes.

We show our results in Table 1. The radii have been discussed in the previous section. The reddenings are an average of various determinations as summarized by Feltz and McNamara (1980). The absolute magnitudes were obtained by two different methods. Firstly, by using the slopes of the colour-surface brightness relations, determined in the previous section, and the relation

$$M = M_{\odot} + a\,[C - C_{\odot}] - 5 \log R/R_{\odot} \qquad (4)$$

where M and C are the absolute magnitude and unreddened colour index respectively. A second method is to obtain T_e from $(B-V)_0$, using the colour-temperature calibrations of Kraft (1961) and Pel (1978), and then the relation

$$M = M_{\odot} - 2.5 \log\,[B(T)/B(T_{\odot})] - 5 \log R/R_{\odot} \qquad (5)$$

where B(T) is the Planck function. This should be a good approximation at the infrared wavelengths. The two methods agree closely which suggests the slopes of the colour-surface brightness relation are reasonable. This is reassuring for the accuracy of the method of section 2.

Figure 1. The radius variation of T Vul derived from equation 2 (solid line) and from inverting equation 1 with the best-fit values of a, b and c (• = using V, B-V and x = using K, V-K).

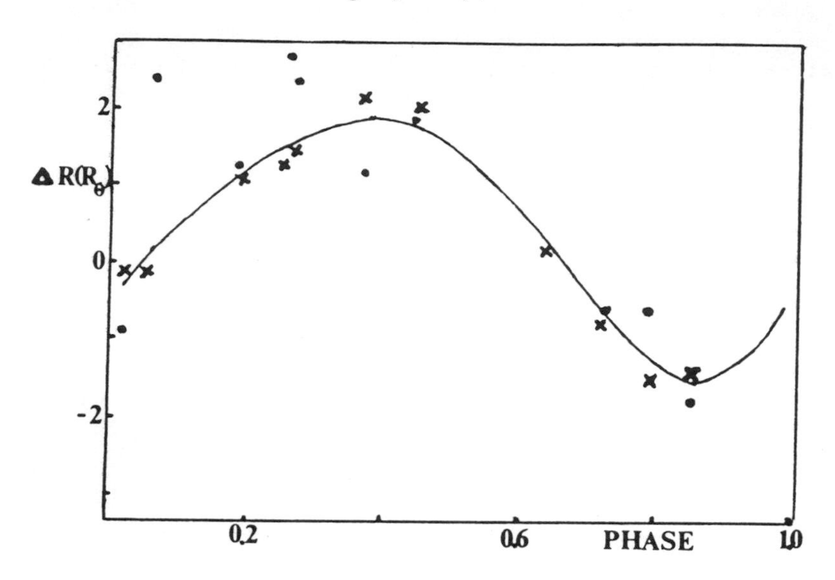

In Figure 2 we compare our results with two recent determinations of the Period-Luminosity relation. Stothers (1983) and McGonegal et al. (1983) used a cluster main-sequence fitting method to obtain their calibration. Their calibration is thus completely independent of ours and the good agreement between the two is encouraging. In the near future we hope to obtain photometry of several more Cepheids to test this agreement more fully.

Table 1

	T Vul	U Vul	T Mon
$R(R_\odot)$	37.70 ± 2.50	63.40 ± 5.00	147.50 ± 14.00
$(B-V)_0$	0.68 ± 0.02	0.82 ± 0.05	1.10 ± 0.05
E(B-V)	0.12 ± 0.01	0.65 ± 0.04	0.33 ± 0.04
T_e	5800 ± 100	5440 ± 150	4900 ± 150
$<M_V>$	- 3.33 ± 0.23	- 4.06 ± 0.29	- 5.54 ± 0.35
$<M_H>$	- 4.47 ± 0.16	- 5.49 ± 0.19	- 7.14 ± 0.26
Log P	0.646	0.902	1.432

Note: R, $(B-V)_0$ and T_e are at phase ϕ = 0.7 (minimum light)

Figure 2. Period-Luminosity relations for galactic Cepheids in the optical and infrared. • = Stothers 1983 (V), McGonegal et al. 1983 (H), x = this work.

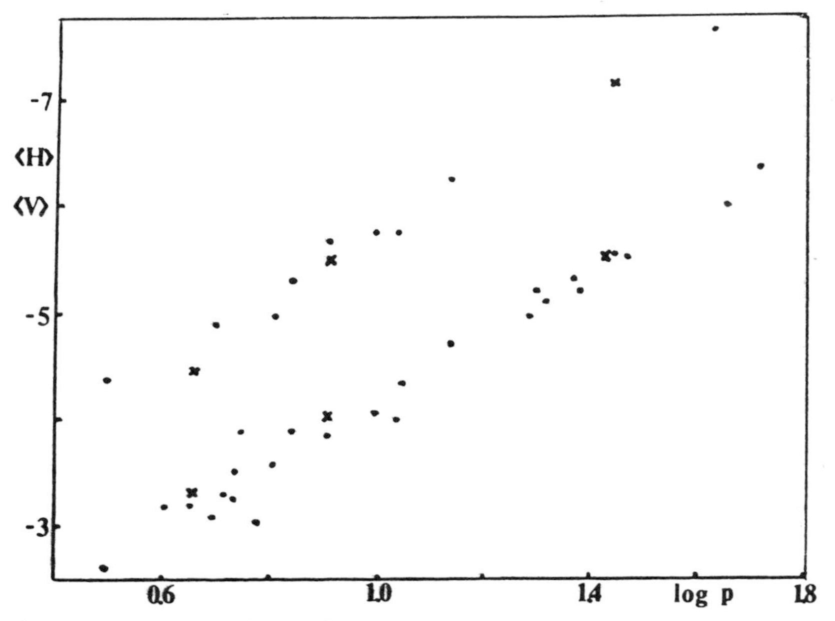

4. Distance to the LMC.

Fitting our results to the slopes of existing optical and infrared Period-Luminosity relations for LMC Cepheids (Martin et al 1979, McGonegal et al 1982) we find distance moduli of $V-M_V = 18.85 \pm 0.15$ and $h-H = 18.55 \pm 0.14$. For $A_V/A_H = 7 \pm 1$ (Jones and Hyland 1980) and two distance moduli converge to 18.50 ± 0.10 for $A_V = 0.35$. This higher value for the reddening confirms the estimate of de Vaucouleurs (1978), however, the distance modulus is approximately 0.2 larger than he derives from the average of various indicators.

5. P-L-C Relation.

A recent determination of the P-L-C relation is due to Martin et al. (1979) who apply a maximum-likelihood technique to the unreddened optical data on LMC Cepheids. They find

$$\langle V \rangle = -3.80 \log P + 2.70 (\langle B \rangle - \langle V \rangle) + 16.41 \quad (6)$$

This is incompatible with the results of McGonegal et al. (1982) and ourselves. McGonegal et al. (1982) find for the LMC Cepheids

$$\langle H \rangle = -3.10 \log P + 15.90 \quad (7)$$

We find, from Section 2, the slopes of the colour-surface brightness relation for V, B-V and H, B-V. Thus we may write

$$\begin{aligned} \langle V \rangle &= -5 \log R/R_\odot + 2.20 (\langle B \rangle - \langle V \rangle) + 3.40 \\ \langle H \rangle &= -5 \log R/R_\odot + 0.50 (\langle B \rangle - \langle V \rangle) + 3.17 \end{aligned} \quad (8)$$

where we have $V_\odot = 4.83$, $H_\odot = 3.48$ and $(B-V)_\odot = 0.65$. Combining (7) and (8) gives

$$\langle V\rangle = -\ 3.10 \log P + 1.70\ (\langle B\rangle - \langle V\rangle) + 16.13 \qquad (9)$$

The discrepancy between (6) and (9) is interpreted as partial confirmation of the arguments raised against the P-L-C (e.g., Stift 1982).

References

Balona, L.A., (1977), M.N.R.A.S., 178, 231.
Feltz, K.A., and McNamara, D.H., (1980), P.A.S.P., 92, 609.
Fernley, J.A., Jameson, R.F., Sherrington, M.R., (1984), M.N.R.A.S. (in press)
Jones, T.J., and Hyland, A.R., (1980) M.N.R.A.S., 192, 359.
Karp, A.H., (1975), Ap.J., 201, 641.
Kraft, R.P., (1961), Ap.J., 134, 616.
Lust-Kulka, R., (1954), Z. Ast., 33, 211.
Martin, W.L., Warren, P.R., Feast, M.W., (1979), M.N.R.A.S., 188, 139.
McGonegal, R., McLaren, R.A., McAlary, C.W., Madore, B.F., (1982) Ap.J., 257, L33.
McGonegal, R., McLaren, R.A., McAlary, C.W., Madore, B.F., (1983), Ap.J., 269, 641.
Pel, J.W., (1978) Astr. Ap., 62, 75.
Sanford, R.F., (1951), Ap. J. 114, 331.
Sanford , R.F., (1956), Ap. J. 123, 201.
Stift, M.J., (1982), Astr. Ap. 112, 149.
Stothers , R.B., (1983), Ap. J., 274, 20.
de Vaucouleurs, G., (1978), Ap. J., 223, 730.
Wallerstein, G., (1972), P.A.S.P., 84, 656.
Wisniewski, W.Z., and Johnson, H.L., (1968), Lunar and Planetary Lab. Comm., No. 112.

NEW EVIDENCE FOR MASS LOSS IN CLASSICAL CEPHEIDS

M.J. Stift
Institut für Astronomie
Türkenschanzstr. 17
A-1180 Wien, Austria

I. Introduction

The question of apparent mass anomalies in classical cepheids was first brought up by Christy (1968) and Stobie (1969), but 15 years later there is still no definite picture concerning the reality and the possible cause of these mass anomalies. The masses obtained from application of standard evolutionary theory were always sensibly larger than the masses derived from pulsation theory using both linear and non-linear codes. Since for various reasons few people have accepted the idea that mass loss could play an important role in cepheids a number of elaborate scenarios have been proposed to account for the mass discrepancies. Among these are helium enriched outer layers and tangled magnetic fields. It is difficult, however, to see how significant mass loss can be avoided during the evolution of the more massive cepheids. In fact, practically all supergiants lose mass over the whole HR diagram, a process frequently manifesting itself in photometric micro-variability. Little hope can be placed in attempts to solve the problem by means of improved determinations of the physical parameters of cepheids; intrinsic colours, luminosities, radii, effective temperatures, and the width of the instability strip have been disputed for years with no definite results yet. Only independent observational evidence will make it possible to confirm - or reject - the mass anomalies. On account of the large number observed and because of the fairly complete sample they represent, the cepheids in the LMC, SMC and in our Galaxy are best suited for this kind of investigation.

II. The frequency-period distribution of classical cepheids

Becker et al. (1977) have carried out a detailed study of the frequency-period distribution of cepheids in several galaxies, comparing observations and standard evolutionary calculations (i.e. assuming an atmosphere in hydrostatic equilibrium and not including the effects of mass loss, convective overshooting and turbulent diffusion). Combining these results with linear adiabatic pulsation theory, Becker et al. were able to determine the period-luminosity relation and the frequency-period distribution as a function of metallicity Z and of helium content Y. Satisfactory agreement was found between theory and observations, but 2 major problems turned up: the predicted width of the frequency-period distribution was too small and the predicted absolute number of cepheids with periods in excess of 10 days was also too small. Realizing that their evolutionary results combined with the initial mass function of Salpeter (1955) were incompatible with the observations, Becker et al. postulated a two-component birthrate-function; the second, much more recent component was expected to yield 10 times more massive stars than the first, old component. We note that Lequeux (1983) does not find any indication for such a birthrate function in the Magellanic Clouds.

A way to reconcile the IMF of Salpeter with the observed frequency of luminous cepheids involves substantial mass loss. In general, cepheids having suffered substantial mass loss in previous stages of their evolution are overluminous for their mass; they thus oscillate at lower frequencies - for a given luminosity - than canonical cepheids. This leads to an increase in the number of long-period cepheids compared to the canonical case. A similar effect on the frequency-period distribution is due to convective overshooting (Matraka et al. 1982) and probably also due to turbulent diffusion (Schatzman & Maeder 1981).

Mass loss leads to an even more dramatic enhancement of the number of long-period cepheids by way of increased lifetimes in the cepheid instability strip. Evolutionary calculations including mass loss (Maeder 1981) show that for large masses lifetimes in the cepheid region may increase by a factor 5-50. The cause for this increase is open to straightforward interpretation and is not tied to obscure numerical details. Figs. 2 to 4 of Maeder (1981) show that for moderate mass loss the "horn" - the region in the HR diagram occupied by core helium burning stars - is displaced towards the red compared to the zero mass loss case. Whereas for Maeder's case A (no mass loss) the "horn" joins the giant branch at approximately $60M_{\odot}$, this mass drops to $30M_{\odot}$ for case B (moderate mass loss). The loops which are limited to masses below $15M_{\odot}$ for case A are predicted to extend up to at least $30M_{\odot}$ for case B. It is evident that this shift in the position of the horn combined with looping at higher masses will greatly affect the relative numbers of long-period cepheids.

III. Estimated cepheid mass loss rates

Based on this mass loss scenario we shall estimate the appropriate mass loss rates for Galactic and Magellanic cepheids. Almost universal agreement has emerged that there is some positive correlation between metallicity and mass loss rate for a given domain in the HR diagram. This implies that mass loss should be smallest in the SMC, somewhat larger in the LMC and most important in our Galaxy. We have chosen the following mass loss rates:
a) SMC - slightly less than case B
b) LMC - about case B
c) Galaxy - between case B and case C
This particular choice can be justified by the relative number of long-period cepheids, by the position of the maximum of the frequency-period distribution, by the longest periods encountered, and finally by the mean metallicity of the galaxies in question. Extensive surveys of Magellanic Cloud cepheids by the Gaposchkins (1966, 1971) have revealed at least 10 cepheids with periods in excess of logP=1.8 whereas

canonical theory predicts about 10 times less cepheids for this period range. Maeder's case B mass loss leads to agreement between the observed and the theoretical relative numbers of long-period cepheids in the Magellanic Clouds; for our Galaxy a somewhat higher mass loss rate appears appropriate since no cepheid with a period in excess of 50 days has yet been detected. At the same time the mean metallicity of our Galaxy is sensibly higher than the respective mean metallicities of the Magellanic Clouds (Harris 1983). Because mass loss enhances the effect of metallicity in the HR diagram, viz. the position of the "horn", it also affects the position of the maximum of the frequency-period distribution and the half width of this distribution. It is thus probable - details have to be confirmed by calculations - that mass loss can explain the observed frequency-period distribution without resorting to exotic IMFs.

Let us note that the scenario sketched above is in accord with observational data concerning Wolf-Rayet and red supergiant luminosities. Apparent magnitudes of single WR stars in the LMC range between $V=12\overset{m}{.}0$ and $V=16\overset{m}{.}5$ (Breysacher 1981). In agreement with theory which predicts a sharp drop in luminosity during the later WR stages (Maeder 1983), most low-luminosity WR stars are of spectral type WNE and WC. The brightest red supergiants are found at about $V=11^{m}$ whereas the most massive cepheids are observed near $V=12^{m}$. This lends additional support to Maeder's scenario of WR stars repre senting a post-red-supergiant stage of stellar evolution. The lack of single WR stars in the SMC indicates somewhat lower mass loss rates; uncertain WR luminosities in our Galaxy make a similar comparison with cepheids rather hazardous.

As to the other hypotheses mentioned above which have been advanced for the purpose of resolving the mass anomalies, we do not consider them particularly promising. For a more detailed discussion including overshooting and turbulent diffusion we refer to Stift (1984).

IV. The consequences for the PL and the PLC relations

Whatever effect is responsible for the disagreement between observations and the canonical theory of cepheids, all possible explanations imply an internal structure of the cepheids sensibly different from the standard case. Taking into account the cosmic dispersion of metallicity, mass loss rates, and perhaps the degree of overshooting, we expect the pulsational characteristics of cepheids to be not as uniform as claimed by Sandage & Tammann (1969) or by Martin et al. (1979). Whereas there exist period-luminosity and period-luminosity-colour relations for cepheids in the same crossing, showing identical helium and metal abundances and being neither affected by mass loss nor by convective overshooting, this is no longer true for a real-life cepheid sample. Particularly the PLC relation loses its meaning once non-canonical effects become of the same order as the effects of finite strip width. That this is indeed the case has been shown by Stift (1982) and by Fernie & McGonegal (1983) who estimate the strip width at less than $\Delta(B-V)=0\overset{m}{.}20$ at constant period. The size of the effects of a cosmic dispersion in abundances and mass loss rates, and of the relative time spent in different crossings can be estimated from the interpolation formulae given by Iben & Tuggle (1972) and by Becker et al. (1977); even for a relatively small scatter in abundances and a moderate percentage of time spent in 1st, 4th and 5th crossings, the theoretical colour coefficient of the PLC relation drops from the canonical $\beta=2.7$ to $0<\beta<2$. Chance selection effects due to a small cepheid sample and a restricted period range become very important, leading to a zero-point error up to $0^m.5$. On the other hand, mean magnitudes at standard period derived with the help of the PL relation are much less affected by the above-mentioned effects and should preferably be used for distance determinations - for a more detailed discussion see Wayman et al. (1984).

Acknowledgement: I should like to thank Prof. A. Maeder for several enlightening discussions.

References

Becker, S.A., Iben, Icko, Jr., Tuggle, R.S. 1977, Astrophys. J. 218, 633.
Breysacher, J. 1981, Astron. Astrophys. Suppl. 43, 203.
Christy, R.F. 1968, Quarterly J. Royal Astron. Soc. 9, 13.
Fernie, J.D., McGonegal, R. 1983, Astrophys. J. 275, 732.
Harris, H.C. 1983, Astron. J. 88, 507.
Iben, Icko, Jr., Tuggle, R.S. 1975, Astrophys. J. 197, 39.
Lequeux, J. 1983, in Structure and Evolution of the Magellanic Clouds, IAU Symposium No. 108, S. van den Bergh and K.S. de Boer, Eds., D. Reidel, p. 67.
Maeder, A. 1981, Astron. Astrophys. 102, 401.
Maeder, A. 1983, Astron. Astrophys. 120, 113.
Martin, W.L., Warren, P.R., Feast, M.W. 1979, Monthly Notices Roy. Astron. Soc. 188, 139.
Matraka, B., Wassermann, C., Weigert, A. 1982, Astron. Astrophys. 107, 283.
Payne-Gaposchkin, C.H. 1971, Smithsonian Contr. Astrophys., no. 13.
Payne-Gaposchkin, C.H., Gaposchkin, S. 1966, Smithsonian Contr. Astrophys., Vol. 9.
Salpeter, E.E. 1955, Astrophys. J. 121, 161.
Sandage, A., Tammann, G.A. 1969, Astrophys. J. 157, 683.
Schatzman, E., Maeder, A. 1981, Astron. Astrophys. 96, 1.
Stift, M.J. 1982, Astron. Astrophys. 112, 149.
Stift, M.J. 1984, submitted to Astron. Astrophys.
Stobie, R.S. 1969, Monthly Notices Roy. Astron. Soc. 144, 461.
Wayman, P.A., Stift, M.J., Butler, C.J. 1984, Astron. Astrophys. Suppl. 56, 169.

CLASSICAL CEPHEIDS: PERIOD CHANGES AND MASS LOSS.

H. Deasy
Dunsink Observatory, Castleknock, Dublin 15, Ireland.

INTRODUCTION.

The period is an ideal parameter for monitoring minute changes in the structure of a star passing through the instability strip, as it can be measured with an accuracy of up to one part per million. The view of Parenago (1956) that only abrupt period changes occur in cepheids is no longer prevalent, and it is generally accepted that random period changes are superposed on the secular variation due to evolution. One possible mechanism for the random fluctuations in period or phase is convection or semiconvection, which Sweigert & Renzini (1979) showed could account for the period changes of RR Lyrae stars. Other mechanisms include the influence of binary companions and mass loss. The latter mechanism forms the basis for a separate study involving the use of IUE spectra to search for evidence of matter being ejected from cepheids.

PERIOD CHANGES

Szabados (1977,1980,1981) found that for periods under 10 days, about 15% of period changes in galactic classical cepheids were secular, while for longer periods this fraction was nearer to 80%. In the present study, estimates for the mean fractional rate of change of period per day, d/dt(lnP), with errors, have been derived for 112 Magellanic Cloud cepheids, 81 in the large and 31 in the small cloud, using Dunsink Observatory observations (C.J. Butler (1976,1978) and Wayman et al. (1984)) in conjunction with Harvard (Payne-Gaposchkin & Gaposchkin (1966) and Payne-Gaposchkin (1971)) and South African Astronomical Observatory observations (Martin et al. (1981)). These estimates correspond, approximately, to the intervals 1940-1966 (denoted by H1), 1966-1976 (denoted by H2) and 1940-1976 (combining H1 and H2 and denoted by H3).

From Figure 1 (a), it is apparent that the larger period changes are concentrated at longer periods (P > 10 days). In Figure 1 (b), for galactic cepheids, this trend is also apparent. Figure 1 (c) uses the evolutionary tracks of Becker, Iben & Tuggle (1977) and Hoffmeister (1967) to construct a plot comparable with Figures 1 (a) and (b). The symbols correspond to crossings of the instability strip, with symbol area proportional to the duration of the crossing and y ordinate equal to the average daily fractional period change over that crossing. It is seen that the ratio of the total area of the symbols corresponding to the larger period changes relative to that of the smaller changes

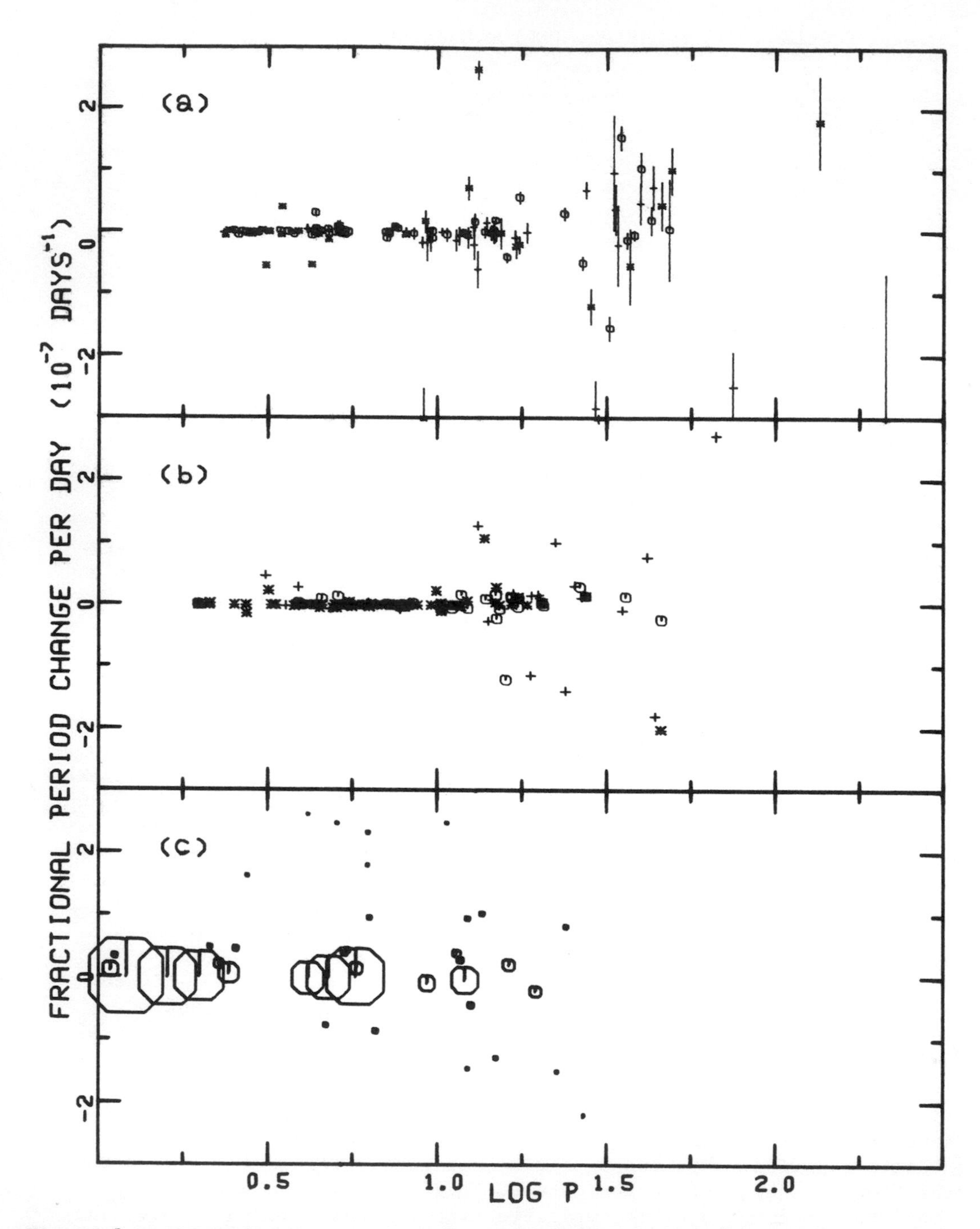

Figure 1.: Mean fractional period change per day for (a) Magellanic Cloud cepheids (using H3), (b) Galactic cepheids, where circles, asterisks, crosses represent data from Parenago (1956), Szabados (1977,1980,1981) and Erleksova & Irkaev (1980), respectively., (c) Evolutionary model calculations (see text for details).

increases toward longer periods (>10 days), corresponding to the greater number of large observed period changes at these periods in Figures 1 (a) and (b). The low cut-off period of the models prevents comparison for periods longer than 30 days.

In Figure 2, the two estimates, H1 and H2, are compared. The LMC cepheids show some positive correlation, while the estimates for SMC cepheids tend to be anticorrelated, indicating non-evolutionary period changes. Comparing these results with those of Szabados (1977,1980,1981), who found a larger fraction of galatic cepheids with secular period changes, it seems that the frequency of erratic changes increases in the order Galaxy-LMC-SMC. This suggests a possible link with metal abundance.

MASS LOSS

Schmidt & Parsons (1984) have studied the Mg II h and k profiles of five cepheids using high dispersion IUE spectra. For several stars they found features corresponding to outflow velocities of the order of the escape velocities. Thus it would seem that mass loss may be occuring in cepheids, driven, possibly, by the pulsation (see Willson & Hill (1979)). A program is under way to search for evidence of such pulsation-related mass loss in classical cepheid variables. High dispersion IUE spectra have been taken of l Car and of two binaries, S Mus and V810 Cen. Preliminary analysis shows no evidence for V<R reversal in the Mg II h and k lines for l Car. For the binaries, it is possible that ejecta from the cepheids will show up in absorption in the ultraviolet spectra of their blue companions, as in the pioneering study of Alpha Herculis by Deutsch (1956).

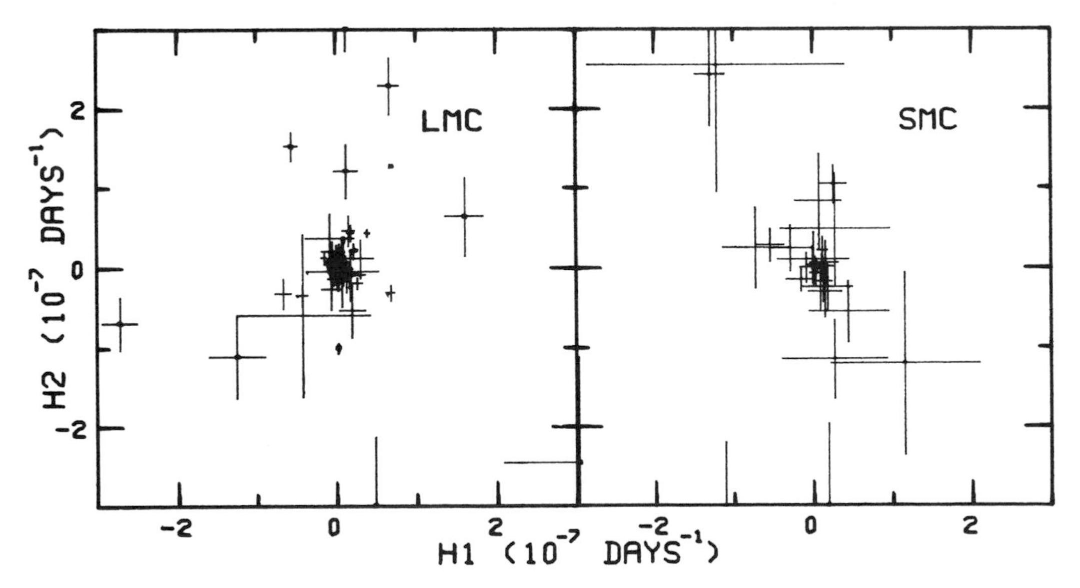

Figure 2: H2 against H1 for LMC and SMC cepheids.

Comparison with previous IUE spectra should determine if there is a pulsation-related component to the mass loss.

CONCLUSIONS

Both abrupt and secular period changes occur in cepheids, with larger changes more prevalent at longer periods. This trend is in accordance with the predictions of stellar evolution models, as is the size of the changes. The fraction of period changes which are irregular increases in the order Galaxy - LMC - SMC, suggesting a link with metallicity. Mass loss, which is one possible cause of the period fluctuations, may be detectable using high dispersion IUE spectra.

REFERENCES

C.J. Butler (1976), Astron. Astrop. Supp. 24,299.
C.J. Butler (1978), Astron. Astrop. Supp. 32,83.
A. Deutsch (1956), Ap.J., 123,210.
G. Erleksova & B. Irkaev (1980), in abstracts from the 5th Eur. Regional Meeting on Variability in Stars and Galaxies (Liege), B.1.8.
E. Hoffmeister (1967), Z. Ap. 65, 194.
W. Martin, Y. Thomas, B. Carter & H. Davies (1981), S.A.A.O. Circ. 2,31.
P.P. Parenago (1956), Perem Zvezd. 11,236.
C. Payne Gaposchkin (1971), Smiths. Contr. Astrop. No.13.
C. Payne Gaposchkin & S. Gaposchkin (1966), Smith. Contr. Astrop. No.9.
E.G. Schmidt & S.B. Parsons (1984), Ap.J. 279,202.
A. Sweigert & A. Renzini (1979), Astron. Astrop. 71,66.
L. Szabados (1977), Mitt. Sternw. Ungar. Akad. Wiss no.70.
L. Szabados (1980), Mitt. Sternw. Ungar. Akad. Wiss no.76.
L. Szabados (1981), Mitt. Sternw. Ungar. Akad. Wiss no.77.
P.A. Wayman, M.J. Stift & C.J. Butler (1984), Astro. Astrop. Supp. 56,169
L.A. Willson & S.J. Hill (1979), Ap.J. 228, 854.

DUPLICITY, MASS LOSS AND THE CEPHEID MASS ANOMALY

G. Burki
Geneva Observatory
CH 1290 Sauverny
Switzerland

The rate of binaries among cepheid stars is 25 - 35% (Burki, 1984a), a value which is in agreement with the rate among F-M supergiants (31-38%) obtained by Burki and Mayor (1983) from a radial velocity survey.

For binary cepheids, the Wesselink radius R_W is affected by the light of the companion. Balona (1977) found that R_W is too small if the companion is a blue main sequence star and too large if it is a red giant star. This effect has been quantitatively examined by Burki (1984a). For example, it was found that, in the case of a cepheid of period P = 32d with a companion fainter by 2 mag., R_W is 70% of its correct value if we are dealing with a main sequence companion and 130% in the case of a companion redder by 0.6 in B-V. The bias on R_W becomes negligible when the companion is fainter than the cepheid by more than 4 mag.

Following Cox (1979), four basic equations can be used: The definition of Te: log L = f(R,Te); the mass-luminosity law: log L = g (M,Y,Z); the period-density relation: log Q = h(P,M.R); the variation of Q: log Q = k(Te,R,M,L). The relations g and k result from model calculations. The theoretical mass M_{Th} and radius R_{Th} are obtained by resolving equations f,g,h,k for given values of P and T_e. The Wesselink mass M_W is deduced from equations f,h,k, the quantities P,T_e and the Wesselink radius R_W being known.

Table 1: Mean values of the ratio R_W/R_{Th} for the cepheids in Table 6 of Cox (1979)

	P < 10d	P > 10d
Single cepheids	1.01 ± 0.10 (22*)	0.83 ± 0.09 (9*)
Binary cepheids	0.92 ± 0.16 (17*)	0.86 ± 0.09 (4*)
All	0.97 ± 0.13 (39*)	0.84 ± 0.08 (13*)

Table 1 gives the mean values of the ratio R_W/R_{Th} for the cepheids listed by Cox (1979). We see that:

1) R_W and R_{Th} are in agreement (ratio equal to 1) in the case of single cepheids with $P < 10$ d.
2) In the case of binary cepheids with $P < 10$d the ratio is not too far from unity but the dispersion is large (0.16): R_W is larger or smaller than R_{Th}, depending on the type of companion.
3) In long period cepheids, R_W is smaller than R_{Th} by about 15%, for both single or binary cepheids. For these stars, it is suggested that the discrepancy is due to the mass loss process, which must be taken into consideration for the calculations of evolutionary models of massive stars.

Indeed, it is now an established fact that the evolution of massive stars is modified by the mass loss process (Maeder, 1980). Burki (1984a) has used the evolutionary log M vs. log L diagram in order to show that preliminary calibrations, derived from the stellar models with mass loss by Maeder, can resolve the inconsistency between M_{Th} and M_W (or R_{Th} and R_W). This result can also be shown in the following way:

Lovy et al. (1984) have determined the pulsation periods of the supergiant models by Maeder, applying the classical linear adiabatic theory. They found the following relation for the fundamental radial mode:

$$\log P = 0.688 \log L - 3.918 \log Te + 13.237 \qquad (1)$$

By using further the definition of T_e ($L \sim R^2 Te^4$) and the location of the instability strip (Cogan, 1978), a theoretical period-radius relation can be derived for the long period cepheids:

$$\log R = 0.68 \log P + 1.14 \qquad (2)$$

Figure 1 shows the period-radius relation for all single Pop I cepheids that have a Wesselink radius determination. The theoretical relations, based on the models with $\dot{M}$ for the long period cepheids (equation 2) and without $\dot{M}$ for the cepheids with $P < 10$d (Cogan, 1978; Fernie, 1984) are also shown, as well as the observational log P-log R relations for cepheids with $P < 10$d and $P > 10$d, obtained by linear regressions. The vertical width of these relations correspond to twice the residual standard deviation in log R.

We see that:

1) In the case of cepheids with $P < 10$d, the agreement between observations and theory is quite remarkable (see Burki, 1984b).
2) In the case of cepheids with $P > 10$d, the theoretical relation based on models with $\dot{M}$ is in satisfactory agreement with the

observations. Note that the theoretical relation deduced from models without $\dot{M}$ would be in poorer agreement with these long period cepheids.

Of course, this comparison between observations and recent stellar models is only preliminary and the following remarks are to be made: i) A different parametrization for the mass loss rate in the models would modify the theoretical relations (1) and (2) ; ii) the observational log P - log R relation for cepheids with P > 10d is based merely on 9 stars ; iii) dividing the cepheids into two groups, with a limit of P =10d, does not have a strong physical significance.

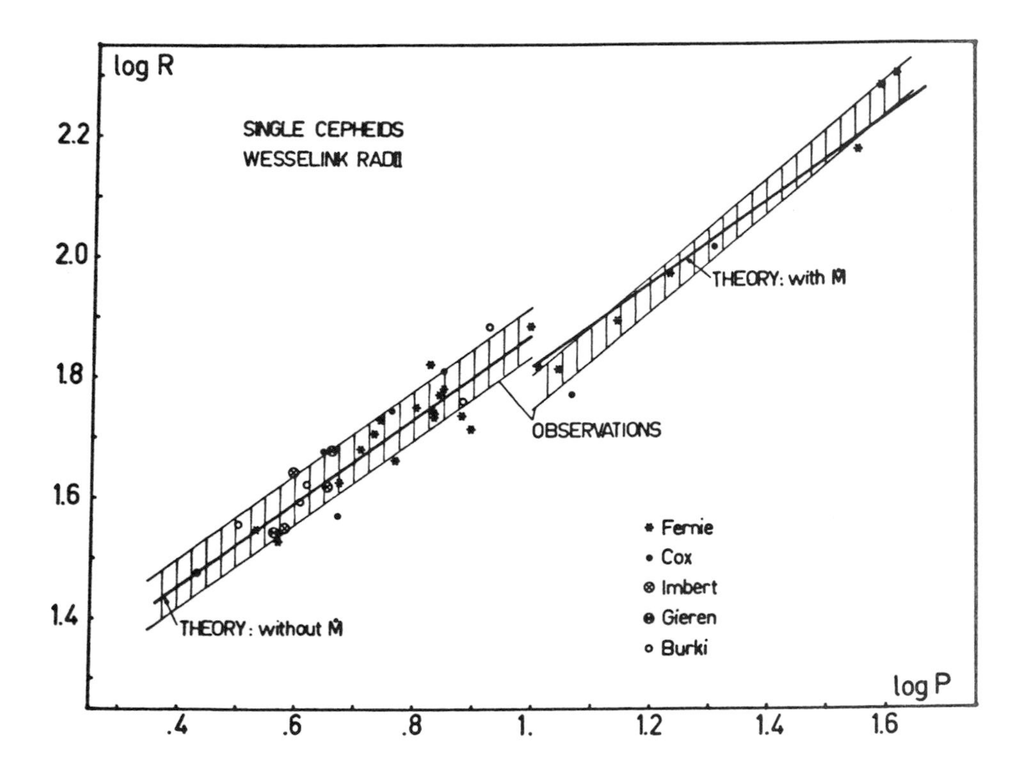

Figure 1: Period-radius relation for the single classical cepheids having a Wesselink determination of the radius. The radii come from Fernie (1984), Cox (1979), Imbert (1981, 1983, 1984), Gieren (1982), Burki (1984b), Burki and Benz (1982). The theoretical relations are from Cogan (1978) and Fernie (1984) for the cepheids with P < 10d and from equation (2) for the long period cepheids.

However, this preliminary result is encouraging, as it backs up a number of other tests made previously. It further brings out the importance of the mass loss process for the evolution of massive stars. The study of long period cepheids must take into account the effect of mass loss.

References

Balona, L.A.: 1977, Monthly Not. R. Astron. Soc. 178, 231
Burki, G.: 1984a, Astron. Astrophys. 133, 185
Burki, G.: 1984b, this colloquium
Burki, G., Benz, W.: 1982, Astron. Astrophys. 115, 30
Burki, G., Mayor, M.: 1983, Astron. Astrophys. 124, 256
Cogan, B.C.: 1978, Astrophys. J. 221, 635
Cox, A.N.: 1979, Astrophys. J. 229, 212
Fernie, J.D.: 1984, preprint
Gieren, W.: 1982, Astrophys. J. 260, 208
Imbert, M.: 1981, Astron. Astrophys. suppl. 44, 319
Imbert, M.: 1983, Astron. Astrophys. suppl. 53, 85
Imbert, M.: 1984, preprint
Lovy, D., Maeder, A., Noëls, A., Gabriel, M.: 1984, Astron. Astrophys., in press
Maeder, A.: 1980, Astron. Astrophys. 92, 101

DUPLICITY AMONG THE CEPHEIDS IN THE NORTHERN HEMISPHERE

L. Szabados
Konkoly Observatory
of the Hungarian Academy of Sciences
H-1525 Budapest, P.O. Box 67
Hungary

Cepheid variables in binaries are important from various points of view. These objects can in some cases provide direct information about the physical parameters of the system, can be used as tracers of stellar evolution, and the effect of the companions may influence the form of various relations (e.g. P-L-C, P-R) derived for Cepheids. While the first two advantages mentioned concern individual stars, the third involves the question of the frequency of binaries among the Cepheids. Systematic searches for binaries containing this kind of variable resulted in increasingly higher frequency of incidence: 2% (Abt 1959), 15% (Lloyd Evans 1968), >20% (Madore 1977), 25% (Pel 1978), 20%-40% (DeYoreo & Karp 1979), 35% (Madore & Fernie 1980). It was only in the early eighties that this trend ceased. The recent determinations of the incidence of binaries among the Cepheids are: 20%-40% (Gieren 1982), 18% (lower limit, Lloyd Evans 1982), 25% (Russo 1982), 25%-35% (Burki 1984). At the same time we have been going over to a qualitative era from the quantitative one, i.e. very thorough studies are now available on some individual cases of binary Cepheids (e.g. McNamara & Feltz 1981; Coulson 1983; Evans 1983; Böhm-Vitense et al. 1984).

The aim of this paper is not only to give an estimation of the percentage of binaries in a previously unused sample but also to recommend for further analysis more than twenty Cepheids suspected of having a companion.

The sample consists of the classical Cepheids whose O-C diagram has recently been constructed (Szabados 1977, 1980, 1981 & 1983). Altogether 89 northern Cepheids ($\delta > 0^{\circ}$) with B magnitude at light minimum brighter than $12^{m}.5$ were investigated. All kinds of evidence concerning the duplicity of the programme stars were then collected including spectroscopic evidence (radial velocity, IUE spectrum, etc.), photometric evidence (loops and location in two-colour diagrams, phase difference between the light and colour curves), and evidence based on the O-C diagram (periodic variations of the O-C residuals, rejump of the period). Strictly speaking, the rejump of the period ("stepwise" O-C curve) cannot be considered as evidence but rather a phenomenon observable exclusively in those Cepheids which show other signs of duplicity.

Table 1 contains a summary of the data about the duplicity of the northern Cepheids, where both the definite and the possible binaries are listed. In column (1) the names in parentheses are Cepheids in possible binaries. Column (2) gives the references of the various studies concerning the duplicity of the given Cepheid. Here, figures in parentheses denote those references where the binary nature of the programme star is not borne out. The key to the second column is in column (3). In addition to the 9 definite cases (10 per cent), 21 other Cepheids are suspected of having a companion (24 per cent). The well known binaries α UMi and CE Cas were not included in the programme. Although this percentage supports the reality of the earlier determinations, no numerical conclusion on the incidence of binaries among the Cepheids can be drawn.

Table 1

(1) Name	(2) References
FF Aql	*1,2,(10),11,(12),13,28,29*
(FM Aql)	*16,28*
(FN Aql)	*5,28*
η Aql	*(12),(13),18,(28)*
(RT Aur)	*2,(8),(10),11,(13)*
(SY Aur)	*(8),11,17*
(YZ Aur)	*(8),16,17,32*
(AN Aur)	*(8),16,17,32*
RW Cam	*3,8,16,17,22,24*
(BY Cas)	*12,17*
(DD Cas)	*16,17,31*
(X Cyg)	*(7),(8),16,17,32*
SU Cyg	*16,17,21,22,30,31*
(SZ Cyg)	*12,13,16,31*
(VX Cyg)	*16,17*
(VY Cyg)	*12,13,16,17*
(BZ Cyg)	*12,17*
(DT Cyg)	*13,17,30*
(V 386 Cyg)	*12,16,17*
(V 532 Cyg)	*16,30*
V 1334 Cyg	*25,27*
(W Gem)	*(8),11,17,22,(28)*
(RZ Gem)	*16,17*
T Mon	*2,4,(12),(13),17,19,28,32*
(CV Mon)	*16,28,31*
(RS Ori)	*16,17,(28),31*
SV Per	*3,8,14,15,16,17,22,32*
AW Per	*2,6,13,16,17,20,22,23,26,31*
S Sge	*2,9,12,13,17,28*
(SZ Tau)	*16,30*

(3) Key to references

1	Abt (1959)
2	Balona (1977)
3	Böhm-Vitense et al. (1984)
4	Coulson (1983)
5	Dean (1977)
6	Evans (1983)
7	Evans (1984)
8	Harris (1981)
9	Herbig & Moore (1952)
10	Hutchinson (1977)
11	Janot-Pacheco (1976)
12	Kurochkin (1966)
13	Lloyd Evans (1968)
14	Lloyd Evans (1982)
15	Lloyd Evans (1984)
16	Madore (1977)
17	Madore & Fernie (1980)
18	Mariska et al. (1980 a)
19	Mariska et al. (1980 b)
20	McNamara & Chapman (1977)
21	McNamara & Feltz (1981)
22	Mianes (1963)
23	Miller & Preston (1964 a)
24	Miller & Preston (1964 b)
25	Millis (1969)
26	Oosterhoff (1960)
27	Parsons (1981)
28	Pel (1978)
29	Plaut (1934)
30	Szabados (1977)
31	Szabados (1980)
32	Szabados (1981)

The uncertainty arising from the possible binaries can only be diminished by further observations. In particular, spectroscopic observations are needed to this end since almost all the suspected northern Cepheid binaries lack this kind of observational data.

Figure 1 shows the frequency of binary Cepheids as a function of the logarithm of period (n_{BC} = number of binary Cepheids, n_C = number of Cepheids). The upper (solid) curve is both for the definite and possible binaries, the lower (dashed) curve is based on the definite cases only. The trend towards higher duplicity with longer pulsation period found by Madore & Fernie (1980) cannot be seen in this figure. Similarly, Burki & Mayor (1983) and Burki (1984) could not confirm this trend. Again, many-sided investigations of suspected binaries are needed to reduce the selection effect that may be present.

Figure 1

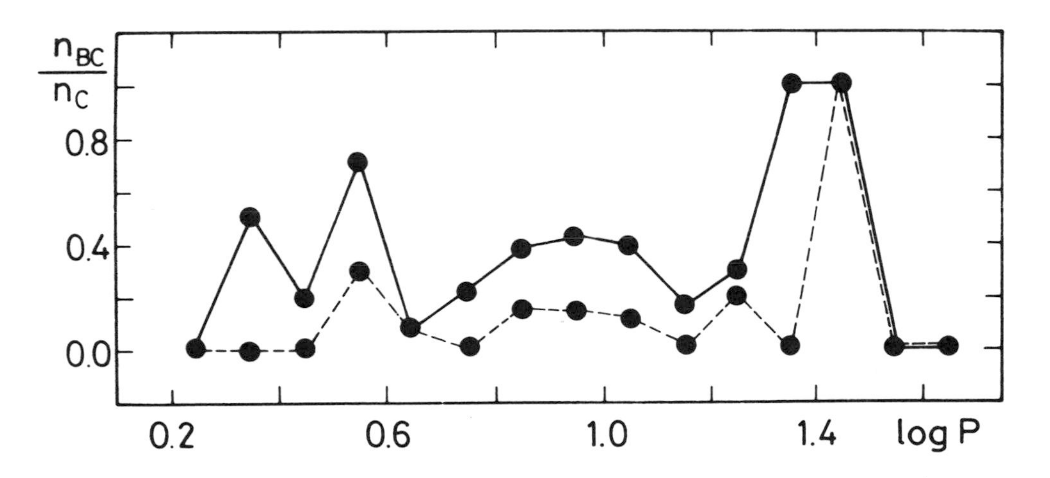

References:

Abt, H.A. (1959). Astrophys. J., 130, 769.
Balona, L.A. (1977). Mon. Not. R. astr. Soc., 178, 231.
Böhm-Vitense, E., Borutzki, S. & Harris, H. (1984). preprint.
Burki, G. (1984). Astron. Astrophys., 133, 185.
Burki, G. & Mayor, M. (1983). Astron. Astrophys., 124, 256.
Coulson, I.M. (1983). Mon. Not. R. astr. Soc., 203, 925.
Dean, J.F. (1977). MNASSA, 36, No.1, 3.
DeYoreo, J.J. & Karp, A.H. (1979), Astrophys. J., 232, 205.
Evans, N.R. (1983). Astrophys. J., 272, 214.
Evans, N.R. (1984). preprint.
Gieren, W. (1982). Astrophys. J. Suppl., 49, 1.
Harris, H.C. (1981). Astron. J., 86, 707.
Herbig, G.H. & Moore, J.H. (1952). Astrophys. J., 116, 348.
Hutchinson, J. (1977). B.A.A.S., 9, 357.

Janot-Pacheco, E. (1976). Astron. Astrophys. Suppl. Ser., 25, 159.
Kurochkin, N.E. (1966). Perem. Zvezdy, 16, 10.
Lloyd Evans, T. (1968). Mon. Not. R. astr. Soc., 141, 103.
Lloyd Evans, T. (1982). Mon. Not. R. astr. Soc., 199, 925.
Lloyd Evans, T. (1984). Observatory, 104, 26.
Madore, B.F. (1977). Mon. Not. R. astr. Soc., 178, 505.
Madore, B.F. & Fernie, J.D. (1980). P.A.S.P., 92, 315.
Mariska, J.T., Doschek, G.A. & Feldman, U. (1980 a). Astrophys.J., 238,L87.
Mariska, J.T., Doschek, G.A. & Feldman, U. (1980 b). Astrophys.J., 242,1083.
McNamara, D.H. & Chapman, R. (1977). P.A.S.P., 89, 329.
McNamara, D.H. & Feltz, K.A.Jr. (1981). In: "Effects of Mass Loss on Stellar Evolution". eds. C. Chiosi & R. Stalio, p. 389. D. Reidel. Dordrecht.
Mianes, P. (1963). Ann. d'Astrophys., 26, 1.
Miller, J. & Preston, G.W. (1964 a). Astrophys. J., 139, 1126.
Miller, J. & Preston, G.W. (1964 b). P.A.S.P., 76, 47.
Millis, R.L. (1969). Lowell Obs. Bull., No. 148.
Oosterhoff, P.Th. (1960). P.A.S.P., 72, 313.
Parsons, S.B. (1981). Astrophys. J., 247, 560.
Pel, J.W. (1978). Astron. Astrophys., 62, 75.
Plaut, L. (1934). Bull. astr. Inst. Neth., 7, 181.
Russo, G. (1982). In: "Binary and Multiple Stars as Tracers of Stellar Evolution". eds. Z.Kopal & J. Rahe, p. 23, D. Reidel, Dordrecht.
Szabados, L. (1977). Mitt. Sternw. ung. Akad. Wiss., Budapest, No. 70.
Szabados, L. (1980). Mitt. Sternw. ung. Akad. Wiss., Budapest, No. 76.
Szabados, L. (1981). Commun. Konkoly Obs. Hung. Acad. Sci., Budapest, No. 77.
Szabados, L. (1983). Astrophys. Space Sci., 96, 185.

A SEARCH FOR CEPHEID BINARIES USING THE CaII H AND K LINES

Nancy Remage Evans
Department of Astronomy and David Dunlap Observatory
University of Toronto, Toronto, Canada

Abstract. A survey of 24 classical Cepheids has been made to search for blue companions using the CaII H and K lines. It is shown that this technique can detect an early A companion for a typical Cepheid. A blue companion of SU Cas was discovered and upper limits for the companions for a number of previously suspected binaries were established.

Introduction

Miller and Preston (1964) demonstrated that for a Cepheid which has a blue companion the line depths of the CaII H and K lines are reversed as compared those of with single supergiants, where K is stronger than H. The light from the companion fills in the K line, but since Hε is close to H, the strong Hε absorption of the hot star results in a deeper total absorption in that region than in the center of the K line.
A study has been undertaken to explore the limits of this technique and to survey a list of Cepheids suspected of having companions. Several factors complicate such an analysis. Since Cepheids vary typically from mid-F to late-G during their cycles, at some phases H and K lines will have central emission. Interstellar absorption lines can also complicate the interpretation of the line cores. For companions later than early A stars, their CaII lines will be strong enough to dilute the effect.

Observations

Spectra with a reciprocal dispersion of 12 Å/mm were obtained with the Cassegrain spectrograph of the 1.9 m telescope at the David Dunlap Observatory. Exposure times were adjusted to obtain a good exposure at 4000 Å on IIaO plates. The plates were scanned on the PDS microdensitometer at the observatory and calibrated by means of spot sensitometer plates. For comparison a series of spectra of nonvariable supergiant standards was obtained covering the same spectral range as the Cepheids. In addition spectra of several nonvariable supergiants known to have blue companions were obtained (μ Per G0Ib +, 22 Vul G2Ib +, and 58 Per G8II + B). Although emission is clearly present in these profiles, the reversal in the line strengths is evident as late as G8, if a companion is bright.

Cepheids

The following list summarizes the Cepheids surveyed and (C) indicates the detection of a companion, (N) indicates not yet reduced: HR 7308, SU Cas (C), TU Cas, DT Cyg, SZ Tau, HR 8157 (C), Alpha UMi, RT Aur, SU Cyg (C), T Vul, FF Aql, Delta Cep, Eta Aql (C),

HR 690, W Gem, HR 9250 (N), S Sge, ζ Gem, TT Aql, X Cyg, Y Oph, T Mon (N), HD 161796, and 89 Her. The only new blue companion which was discovered was that of SU Cas. It was later confirmed by the author on IUE spectra to be approximately spectral type A0. The companion to η Aql which was a complete surprise on IUE spectra (Mariska, Doschek, and Feldman, 1980) would have been discovered by this survey. During the course of this survey a number of stars were observed with IUE, and the CaII and IUE results are in agreement. Both the full amplitude and small amplitude pulsators are mixtures of single and double stars.

What limits can be placed on an undetected companion? The companion of Eta Aql produces an easily discernable effect on the profiles. If we adopt the temperature Mariska, Doschek, and Feldman determined (9500°K), the companion is an A0V star. We estimate that a companion contributing half as much light at 3900 Å should be detectable. This corresponds to an A3V star for a typical star on the list. This limit is coincident with the limitation imposed by the strengthening of the Ca lines in main sequence stars.

Unusual Line Profiles

The H and K profiles confirm that the wings of HD 161796 are a very good match for those of 89 Her, in agreement with the F3Ib spectral type of Fernie and Garrison (1984), however the line cores are markedly different. For 89 Her the cores are blue shifted with respect to weaker features. This is not true for HD 161796, implying that in this respect the atmospheres are quite different.

For HR 7308, there is always light in the line cores, although the strength of the profiles corresponds to a spectral type no later than F7. This is not true for other cepheids.

References

Fernie, J. D. and Garrison, R. F. 1984, preprint.

Mariska, J. T., Doschek, G. A., and Feldman, U. 1980, Ap. J. Lett., 238, L87.

Miller, J. and Preston, G. 1964, Ap. J., 139, 1126.

ON THE BINARY NATURE OF 89 HERCULIS

A. Arellano Ferro
Department of Astronomy, University of Toronto, and
Instituto de Astronomía, Universidad Nacional Autónoma de México.

ABSTRACT. The radial velocities of 89 Herculis (89 Her) between 1977 and 1981 show a clear periodicity of about 285 days. This periodic variation is interpreted as the orbital variation of 89 Her around an unseen companion. From the orbital elements no support is found for a low mass ($M \sim 2M_{\odot}$) for 89 Her, but rather a high mass ($13 < M/M_{\odot} < 24$) is preferred.

DISCUSSION

The high galactic latitude run-away yellow supergiant variable 89 Her, has attracted great interest in the past due mainly to its apparent peculiar position off the galactic plane and to its capricious variational behaviour (e.g., Fernie 1981). The star pulsates with a period of about 63.5 days (Arellano Ferro 1984) and has been observed to interrupt its rhythmic variation for several months (Fernie 1981). Its place of formation is controversial. It may have been formed out of the galactic plane [although it has solar composition (Searle, Sargent & Jugaku 1963)]. On the other hand, it may have traveled from the plane during its lifetime. It is of obvious importance then, to know its mass (age) and luminosity (distance to the plane).

A period analysis on the 1977-81 radial velocities revealed two clear periodicities; one of about 60 days, which is due to the pulsational variation, and the other of about 285 days which is interpreted as an orbital period.

Table 1. Several solutions for the 89 Her system.

Solution	$M_{89}/M_{\odot}$	$M_2/M_{\odot}$	$a(R_{\odot})$	i
1	20	15	596	3.5°
2	20	2	510	19°
3	2*	15	413	2°
4	2*	2	144	6°

* $R_{89} \sim 130\ R_{\odot}$ (Fernie 1981)

The orbital solution is:

$\gamma = -27.8 \pm 0.2$ km s^{-1} $K = 2.8 \pm 0.2$ km s^{-1}

$e = 0.13 \pm 0.08$ $\omega = 338° \pm 11°$

$T = 244\ 3970.0 \pm 9.6$ $P = 285.8 \pm 41$ days

and

$$f(M) = (M_2 \sin i)^3/(M_{89} + M_2)^2 = 6.063\times10^{-4} \pm 1.485\times10^{-4} M_\odot$$

$$a_{89} \sin i = 1.074\times10^7 \pm 0.089\times10^7 \text{ km.}$$

This solution fits the observations very well. If $-9 < M_V < -6$ then $13 < M/M_\odot < 24$ (see Arellano Ferro 1984) and $109 < R / R_\odot < 314$. In order to progress on the mass controversy (Fernie 1981; Bond et al. 1984) let us assume 20 $M_\odot$ and 2 $M_\odot$ for 89 Her. Given the above elements we find solutions 1 to 4 in Table 1. Solution 3 may be ruled out as it is difficult to understand how a 2 $M_\odot$ star can dominate over a 15 $M_\odot$ star and how such a system could have originated. In solution 4 the separation and the stellar radius are nearly equal and this solution may be rejected. Therefore the low mass case receives no support. Attention is called to the similarity of the 89 Her system to the 20 $M_\odot$ star BL Tel system which also has high galactic latitude and has been classified as a run-away star (Wing 1963, Feast 1967).

REFERENCES

Arellano Ferro, A. (1984). Proceedings of this symposium.

Bond, H.E., Carney, B.W. & Grauer, A.D. (1984). Publs. astr. Soc. Pacific, **93**,367.

Feast, M.W. (1967). Mon. Not. R. astr.Soc., **135**, 287.

Fernie, J.D. (1981). Astrophys. J., **243**, 576.

Searle, L., Sargent, W.L.W. & Jugaku, J. (1963), Astrophys. J., **137**, 268.

Wing, R.F. (1963). Mon. Not. R. astr. Soc., **925**, 188.

NON-CEPHEIDS IN THE INSTABILITY STRIP

William P. Bidelman
Warner and Swasey Observatory, Case Western Reserve Univ.
Cleveland, Ohio 44106

The question of whether non-Cepheids populate the cepheid instability strip has been the subject of much recent interest. Several investigators (see especially Fernie and Hube 1971 and Fernie 1976) have concluded that indeed there are an appreciable number of objects not presently recognized as Cepheids within the strip, though both the size and position of the strip and the placement of supergiants in the H-R diagram are subject to considerable uncertainty. In view of the interest in this matter, I have attempted to shed some light on this situation by simply considering the nature of the stars spectroscopically classified as of high luminosity in spectral classes F and G. For this one needs, of course, a complete sample of the relevant stars, which fortunately is now to some extent available.

A complete listing of all class Ib and brighter supergiants bright enough to be contained in the Henry Draper Catalogue south of $\delta = -53^{o}$ is given by Houk, Hartoog, and Cowley (1976), the data being taken from the first volume of the Michigan spectral catalogue (Houk and Cowley 1975). This region of course, includes much of the southern Milky Way, though only a small part of the whole sky.

From Table II of Houk et al.(1976) counts have been made of all known Cepheids (not all of which are so identified in their paper) and of all non-Cepheid supergiants within two different spectral ranges. The results are as follows:

Type	Spectral Range	
	F2-G5	F5-G2
Cepheids	18	13
non-Cepheids	53	29
total	71	42

Though the numbers are small, the percentages of Cepheids are 25% and 31% for the two spectral ranges. The narrower range is perhaps closer to the truth since the spread in the assigned spectral types is actually quite small. In making the counts it was assumed that several controversial variable supergiants (V382 Car = HR 4337, 0^{1} Cen, V810 Cen = 4511, and V766 Cen = HR 5171) were not Cepheids.

It is probably unwarranted to draw any very firm conclusions from these data, for at least the following reasons: First, the assignment of similar spectral types and luminosity classes does not necessarily mean that the variables and non-variables are actually in the same positions in the H-R diagram. And, secondly, some, perhaps many small-range Cepheids may await discovery in the southern sky. Nevertheless the data taken at face value would appear to indicate that there are at least as many non-Cepheids in the instability strip as Cepheids.

Further information could be gleaned from Dr. Houk's classifications if detailed counts of somewhat intrinsically fainter normal stars were available. In the same region of the sky she has classified an additional 24 Henry Draper Catalogue Cepheids as of luminosity classes Ib-II or II--almost all among the F's--but the number of normal supergiants of these luminosity classes is also no doubt substantially larger than the number of Ib stars. It will be of interest to see the results of Dr. Houk's classifications in other parts of the sky. In the meantime detailed study of the apparently non-variable supergiants of the southern sky might be well worthwhile.

References

Fernie, J.D. 1976, Publ. Astron. Soc. Pac. 88, 116.
Fernie, J.D. and Hube, J.O. 1971, Astrophys. J. 168, 437.
Houk, N. and Cowley, A.P. 1975, U. of Michigan Catalogue of Two-Dimensional Spectral Types for the HD Stars, Vol. 1 (Ann Arbor: U. of Michigan Dep't of Astronomy).
Houk, N., Hartoog, M.R., and Cowley, A.P. 1976, Astron. J. 81, 116.

Cepheid-like Supergiants in the Halo

D. D. Sasselov
Department of Astronomy, University of Sofia
Department of Astronomy and National Astronomical Observatory, Bulgarian Academy of Sciences

Abstract. The newly proposed UU Her-type stars are discussed; their main features being long-period variability and high galactic latitude. It is unlikely that the UU Her stars originated in the plane of the galaxy as they are now located more than a kiloparsec above it. An alternative explanation of their normal Population I characteristics must be looked for.

In the past several years, there has been considerable interest in the so-called Cepheid-like supergiants. One intent of these studies has been to find very long-period Classical Cepheids. However, it has gradually become evident that there also exist variable supergaints blueward of the instability strip. The star UU Her is a good example of such a variable, yet, for many years, little attention has been paid to it. However, now it seems that pulsation to the left of the blue radial pulsation edge can be explained theoretically (e.g., Shibahashi & Osaki 1981, predict observable low-harmonic non-radial modes in these regions). Thus the similar variability of many supergiants of different status might be the result of a single pulsation mechanism. Such stars could be considered as part of a wider class (Percy 1980); however we then lack overall homogeneity. Accordingly, we might expect that subtle differences in the variability of Cepheid-like supergiants should exist; and as more details of their photometric behaviour become known, this appears to be the case.

In that context comes our suggestion that there exists a small group of Cepheid-like supergiants sharing similar variability, and having other properties in common, such as their galactic distribution. We have called these stars UU Herculis stars (Sasselov 1983). They seem to be F0-F7 supergiants of roughly normal composition yet located at high galactic latitudes - an apparent contradiction in terms (Sasselov 1984). The main features of their semi-regular variability are:

- small amplitudes (0.1 to 0.6 mag) and long periods (40 to 100 days);
- pulsation mode switching: two (or three) distinct alternating modes (within the above period range), switching semi-regularly from one to the other, with a shorter interval of erratic fluctuations in between;
- short standstills: unpredictable abrupt ceasing of pulsation for a couple of months.

Nonradial pulsations seem to offer the least contradictory explanation but evidence is inconclusive so far.

At present the established UU Her stars are five in number and there is a list of some 33 suspected, as well. We tend to widen the frames of the term "variability type", paying particular attention to the contradiction between their Population I parameters and high galactic latitudes.

Concerning the status of the UU Her stars: Are they really normal Population I supergiants, or are they post-AGB stars of the old-disk population masquerading as normal? The latter explanation has been suggested recently by Bond, Carney & Grauer (1984) who call them 89 Her variables. Another suggestion comes from the work of Belyakina et al. (1984) who find many similarities between the PU Vul-phenomenon and the UU Her stars. Following this line of reasoning there is a interesting analogy with the discussion about the status of the high-latitude B-stars. The initial suggestion that they are subluminous and "mimic" normal Population I parameters seems now to be inconsistent with the many detailed analyses of Keenan et al. (1982), Keenan & Lennon (1984), Tobin & Kaufmann (1984) and other. Similarly normal A-type stars have been found towards the South Galactic Pole by Rodgers et al. (1981).

So, our further discussion of the UU Her stars depends critically upon the reliability of their spectroscopic and photometric analyses. We prefer as most reliable the parameters derived from detailed studies of the two UU Her stars in binary systems, namely BL Tel and 89 Her. For them normal Population I parameters and solar abundances have been obtained (Table 1). As regards the other UU Her stars, HD 161796 seems to have solar abundances (see Fernie & Garrison, 1984); and HD 112374 is moderately (Luck et al., 1983) or slightly (Sasselov & Kolev, 1984) metal-poor. The latter fact perhaps should not lead automatically to a low-mass interpretation, as there exist young massive F-supergiants in the plane which are also slightly metal deficient (e.g. i Car, analysed by Boyarchuk & Lyubimkov, 1984). on the other hand, it should be realised that we are quite likely to come upon some low-mass old stars among the suspected UU Her stars, as is the case with HD 46703 (Bond et al., 1984). However, it seems that generally the low mass alternative is preferred mainly because of the problems arising from the high galactic latitudes of these stars. Otherwise, we should have to admit that star formation is possible in the halo. Indeed, this unorthodox possibility does not appear so exotic at present. Dyson & Hartquist (1983) showed recently that OB-stars may well be formed in the halo by the collision of cloudlets within intermediate and high-velocity clouds. Their quantitive calculations are based on parameters derived from radio observations. Moreover now that there is evidence for normal B-stars in the halo (as mentioned above) such early-type objects might be the predecessors of the UU Her stars.

Following the theoretical results of Dyson & Hartquist (1983) one should expect to find in the halo a very sparse population of early-type stars, occasionally forming in wide groups of several stars each.

Perhaps such might be the concentration of F-type supergiants (MK-classification) we have noted around SA 11, 27, 28 and 31. They comprise 5 to 7 stars each, of 8th to 10th magnitude in an area of 4 to 6 deg^2. This gives rough dimensions of about 200 to 400 pc. Thus they seem to resemble some peripheral isolated OB-groups in the nearby galaxy M33. However a detailed study of these stars is necessary before discussing the structure of the apparent concentrations they form.

In summary, we suggest the existence of a group of Cepheid-like supergiants - UU Her stars - which pose two major problems. First of all, they exhibit a quite specific variability with some unusual properties such as pulsational mode switching and standstills. On the other hand, they are at high galactic latitudes being either tracers of recent occasional star formation in the halo, or perhaps post-AGB stars - both alternatives being interesting. Or it may be so, that stars of both status exist but are not well separated yet.

We would like to conclude here, leaving the questions open, being only convinced that the UU Her stars deserve our special attention in the future.

The author is much indebted and grateful to Dr. J.R. Percy and Dr. J.D. Fernie, and to the Organizing Committee, for their cordial invitation and the grant which enabled him to attend IAU Colloquium 82. It is a pleasure to express warm thanks also to Prof. N.S. Nikolov and to the Faculty of Physics, University of Sofia, for their encouragement and support.

References

Belyakina, T., Bondar, N. & Chochol, D. et al. (1984). Astron. Astrophys. 132, L12.

Bond, H., Carney, B. & Grauer, A. (1984). Publ. Astron. Soc. Pacific 96, 176.

Boyarchuk, A. & Lyubimkov, L. (1984). Astrofizika 20, 85.

Dyson, J. & Hartquist, T. (1983). M.N.R.A.S. 203, 1233.

Fernie, J. & Garrison, R. (1984), preprint.

Keenan, F., Dufton, P. & McKeith, C. (1982), M.N.R.A.S., 200, 673.

Keenan, F. & Lennon, D. (1984), Astron. Astrophys. 130, 179.

Luck, R., Lambert, D. & Bond, H. (1983), Publ. Astron. Soc. Pacific 95, 413.

Percy, J. (1980), Journal AAVSO 9, No. 2, 64.

Rodgers, Harding & Sadler (1981), Astrophys. J., 244, 912.

Sasselov, D. (1983), I.A.U. Inf. Bull. Var. Stars, No. 2387.

Sasselov, D. (1984), Astrophys. Sp. Sci., in print.

Sasselov, D. & Kolev, D. (1984), in preparation.

Shibahaski, H. & Osaki, Y. (1981), Publ. Astron. Soc. Japan 33, 427.

Tobin, W. & Kaufmann, J. (1984) M.N.R.A.S. 207, 369.

Table 1

	89 Her	BL Tel(F)
MK	F2 Ib	F5 Ia
<B-V>	0.25	0.37
T_{eff}	7000°K± 100	6700°K± 100
M_v	-6.8	-7.2
$R/R_\odot$	210 ± 100	200 ± 40
$M/M_\odot$	13 - 24	14 - 23
Z	1100 pc	-2900 pc
V_r	-27 km/sec	+92 km/sec
Abundance	solar	solar
Binary type	SB (?)	SB, EB
Source	Fernie (1981) Arellano Ferro (1984)	van Genderen(1982)

Fernie, J. (1981), Astrophys. J. 243, 576.
Arellano Ferro, A. (1984), P.A.S.P., preprint.
van Genderen, A. (1982), Astron. Astrophys. 105, 250.

THE QUASI-CEPHEID NATURE OF RHO CASSIOPEIAE

John R. Percy and David Keith
David Dunlap Observatory, Department of Astronomy
University of Toronto, Toronto, Canada M5S 1A1

Rho Cassiopeiae (HR 9045, HD 224014) is a bright yellow supergiant (F8pIa). If it is a member of the association CAS OB5, it has an Mv of about -9.5, and is therefore one of the most luminous yellow supergiants known (Humphreys 1978).

The brightness of ρ Cas varies in two ways: (i) it undergoes semi-regular cycles with amplitudes of about $0^m.2$ and time scales of about a year, perhaps due to pulsation, and (ii) in the 1940's, it faded by more than a magnitude for 660 days (Gaposchkin 1949); as a result, it is often classified as "R CrB?". Although various writers have pointed to a dissimilarities between ρ Cas and the R CrB stars, it should be noted that R CrB itself undergoes semi-regular cycles as well as fadings. Thus the scientific interest of ρ Cas lies in the nature of its variations, their possible relationship to each other and to the mass loss and extreme luminosity of the star.

To follow the variations in brightness of ρ Cas, it would be desirable to have many decades of continuous photoelectric photometry. Such photometry, unfortunately, has been sporadic. Visual observations, on the other hand, have been made almost continuously by the British Astronomical Association (BAA) and the American Association of Variable Star Observers (AAVSO). These observations are numerous and accurate enough so that they can be combined into 30-day means whose formal standard deviations are typically $0^m.03$ - comparable with those of photoelectric observations. The BAA observations have already been published (Bailey 1978). In this paper, we discuss these observations and (in a preliminary way) the AAVSO observations, particularly as they relate to the Cepheid nature of the star.

We have compared the BAA and AAVSO 30-day means with each other and with available photoelectric photometry as listed by Bailey (1978) and Arellano Ferro (1983). The 30-day means are of comparable accuracy. There is an offset m_v - V of about $0^m.30$ between the visual and the photoelectric observations. Bailey (1978) notes that the BAA means are on a system which is fainter by about $0^m.1$ than V magnitudes, and he attributes the rest of the offset to the colour difference between the comparison stars (which are blue) and the variable star (which is red). In fact, we find some preliminary evidence that the offset depends on the instantaneous colour of the star and on the individual visual observer. This could also be understood in terms of the difference in colour sensitivity between the V filter and the individual human eye.

In principle, the offset can be corrected for: to first order by applying a constant correction to the visual means, or to second order by applying a colour-dependent correction. Conversion of m_v to V magnitudes, however, should not be done without considerable thought.

The motivation for using the visual data was to investigate the power spectrum. We therefore wished to reassure ourselves that there were no systematic seasonal effects caused by the visual observing methods. Bailey (1978) applied various tests to the BAA data, and found no such effects. We have applied other tests to the BAA and AAVSO data: (i) we find no obvious correlation between the time of year and the occurrence of maxima or minima (ii) there is no peak in the power spectrum at a period of one year (iii) the light curve obtained by folding the data with a period of one year has insignificant amplitude and (iv) the observations follow the photoelectric observations (e.g. Arellano Ferro 1983) well.

The BAA (1964-1976) and AAVSO (1975-1983) observations both show semi-regular cycles of amplitude about $0^m.2$ and time scale about a year. We determined the power spectra of each data set using Deeming's (1975) and Scargle's (1982) methods for unequally-spaced data, and using the standard Fast Fourier Transform (FFT) method for equally-spaced data. The 30-day means are equally-spaced, and the FFT proved to be quite adequate. The BAA data show a mildly significant (P=0.9) peak in the power spectrum at 275 $\pm$ 25 days. The AAVSO data show a less significant peak at the same period.

We have compared this period with that predicted by infrared (J and K) period-luminosity relations for Cepheids in clusters and associations (Welch 1983), assuming that ρ Cas is a member of CAS OB5 and has a distance modulus (12.0) and absorption (Av = 2.1) consistent with such membership (Humphreys 1978). The star lies within $0^m.2$ of these P-L relations. This, together with the position of the star near the Cepheid instability strip, suggests that it is related to the Cepheids. It is also similar to the "Leavitt variables" in the Magellanic Clouds, described by Grieve *et al*. elsewhere in these proceedings.

Acknowledgements. We thank the AAVSO observers and staff (especially Dr. Janet A. Mattei) for providing their observations, and the Natural Sciences and Engineering Research Council of Canada for a research grant.

Arellano Ferro, A. (1983). Unpublished Ph.D. thesis, Univ. of Toronto.
Bailey, J. (1978). J. Brit. astron. Assoc. 88, 397-401.
Deeming, T. J. (1975). Astrophys. Space Sci. 36, 137-158.
Gaposchkin, S. (1949). Harvard Bull #919, 18-19.
Humphreys, R.M. (1978). Astrophys. J. Suppl. 38, 309-350
Scargle, J.D. (1982). Astrophys. J. 263, 835-853.
Welch, D.L. (1983). Unpublished M.Sc. thesis, Univ. of Toronto.

ON THE BRIGHTNESS AND PULSATIONAL PROPERTIES OF YELLOW SUPERGIANTS

A. Arellano Ferro
Department of Astronomy, University of Toronto, and Instituto de Astronomía, Universidad Nacional Autónoma de México.

ABSTRACT. Most of the existing data in the literature is used to search for the periodicity of five yellow supergiant variables. The pulsational mode and Q-values are discussed, being the luminosity the key parameter for their accurate determination. An attempt to find the spherical harmonic ℓ from the light and colour variation phase shift is carried out.

PERIODICITY

The periodicity in these stars is not strict. The characteristic time of variation has been found for the five yellow supergiants in Table 1, using virtually all existing observations in the literature. Photometry and radial velocities have been studied when available. In the case of the eclipsing binary ε Aur, data out of the eclipse have been preferred to avoid contamination by the companion. Period changes were detected (ε Aur). A new period switch was found for HD161796, from 43 days in 1981 to 62 days in 1982 (details will be published in a forthcoming paper).

Table 1. Pulsational properties of five yellow supergiants.

Star	P (days)	Epoch	M_V	$M/M_\odot$	Q	ℓ
HD161796	62,43	1979-82	-8 — -9	14-24	0.047-0.071	0
89 Her	63.5	1977-78 1980-81	-6.7 — -9	13-24	0.056-0.190	?
ε Aur	123, 160	1927-30 1931-61	-8.7	16	0.160	$\sim$2 or odd
V509Cas	385	1978-81	-9.5 — -9.1	26-30	0.066-0.087	$\geq$ 4
ρ Cas	483	1964-68 1979-81	-9.5	29	0.180	$\sim$2 or odd

PULSATION CONSTANT Q

Theoretical calculations show that Q-values for luminous stars with convective envelopes range between 0.04 and 0.06 for radial fundamental mode pulsations (Takeuti 1979). Larger Q-values are normally interpreted as non-radial oscillations (Maeder 1980). Empirical Q-values were obtained using equation (2) in Maeder (1980), where Q is a function of P,M,M_V and T_e. T_e was estimated from the spectral type; M from theoretical models with mass loss (Maeder 1980). M_V is then the key parameter for the determination of Q. For ε Aur,V509 Cas and ρ Cas, M_V is known from their membership in OB associations (Stothers 1971; Humphreys 1978). They are non-radial pulsators. For HD161796 and 89 Her M_V is uncertain. However, arguments such as period ratios (HD161796) and capricious phasing changing from season to season between V, B-V and radial velocities (89 Her),seem to indicate that HD161796 is a radial pulsator (Fernie 1983) and 89 Her is a non-pure radial pulsator.

SPHERICAL HARMONIC ORDER ℓ

The spherical harmonic order ℓ (ℓ = 0 for radial modes) can be estimated from the phase shift between V and B-V variations (Balona & Stobie 1979;1980). For the long period yellow supergiants, the main difficulty is that simultaneous observations seldom exist and cycles are ill-defined. Simultaneous observations by the author (to be publish ed) and some from the literature permit the estimations of ℓ in Table 1. This method leaves much room open for improvement when more appropriate data become available.

REFERENCES

Balona, L.A. & Stobie, R.S. (1979). Mon. Not. R. astr. Soc., **189**, 649.

Balona, L.A. & Stobie, R.S. (1980). Mon. Not. R. astr. Soc., **190**, 931.

Fernie, J.D. (1983). Astrophys. J., **265**, 999.

Humphreys, R.M. (1978). Astrophys. J. Suppl., 38, 309.

Maeder, A. (1980). Astr.Astrophys., **90**, 311.

Stothers, R. (1971). Nature, **229**, 180.

Takeuti, M. (1979). Science Reports, Tohoku University, 62, 7.

SOME REMARKS ON THE UNIQUE CEPHEID HR 7308

N. R. Simon
Department of Physics and Astronomy
University of Nebraska-Lincoln, Lincoln, NE 68588-0111

HR 7308 is an apparently normal Pop. I Cepheid (Percy & Evans 1980; van Genderen 1981) with a constant period P = 1.49 days. However, its amplitude is variable with a range from about 5 to 20 km/sec in radial velocity and 0.05 to 0.30 in visual magnitude. The timescale for this variation is about 1200 days. According to Breger (1981) and Burki, *et al*. (1982; hereafter BMB), the star seems to be pulsating in a single modulated radial mode. Using published Q-values BMB identify this mode as a second or higher overtone. These authors also estimate a radius $R = 34 \pm 5\ R_\odot$ and, based in part upon the observations of van Genderen (1981), a temperature $\log T_e = 3.786 \pm 0.01$ and a gravity $\log g = 2.25 \pm 0.25$.

In the present investigation we have constructed linear nonadiabatic (LNA) pulsation models for HR 7308 using the Lagrangian code described by Aikawa & Simon (1983). These models assume the standard mass-luminosity relation for Pop. I stars, and fix the effective temperature at the value quoted above, namely T_e = 6110K. For masses appropriate to classical Cepheids we rule out fundamental or first overtone pulsation, in agreement with BMB. In addition, however, we are able to effectively discard the third and higher overtones which are linearly stable and thus not expected to exist at finite amplitude. Thus if HR 7308 is a normal Pop. I Cepheid with temperature as given by BMB, *it must be pulsating in the second overtone*.

To reproduce the period of HR 7308 we have constructed an LNA model with the following parameters: $M = 5\ M_\odot$, $L = 1000\ L_\odot$, T_e = 6110K, X = 0.70, Z = 0.02. The periods and growth rates (P,η) for the three lowest modes of this model are (2.40, 2.6E-3), (1.81, 1.8E-2) and (1.45, 3.4E-2). The higher modes (third overtone and up) are stable. One notes that the second overtone approximately matches the period of HR 7308 and has a growth rate which exceeds that of the fundamental by a factor 13 and that of the first overtone by a factor of nearly 2. While it is not possible to directly infer limit cycle characteristics from linear growth rates, experience indicates (e.g., King, *et al*. 1973) that the growth rates exhibited by the present model are at least not unfavorable for second overtone pulsation. In addition, this model has a gravity $\log g = 2.24$ and radius $R = 28\ R_\odot$, both reasonably consistent with the values estimated by BMB.

In another part of their study, BMB Fourier decompose the radial velocities from a number of epochs of HR 7308 and publish the Fourier coefficients up to and including the second order quantities A_2 and ϕ_2. The zeroth order (static) velocity A_0 is found to vary from epoch to epoch with a relatively large range exceeding 1 km/sec. However, it is not clear to what extent this might be due to scattered phase coverage in the observations. The small second order quantities are subject to even greater distortion due to sparse coverage and are thus probably not well determined.

For one particular epoch BMB publish Fourier coefficients for both radial velocity and V magnitude. Using these we have calculated the first order phase shift between light and velocity, obtaining $(\Delta\phi)_1 = -0.24$. This value is normal for Pop. I Cepheids (Simon 1984) but may not confirm the overtone nature of the pulsations of HR 7308. A larger sample of very short period stars will be necessary to make further progress on this point.

Breger (1981) published 87 V magnitude observations of HR 7308 spread over about 3 years. We have selected for Fourier analysis 62 of these points all obtained within 3 months and thus corresponding to a given epoch of the 1200-day amplitude cycle. It was hoped to determine in this way the second-order Fourier quantities ϕ_{21} and R_{21} (Simon & Lee 1981) which have proven so useful in treating the classical pulsating stars. However, the oscillations of HR 7308 are so sinusoidal that the (very small) second order coefficients cannot be determined without extremely fine phase coverage which Breger's observations did not provide. Thus the calculation of ϕ_{21} and R_{21} must await a more extensive set of observed data.

References

Aikawa, T. & Simon, N. R. (1983). Ap. J., 273, 346.
Breger, M. (1981). Ap. J., 249, 666.
Burki, G., Mayor, M. & Benz, W. (1982). Astr. and Ap. 109, 258 (BMB).
King, D. S., Cox, J. P., Eilers, D. D. & Davey, W. R. (1973). Ap. J. 182, 859.
Percy, J. R. & Evans, N. R. (1980). A. J. 85, 1509.
Simon, N. R. (1984). Ap. J., in press.
Simon, N. R. & Lee, A. S. (1981). Ap. J., 248, 291.
van Genderen, A. M. (1981). Astr. and Ap. 99, 386.

KINEMATIC EVIDENCE FOR FUNDAMENTAL MODE PULSATION IN THE SHORT-PERIOD CLASSICAL CEPHEID SU CASSIOPEIAE

D.G. Turner
Saint Mary's University, Halifax, NS, B3H 3C3, Canada

D.W. Forbes
Trent University, Peterborough, Ont., K9J 7B8, Canada

R.W. Lyons
David Dunlap Observatory

R.J. Havlen
N.R.A.O.

The short period and small amplitude of pulsation for the $1^d.95$ Cepheid SU Cas make it an excellent candidate for pulsation in a purely excited mode, and, as summarized by Gieren (1982), there is some evidence from recent radius determinations for this variable which suggests that it is indeed an overtone pulsator. This conclusion considers only the Cepheid's observed characteristics, however, and ignores 3 important additional pieces of information:

1. SU Cas illuminates reflection nebulosity from a nearby dust cloud, whose distance of 258 pc (Turner & Evans 1984) is consistent with fundamental mode pulsation for the Cepheid.

2. One previous radius determination by Milone (1971), making use of data independent of that used in the other studies quoted by Gieren, yields a radius for SU Cas consistent with fundamental mode pulsation.

3. The radial velocity of -6 ±1 km s^{-1} quoted by Gieren (1976) for SU Cas is kinematically consistent with fundamental mode pulsation.

This last result is based upon an unpublished study by Havlen, the data from which have recently been reanalyzed by us with the addition of new photometric observations for B stars in the field of SU Cas. Havlen's original purpose for this study was to search for possible members, by now mostly main-sequence stars of spectral type B8 or later, of the cluster or association in which SU Cas originated. The search included all stars in the HD Catalogue brighter than 9th magnitude, of spectral type B9 or earlier, and lying within $\sim 5°$ of SU Cas, in order to ensure possible detection of such companions out to and beyond the expected distance of SU Cas. A few early A stars near the Cepheid were added by Turner & Evans (1984). The distribution of program stars on the plane of the sky is illustrated in Fig. 1, where filled circles denote stars comparable in age to SU Cas, open circles denote stars much younger than SU Cas, and a filled diamond symbol denotes the one star judged to be older than the Cepheid. Identical symbols are used in Figs. 2 & 3.

Age discrimination was based upon the location of each star in the H-R diagram of Fig. 2, in conjunction with an estimated age for SU Cas (embracing the possibilities of fundamental mode, F, and first harmonic, 1H, pulsation), assuming it to be in the first or higher crossing of the instability strip. The luminosity of SU Cas was derived from the PL relation of van den Bergh (1977), which is consistent with the B star

Figure 1.

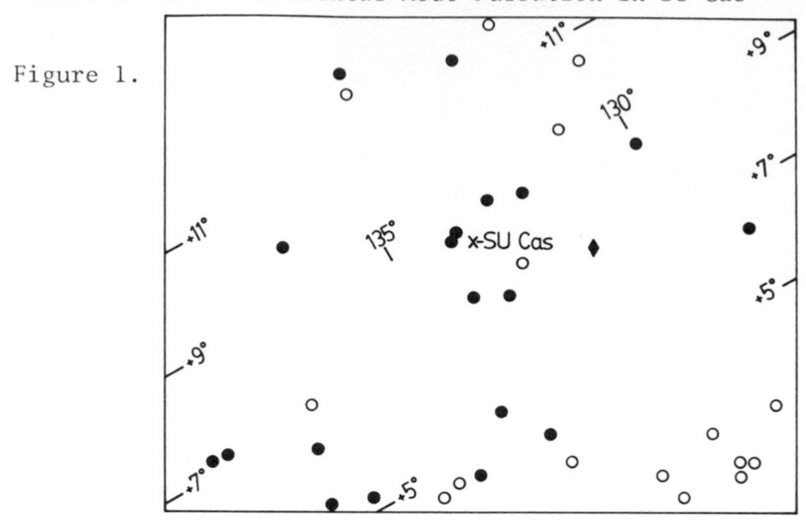

luminosity scale used here. Individual data for the stars in Fig. 2 were derived from broad band UBV photometry, Strömgren and Hβ photometry, and up to 3 blue spectra at a dispersion of 63 Å mm^{-1} for each star. Radial velocities were obtained from Abt & Biggs (1972), Turner & Evans (1984), and/or from radial velocity measures derived from the original plate material of Havlen. The resulting heliocentric velocities are plotted as a function of distance modulus for each star in Fig. 3, where the shaded region denotes the velocity of SU Cas. The main features of Figs. 1 & 3 which pertain to the pulsation mode of SU Cas are summarized below:

1. The majority of stars younger than SU Cas are located within 5° of the galactic plane, which is typical of a group of young field objects unrelated to the Cepheid. Stars comparable in age to SU Cas predominate

Figure 2.

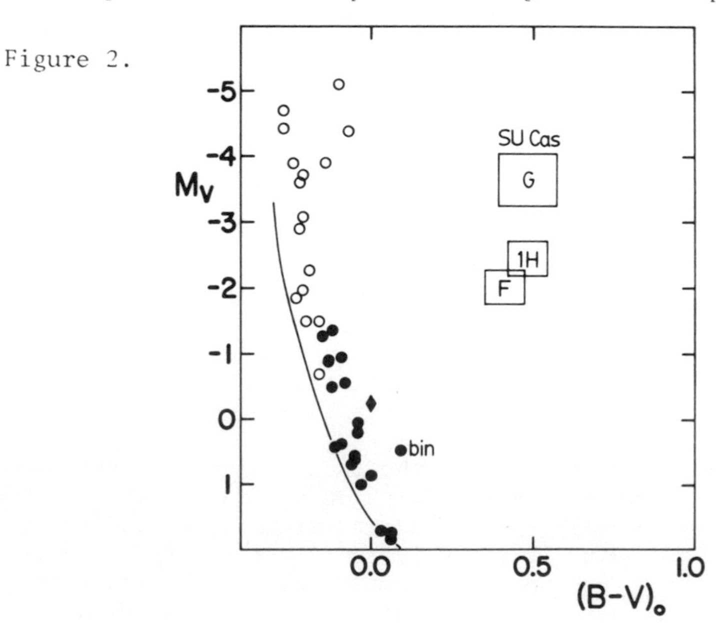

Figure 3.

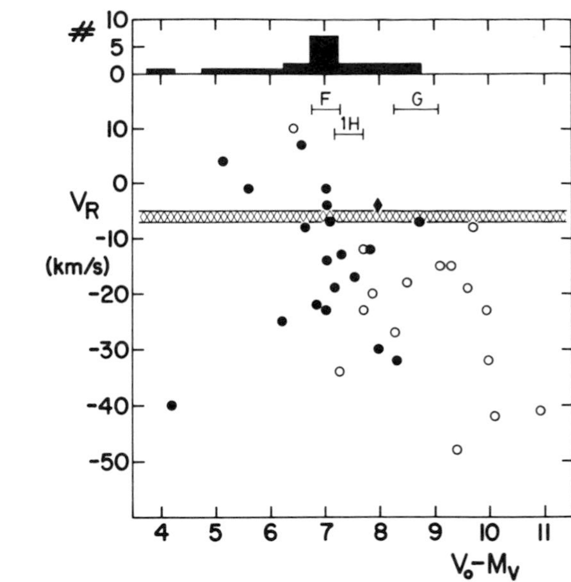

at galactic latitudes similar to that of the Cepheid (+8°.5).

2. As noted in the upper part of Fig. 3, stars comparable in age to SU Cas peak in number at a distance modulus of 7.0, which coincides in distance with the SU Cas BA association discovered by Turner & Evans (1984). This is also the distance at which most of the stars of comparable velocity to SU Cas are found.

3. There is only one star of comparable velocity and age to SU Cas at the distance indicated by the results of Gieren (1982; G in Figs. 2 & 3). However, this star (HD 19856) is almost 6° distant from SU Cas and lies near the edge of the field of Fig. 1.

4. If SU Cas is an overtone pulsator, then it has no nearby companions of similar age and radial velocity.

In general, the kinematic evidence supports the conclusions of Turner & Evans that SU Cas lies at a distance consistent with fundamental mode pulsation. It should be emphasized that this conclusion also rests upon the morphological evidence that SU Cas illuminates reflection nebulosity from the same dust cloud in which are embedded two other members of the association containing the Cepheid. It seems impossible to reconcile all of these arguments with anything other than fundamental mode pulsation for SU Cas, and the suggestion that this Cepheid is an overtone pulsator must therefore be considered untenable.

References.

Abt, H.A. & Biggs, E.S. (1972). Bibliography of Stellar Radial Velocities, New York, Latham Process Corp.
Gieren, W. (1976). Astron. Astrophys. 47, 211.
Gieren, W. (1982). Publ. Astron. Soc. Pacific 94, 960.
Milone, E.F. (1971). Bull. Am. Astron. Soc. 3, 351.
Turner, D.G. & Evans, N.R. (1984). Astrophys. J. 283, in press.
van den Bergh, S. (1977). IAU Colloq. 37, 13.

A STUDY OF THE SHORT-PERIOD CEPHEID EU TAURI

W.P. Gieren
Observatorio Astronómico, Universidad Nacional, and Physics Department, Universidad de los Andes, Bogotá, Colombia

EU Tauri is a classical Cepheid with one of the shortest periods ($2^{d}.1$) known. A Fourier decomposition study of its V light curve by Simon & Lee (1981) revealed a peculiar position in phase-period diagrams, very similar to the star SU Cas which is probably an overtone pulsator (Gieren, 1982). This suggested the possibility that EU Tau might be another galactic overtone pulsator. In order to investigate this question, some 100 new photometric observations of this star on the UBVRI (Johnson) system were obtained with the #2 0.9m telescope of KPNO. Simultaneously, 43 CCD spectra of the star were secured on the Coudé feed telescope of KPNO which were measured for radial velocities using a correlation technique. These velocities have an internal accuracy of better than 0.5 km/s and define a complete velocity curve of EU Tau.

The new photometry shows that a secondary oscillation is present in this star which causes variations of the V magnitude at given phases corresponding to the primary period of about 0.02-0.03 mag. Inspection of the older light curves of the star of Guinan (1972) and Sanwal & Parthasarathy (1974) suggest that this double mode effect of similar size has been present during their observing seasons as well. The surface brightness method of Barnes & Evans (1976) was applied to the new data, and values of

$$R = (18.4 \pm 1.8) R_{\odot}$$
$$d = 746 \pm 74 \text{ pc}$$

for the mean radius and distance of the star were obtained. Using a method based on BVRI photometry given by Fernie (1982), a reddening value of E(B-V) = 0.13 was derived. The star's distance together with this reddening value yield an absolute magnitude of

$$M_{\langle V \rangle} = -1.69 \pm 0.25$$

for EU Tau. This value is about 0.5 mag fainter than the value predicted by recent calibrations of the P-L-C relation. A mean effective temperature of 5970 K obtained from Flower's (1977) $T_e - (B-V)_o$ calibration produces a consistent absolute magnitude of $-1^{m}.75$, using the mean radius obtained from the surface brightness method.

In conclusion, EU Tau is not a single mode overtone pulsator but rather a new double mode Cepheid with a very small amplitude of the secondary oscillation. Its radius is close to the value expected for a normal single mode Cepheid of its period, suggesting that EU Tau has a normal mass for a classical Cepheid, thus not confirming the low-mass hypothesis for double mode Cepheids. However, EU Tau might be slightly sub-luminous when compared to single mode Cepheids of the same period.

ACKNOWLEDGMENTS

I am grateful to the Kitt Peak National Observatory for allocation of observing time there. Special thanks I wish to express to Daryl Willmarth of KPNO for permitting the use of his CCD radial velocity reduction program. Financial aid of the CINDEC of the Universidad Nacional and of COLCIENCIAS are acknowledged.

REFERENCES

Barnes, T.G., and Evans, D.S.(1976). M.N.R.A.S., 174, 489.
Fernie, J.D. (1982). Ap. J., 257, 193.
Flower, P.J. (1977). Astr. Ap., 54, 31.
Gieren, W.P. (1982). Pub. A.S.P., 94, 960.
Guinan, E.F. (1972). Pub. A.S.P., 84, 56.
Sanwal, N.B., and Parthasarathy, M. (1974). Astr. Ap. Suppl., 13, 91.
Simon, N.R., and Lee, A.S. (1981). Ap. J., 248, 291.

ABUNDANCE ANALYSIS OF ALPHA URSAE MINORIS

Sunetra Giridhar
Indian Institute of Astrophysics, Bangalore 560034, India

Abstract. We have derived atmospheric abundances of the bright Cepheid α UMi in order to study the abundance anomalies in different elements. The atmospheric abundance of C, O, Fe-peak elements Ca, Sc, Ti, Cr, Fe and heavier s-process elements Y, Ba, Ce and Sm have been derived using the method of spectrum synthesis. The abundance of carbon is derived using the C I lines in 4700A region, whereas for oxygen, the forbidden line at 6300.311A is employed. Abundances of the Fe-peak elements and s-process elements are obtained by synthesizing selected portions in the wavelength range 4330A - 4650A. The estimates of C/O derived from the present investigation are compared with other Cepheids of similar period. The evolutionary status of α UMi is discussed in the light of these derived abundances.

INTRODUCTION

The atmospheric abundances of various elements and abundance ratios like C/O, $^{12}C/^{13}C$, s/Fe etc. lead us towards a better understanding of the evolutionary processes that take place in stellar interiors. The effect of various processes like mixing between the interior and the atmosphere of a star, mass loss, and enrichment of interstellar medium in different elements caused by massive and intermediate mass stars, could be studied in detail when good estimates of light as well as heavy elements are available.

Atmospheric abundances of Fe-peak elements have been derived for a number of bright Cepheids by many investigators. For the Cepheids belonging to the solar neighbourhood these abundances do not differ considerably from the solar value. The abundances of C, N and O for a sample of 14 Cepheids have been derived by Luck & Lambert (1981; hereafter LL) who reported considerable deficiency of carbon and oxygen accompanied by an enhancement in nitrogen. Though the direction of changes in the abundances reported by LL are in accord with the predictions of evolutionary calculations, the magnitude of the changes is larger than the predictions. LL have suggested the extensive mixing into hydrogen-burning region and ON-processing region caused by more convection or meridianal circulation current as possible mechanisms to explain the observed abundance anomalies.

Here we report the derived atmospheric abundances of C, O and Fe-peak elements Ca, Sc, Ti, Cr & Fe and heavier s-process elements Y, Ba, Ce *and Sm for the bright Cepheid α UMi.*

OBSERVATIONS AND ABUNDANCE ANALYSIS

High-resolution spectra of α UMi were obtained with the 102-cm reflector of Kavalur Observatory. The details of observations are given in Table 1. We have used the forbidden line of O I at 6300.311A to derive the oxygen abundance. We have used a spectrogram with a dispersion of 4.2 A mm^{-1} to measure this weak line of oxygen. At this dispersion one could resolve the O I line and also the Sc II line at 6300.678A into two components. We used C I 4769.997A and 4775.877A to derive the carbon abundances. The gf values for the O I line and C I 4775.87A are taken from Lambert (1978). The gf value of C I 4769.997A is taken from Kurucz & Peytremann (1975). The other line of C I arising from the same multiplet (6) at 4771.72A is very badly blended with Fe I 4771.712A and therefore could not be used. The abundances of Fe-peak elements are derived using several lines in the 4350A-4700A spectral region. The solar gf values derived from inverted solar analysis were used for these metalic lines. Selected portions of the stellar spectrum were synthesized using the model atmospheres selected from the grid of model atmospheres by Kurucz (1979), employing the spectrum synthesis code of Sneden (1974). The atmospheric parameters derived by Parsons (1970) by matching the computed and observed UVBGRI colours were used as first guess to the atmospheric parameter.

Table 1. Journal of Observations, and derived Atmospheric Parameters

Date of observation	Phase	Spectral region	Dispersion A mm^{-1}	T_{eff}	Log g	V_t
12-10-1979	0.880	6300	4.2	6300	2.0	5.3
27-09-1980	0.286	4550	11.3	6000	2.0	5.0
02-11-1980	0.329	4700	11.3	5800	1.8	4.8

Table 2. Elemental abundances in α UMi with respect to the solar value

Element	Z	Lines Max	Lines Min	ϕ=0.286	ϕ=0.329	ϕ=0.880
C	6	2	2		-0.30	
O	8	1	1			-0.05
Ca	20	4	2	-0.25	-0.20	
SC	21	2	2		-0.10	
Ti	22	10	6	+0.05	+0.10	
Cr	24	12	8	+0.15	+0.17	
Fe	26	25	20	-0.05	-0.10	-0.05
Y	39	2	2		-0.10	
Ba	56	1	1	-0.20		
Ce	58	4	2	-0.10	-0.15	
Sm	62	2	2	-0.15	-0.10	

Figure 1. Observed (continuous line) and computed (broken line) spectra of αUMi in different spectral regions: (a) the spectral region containing important lines of BaII and CeII; (b) the spectral region containing the forbidden line of OI and (c) the spectral region containing the lines of CI.

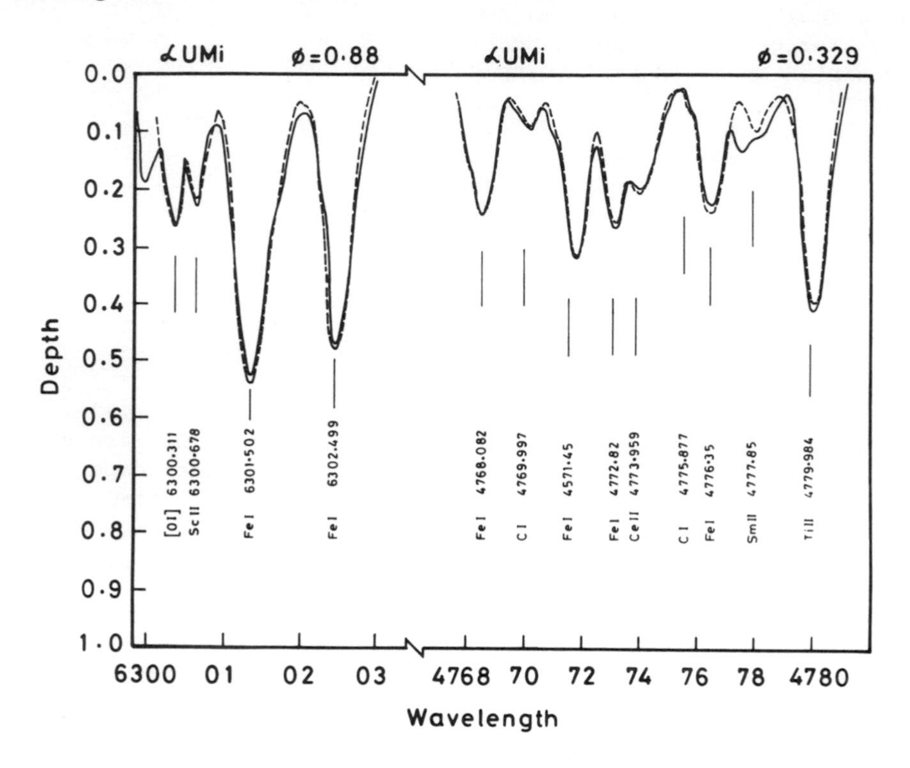

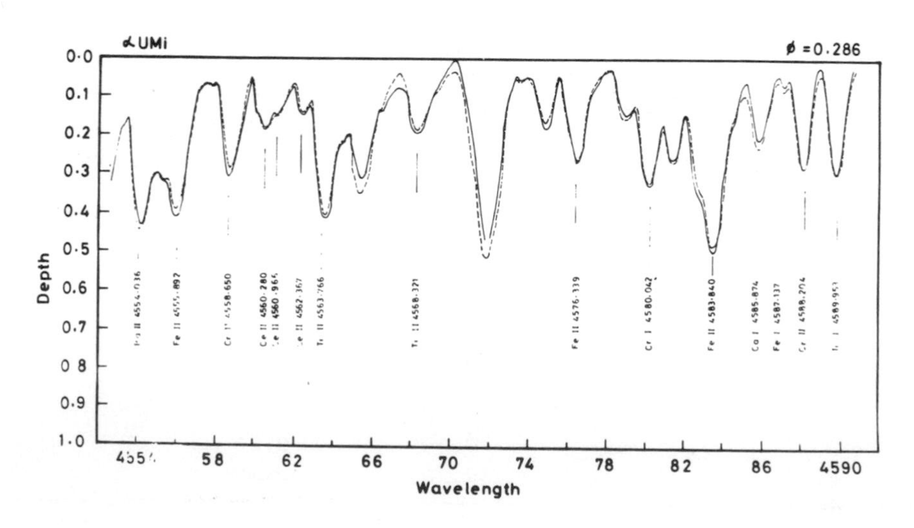

Our method of deriving the final atmospheric parameters and abundances, and justification of assumptions employed in synthesis calculation is given by Giridhar (1983). Spectroscopic gravities derived by requiring neutral and ionised species of the same element leading to the same value of abundances are in good agreement with the gravities calculated using the evolutionary models of Becker et al. (1977). Assuming the second crossing of pulsation strip for α UMi we estimate a mass of 5.91 $M_{\odot}$. Using the period-luminosity relation of van den Bergh (1977) we derived the luminosity, and using the mass derived abc ? we deduced the surface gravity of the star. The value of log g=2.06 derived from evolutionary consideration is in good agreement with the spectroscopic estimate given in Table 1.

RESULTS

The estimates of the abundances of different elements are listed in Table 2. Calcium appears to be underabundant and chromium shows slight enhancement with respect to the solar value. The abundances of Sc, Ti and Fe do not differ significally from the solar value. The s-process elements Y, Ba, Ce and Sm are marginally underabundant and there is no indication of s-process anomalies with respect to iron abundance. Carbon is underabundant by 0.3 dex whereas the abundance of oxygen is not significantly different from the solar value. Figure 1 shows the agreement between the computed and observed spectra.

DISCUSSION

For the stars of intermediate mass passing through the red giant phase, the evolutionary calculations predict the carbon in the stellar atmosphere to be reduced, nitrogen enhanced, while oxygen remains almost unaffected. The magnitude of the changes becomes higher with increasing mass. However, for the sample of LL, the abundances did not show any dependence on period and hence on the mass, and also the observed differences much higher than the predicted ones. Various mechanisms, like extensive mixing of ON-processed material induced by meridianal circulation currents, have been suggested to explain this anomaly. However, it appears that each Cepheid has its own peculiar evolution; i.e, the efficiency of mechanism suggested by LL seems to differ from star to star independent of its period. Nearly solar value of oxygen abundance, with underabundant carbon observed for α UMi could imply an evolution entirely different from the stars from the sample of LL. A study of $^{12}C/^{13}C$ ratio which is a sensitive monitor of CN-processing for a large sample of stars could be a useful step towards understanding the evolutionary differences between individual Cepheids.

REFERENCES

Becker,S.A., Iben,I. & Tuggle, R.S. (1977). Astrophy.J.,218, 633-653
Giridhar,S. (1983). J. Astrophys. Astr. 4, 75-107.
Kurucz, R.L.(1979). Astrophys. J.Suppl. Ser.40, 1 - 340.
Kurucz, R.L., & Peytremann, E. (1975). Smithsonian Astrophys. Sp. Rep. 362.
Lambert,D.L.(1978). Mon. Not.R.Astr. Soc., 182, 249-272.
Luck, R.E. & Lambert, D.L.(1981). Astrophys. J, 245, 1018-1034.
Parsons, S.B. (1970). Warner Swasey Obs. Preprint No.3.
Sneden, C.A. (1974). Ph.D. Thesis, The University of Texas at Austin.
van den Bergh., S. (1977). In Decalages vers le Rouge et Expansion de l' Universe, ed. C. Balkowski & B.E.Westerlund, Paris: CNRS, pp.14 - 38.

CEPHEID EVOLUTION

S. A. BECKER
University of California, Los Alamos National Laboratory,
Los Alamos, NM 87545 USA.

Abstract. A review of the phases of stellar evolution relevant to Cepheid variables of both Types I and II is presented. Type I Cepheids arise as a result of normal post-main sequence evolutionary behavior of many stars in the intermediate to massive range of stellar masses. In contrast, Type II Cepheids generally originate from low-mass stars of low metalicity which are undergoing post core helium-burning evolution. Despite great progress in the past two decades, uncertainties still remain in such areas as how to best model convective overshoot, semiconvection, stellar atmospheres, rotation, and binary evolution as well as uncertainties in important physical parameters such as the nuclear reaction rates, opacity, and mass loss rates. The potential effect of these uncertainties on stellar evolution models is discussed. Finally, comparisons between theoretical predictions and observations of Cepheid variables are presented for a number of cases. The results of these comparisons show both areas of agreement and disagreement with the latter result providing incentive for further research.

1 GENERAL OVERVIEW OF STELLAR EVOLUTION MODELS IN REFERENCE TO CEPHEID VARIABLES.

1.1 The Population I Picture

For purposes of discussion in this section, attention will be directed to results obtained from models of single, non-rotating stars. Population I stars are arbitrarily defined to be those for which $Z \geq 0.005$. Figure 1a shows the evolutionary tracks in the H-R diagram for a number of intermediate-mass stars of composition $(Y,Z) = (0.28, 0.02)$. Intermediate-mass stars are those which ignite He non-degenerately but following core He exhaustion develop electron degenerate carbon-oxygen cores. For the composition depicted in Figure 1a, such stars span the range of approximately 2.25 to 9 $M_\odot$. Stars of mass greater than approximately 9 $M_\odot$ do not develop degenerate carbon-oxygen cores and are called massive stars. Finally, stars of mass less than approximately 2.25 $M_\odot$ are called low-mass stars and these objects develop electron degenerate He cores prior to core He ignition.

In Figure 1, the boundaries of the Cepheid instability strip (as determined by the calculations of Iben & Tuggle, 1975) are represented by three parallel dashed lines which are, going from left to right, the first harmonic blue edge, the fundamental blue edge and the fundamental red edge. Standard pulsation theory argues that whenever a star's evolutionary track lies within the Cepheid strip, the star is unstable to surface pulsations and the star should be recognizable as a Cepheid variable. Observations confirm that the majority of stars in the instability strip are indeed Cepheid variables although some exceptions exist (see e.g. Eggen, 1983 and Bidelman, this conference). The boundaries of the Cepheid instability strip extend to the domains of the massive and low-mass stars not shown in Figure 1.

FIGURE 1. Evolutionary H-R Diagrams of models with Y = 0.28.
In (a) Z = 0.02 and (b) Z = 0.01.

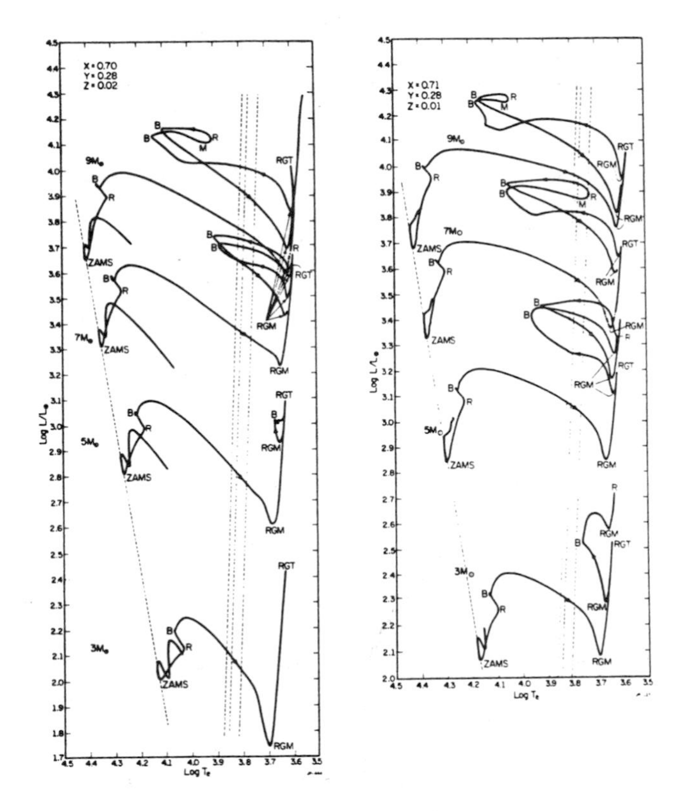

Stellar evolution calculations show that the progenitors of Cepheid variables are main sequence stars, and that to become a Cepheid a star must be in a post core hydrogen-burning phase. Figure 1 shows that a given intermediate-mass star can cross the Cepheid instability strip (depending on its mass or composition) once, thrice, or five times. Massive stars in the range of 9 to 20 $M_{\odot}$ (depending on the mass-loss rate) can also experience up to three crossings (see e.g. Sreenivasan & Wilson, 1978). Stars more massive than 20 $M_{\odot}$ (see e.g. Brunish & Truran, 1982a) as well as stars in the low-mass range (see Mengel et. al, 1979) can undergo at best one crossing. The first crossing occurs for all but the most massive stars during the hydrogen-burning shell phase as the star evolves to cooler temperatures in the H-R diagram on its way to becoming a red giant. The time scale for the first passage through the instability strip ranges from a few x 10^6 yr to about 10^3 yr with the lifetime decreasing with increasing mass. This time scale is approximately on the order of a thermal (Kelvin-Helmholtz) time scale. For example, the 3 $M_\odot$ model of Figure 1a takes about 2 x 10^5 yr to evolve from the fundamental blue edge to the fundamental red edge while a 9 $M_\odot$ model of the same composition takes about 4 x 10^3 yr.

There are effective limits to prevent stars from becoming Cepheids at both the high and low extremes of the stellar mass range. For stars of low enough mass like the sun, the main sequence surface temperatures are cooler than the red edge of the instability strip and consequently, their normal evolutionary tracks do not intercept the instability strip. In fact, low-mass stars which do intersect the instability strip are recognized as a separate class of variables called the δ Scuti stars. Although the boundary between Cepheids and δ Scuti stars is somewhat indistinct, Cepheid variables in practice originate from stars of at least intermediate mass. At the other extreme, stars more massive than about 40 $M_\odot$ appear to lose mass so prodigiously that instead of evolving into red giants their evolutionary tracks reverse before the instability strip and evolve blueward into the domain of the WR stars (deLoore, 1980). This effect limits the brightest Cepheids to periods of no more than about 200 days.

The second crossing of the instability strip occurs during the core helium-burning phase as a star evolves to higher temperatures on the first blue loop in the H-R diagram. Only intermediate-mass stars and massive stars of about 9 - 20 $M_\odot$ exhibit blue loops in their H-R diagrams. The first blue loop arises due to a complicated interplay between the X abundance profile left by the former hydrogen-burning core, the hydrogen-burning shell, the maximum depth reached by the convective envelope during the star's first ascent of the red giant branch, and the nature of stellar envelope solutions (see Schlesinger, 1977 for a review). When the tip of the first blue loop is tangent to the blue edge of the instability strip, passage through the instability strip is driven on a nuclear time scale lasting over several x 10^6 yr. For the composition depicted in Figure 1a, this condition would occur for a model of about 6 $M_\odot$. The lifetime of the second crossing decreases with increasing stellar mass approaching a thermal time scale at larger masses due to the fact that a significant portion of the core helium-burning phase is spent in the vicinity of the first blue loop tip and that as more massive stars are considered the temperature of the blue loop tip increases. In Figure 1a the lifetime of the second crossing is for example about 2.6 x 10^5 yr for the 7 $M_\odot$ model and 3 x 10^3 yr for the 9 $M_\odot$ model.

When it occurs, the second crossing of the instability strip is almost always the longest lived. (It is only in the case of the more massive blue-looping models where all crossings are effectively driven on a thermal time scale that this rule is no longer valid). Consequently, most Cepheids observed are stars undergoing the second passage of the instability strip. In addition, the occurrence of the first blue loop provides an effective lower mass limit to the observed distribution of Cepheids. Only stars which are able to evolve to a high enough temperature to intercept the instability strip during their first blue loop are generally seen as Cepheids. Stars whose first blue loop does not intercept the instability strip can only be observed as Cepheids during their first passage of the strip and these stars due to their shorter lifetimes probably make up only about 10% of the total observed Cepheid population.

The third passage of the instability strip can occur under two different conditions. The most common of the two takes place on the first blue loop near the end of the core helium-burning phase as the star evolves back to the red giant branch. Such is the case for the 7 $M_{\odot}$ models of Figure 1. Massive stars which intercept the Cepheid instability strip only once are also at a similar point in their internal evolution, i.e., they are also near the end of their core helium-burning phase, however, unlike the intermediate-mass stars, they are evolving to the red giant branch for the first time.

The lifetime for the third passage of the instability strip when it occurs near the end of the core helium-burning phase can be as large as just over 10^6 yr for the case of a model whose blue loop tip is tangent to the blue edge of the instability strip. The lifetime of this crossing then deceases with increasing mass until it no longer takes place and the second option for the third crossing is instead in effect. The 7 $M_{\odot}$ model of Figure 1a is an example of the first option for the third crossing and this passage has a lifetime of about 5×10^4 yr which is about one fifth as long as the lifetime of the second crossing but nearly 10 times as long as the lifetime of the first crossing. Generally, the the first option for third passage of the instability strip is the second longest crossing.

The other condition under which a third crossing of the instability strip can occur takes place after core helium exhaustion during the helium-burning shell phase. Such is the case for the 9 $M_{\odot}$ models of Figure 1 and other more massive models which experience blue loops in the H-R diagram. In this case evolution takes place so rapidly in the stellar interior that the model is unable to cross the instability strip until after the helium-burning shell has established itself as the primary energy source for the star. As a result, the second option for the third crossing is the final passage of the instability strip for the stars which experience it. This type of crossing takes place rather rapidly on the order of a thermal time scale lasting approximately 10^3 yr.

Under certain circumstances a second blue loop may occur in the H-R diagram which does intercept the instability strip allowing for two additional crossings of the Cepheid strip. Hoppner et al. (1978) and Becker (1981b) show that this additional loop is due to a complicated interaction between the contracting helium exhausted core, the helium-burning shell and the nature of the stellar envelope solutions. The lifetime of either the fourth or fifth crossing is roughly the same being potentially as large as a few x 10^5 yr for the case where the tip of the second blue loop is tangent to the blue edge of the instability strip. The lifetimes of these two crossings then decrease rapidly with increasing mass until for the more massive models the second blue loop takes place entirely to the left of the instability strip. For example for the 7 $M_\odot$ model of Figure 1a, the lifetimes of the fourth and fifth crossings are each approximately 10^4 yrs. Both the fifth crossing of Cepheid strip and the second condition for the occurrence of the third crossing of the Cepheid strip (see the 9 $M_\odot$ models in Figure 1) take place at the same evolutionary phase, i.e., during the helium-burning shell phase.

To summarize, Population I stars observed to be in the Cepheid instability strip can be undergoing, depending on the circumstances, the hydrogen-burning shell phase, the core helium-burning phase (with an active hydrogen-burning shell) or the helium-burning shell phase. When multiple crossings of the instability strip are possible, the most likely phase to be encountered is the core helium-burning phase with the second crossing being the longest lived of all the passages and the third crossing being the second longest. The cumulative lifetime of the other three crossings (when they occur) is generally small when compared to time spent during the second and third crossings. It is only for the more massive cases (like the 9 $M_\odot$ models in Figure 1) where all crossings of the strip take place on a thermal time scale that the lifetimes of the various crossings became comparable to each other. For stars of sufficiently small or large mass only one crossing of the instability strip is possible. In our galaxy, Cepheids having non-harmonic periods less than 3 days are most likely due to stars which make only one crossing of the Cepheid strip during their hydrogen-burning shell phase. At the other extreme, Cepheids having periods greater than about 30 days are most likely due to massive stars making their single passage of the strip near the end of their core helium-burning phase. The bulk of the Cepheids observed in the Galaxy, however, have periods between these two extremes and the vast majority of these stars should be in the core helium-burning phase. The rarest type of Cepheid would be one which is undergoing the fourth or fifth crossing of the instability strip.

Finally, it should be noted that the evolutionary behavior of a stellar model of a given mass varies as the initial composition is changed (see Becker et al., 1977 and Becker 1981a for an extensive discussion). Figure 1b shows how models of the same masses as used in Figure 1a behave when their metalicity has been reduced to Z = 0.01. For this case, stars having a mass as small as about 4 $M_\odot$ become Cepheids during the core helium-burning phase as opposed to about 6 $M_\odot$ being the lower limit when Z = 0.02. The effect of changing the helium content is shown in Figure 2 where much larger changes in Y are required to produce changes of a similar size as those produced by much smaller changes in Z. By taking into account the composition dependence on a model's evolutionary behavior one can show that the average mass of a Cepheid is about 6 $M_\odot$ for the Galaxy, 4.5 $M_\odot$ for the LMC, and 3.5 $M_\odot$ for the SMC. In general for a model of fixed mass, reducing Z or increasing Y acts to make nearly all evolutionary phases of a model both hotter and brighter.

FIGURE 2. Evolutionary H-R diagrams of models with Z = 0.02. In (a) Y = 0.36, (b) Y = 0.28, and (c) Y = 0.20.

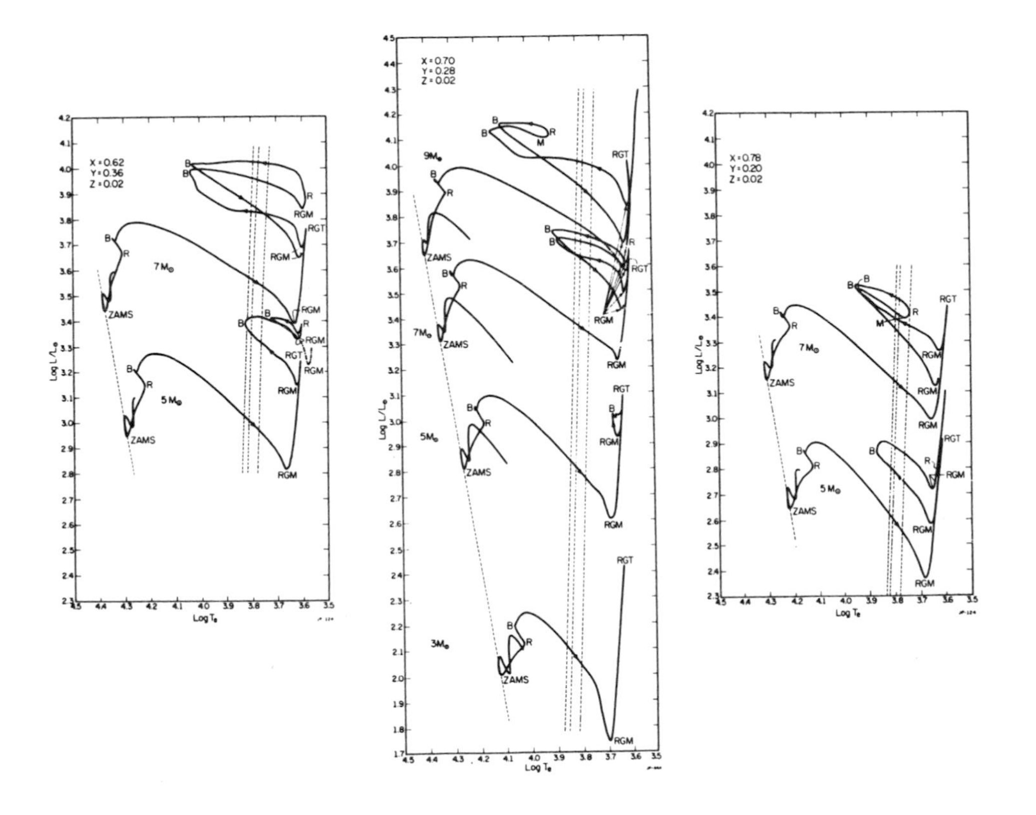

In any case, the progeny of Cepheid variables are later seen as red giants and red supergiants. Intermediate-mass stars evolve onto the asymptotic giant branch (AGB) and undergo helium-shell flashes during the double shell (hydrogen- and helium-burning) phase. The majority of these stars lose their massive envelopes through mass loss and evolve into white dwarfs. A small fraction of the most massive intermediate-mass stars may experience degenerate ignition of the central carbon and become carbon-deflagration supernovae. Massive stars, in contrast, evolve into red supergiants that undergo all the remaining phases of nuclear burning before possibly becoming Type II supernovae.

1.2 The Population II Picture

If single non-coalesced stars of Population II composition ($Z < 0.005$) having masses greater than 2 $M_\odot$ still existed today, such objects would undergo the same behavior as discussed in the previous section and consequently, some of these stars would mimic the behavior of Population I Cepheids. Unfortunately except perhaps in the SMC, no examples of this type of star currently exist although such stars must have existed in the past. Nonetheless as described in the review article by Wallerstein & Cox (1984), Population II Cepheids do exist but they arise due to circumstances that are different from those experienced by Population I Cepheids. Population II Cepheids obey a different period luminosity relation from that of the Population I Cepheids. As a class Population II Cepheids lie in the H-R diagram between the RR Lyrae variables at low luminosities and the RV Tauri and long period variables at high luminosities. Population II Cepheids are broken into three main categories which are BL Herculis (BL Her) stars, the W Virginis (W Vir) stars, and the anomalous Cepheids.

Population II stars having an age near that of the universe can initially be no more massive than 0.8 $M_\odot$ in order to be observed today. Such a star ignites helium under degenerate conditions which leads to the core helium flash and evolution to quiescent core helium-burning on the horizontal branch (see e.g. Figure 1 of Despain, 1981). A portion of the horizontal branch intercepts the instability strip and any stars located at this junction should behave as RR Lyrae variables (see Iben, 1974a for a review of the horizontal branch phase). When helium is exhausted in the center of a horizontal branch star, the star evolves upward in the H-R diagram onto the suprahorizontal branch where the thick helium-burning shell phase is established (see e.g. Iben & Rood, 1970, and Sweigart & Gross, 1976). If while on the horizontal branch the star is located either inside or to the left of the instability strip in the H-R diagram, the post-horizontal branch evolution will cause the star's evolutionary track to intercept the instability strip at which point the star should behave as a BL Her variable (Gingold, 1974 and Iben, 1974b) with a period between 1 - 5 days. Depending on the mass and composition of the star, the evolutionary track may intercept the instability strip more than once with three crossings being the maximum number (see e.g. Figure 7 of Iben & Rood, 1970). When three crossings occur, the third crossing has the longest duration. Unlike Population I Cepheids, BL Her stars originate from a very narrow mass range of about 0.6 $M_\odot \pm 0.05$ $M_\odot$. The total lifetime of this phase of variability lasts up to several x 10^6 yr due to the evolution proceeding on a nuclear-burning time scale.

Eventually a low-mass Population II star evolves off the supra-horizontal branch to the AGB where the hydrogen-burning shell re-establishes itself and the double burning shell phase begins. When the hydrogen envelope is nearly exhausted ($<5 \times 10^{-3}$ $M_\odot$) Gingold (1974) and Schonberner (1979) find that as a result of the last helium shell flash on the AGB one or two loop-like excursions from the AGB are possible for certain models. These excursions can intercept the instability strip causing a star to become a W Vir variable having a period of roughly 10 to 50 days. Even if no loop-like excursion occurs for a particular model, once the hydrogen envelope is essentially exhausted ($< 3 \times 10^{-3}$ $M_\odot$) the star is forced to evolve off the AGB across the H-R the diagram toward the realm of the white dwarfs (see e.g. Figure 3 of Iben, 1982). Such a track results in one guaranteed passage of the instability strip.

Again as is the case for the BL Her stars, the W Vir stars originate from a narrow mass range which is about 0.6 $M_\odot \pm 0.1$ $M_\odot$. The lifetime of this phase of variability can range from about 10^4 yr for the single crossing case or the most optimum loop-like excursion to a few x 10^2 yr for the most transient of the loop-like excursions. Observations of period changes of W Vir stars should help to determine if these loop-like excursions actually occur.

Finally, in dwarf spheroidal galaxies there is another type of Population II Cepheid observed which are known as the anomalous Cepheids because these Cepheids are more luminous at a given period than Cepheids found in globular clusters. Hirshfeld (1980) has shown that extremely metal poor stars ($Z < 5 \times 10^{-4}$) of mass 1.3 to 1.6 $M_\odot$ also lie on something like a horizontal branch during their core helium-burning phase unlike their more metal rich Population II counterparts. This horizontal branch for more massive low-mass stars occurs at a higher luminosity than the traditional horizontal branch, and portions of this new horizontal branch intercept the instability strip (see e.g. Figure 1 of Hirshfeld, 1980). Stars located at this intersection are found to match the observed properties of anomalous Cepheids.

Without invoking recent star formation, the initial mass of stars in Population II systems must be 0.8 $M_\odot$ or less. In order to form stars in the range of 1.3 to 1.6 $M_\odot$ in dwarf spheroidal galaxies, Hirshfeld (1980) argues for coalescence of two stars via binary mass transfer. Wallerstein and Cox (1984) reach the same conclusion based on their analysis of the pulsational masses of anomalous Cepheids. Once formed out of two stars, the coalesced star can have a lifetime in the instability strip of several x 10^7 yr.

To summarize, Population II Cepheids originate from a much narrower and smaller range of masses than Population I Cepheids. Most Population I Cepheids are in the core helium-burning phase while only the anomalous Population II Cepheids of very metal poor systems are undergoing this evolutionary phase. Most Population II Cepheids (i.e., the BL Her and the W Vir stars) are in their post core helium-burning phases prior to their becoming white dwarfs.

2 THEORETICAL UNCERTAINTIES IN THE STELLAR EVOLUTION CALCULATIONS

A theoretical model is only as good as the physics going into it, and therefore uncertainties in how best to model a physical process like convection or uncertainties in the input physics like the nuclear reaction rates can lead to uncertainties in the evolutionary results. In this section a brief discussion of the various modeling problems is presented in terms of how these uncertainties may affect the theoretical predictions for Cepheid variables.

2.1 Convective Overshoot and Semiconvection

Convective overshoot occurs at the boundary of a formally convective region where the kinetic energy of a convective element carries it a finite distance into a region which is formally stable against convection. Mixing may arise between the convective region and a part of its neighboring radiative region due to the finite decay time of overshooting convective elements and the introduction of new material into the radiative region from the overshooting convective elements causing part of the radiative region to become convectively unstable. Semiconvection occurs in a region of variable chemical composition that is marginally unstable to convection. In such a case convective mixing is only able to partially mix the region before convective stability is established. When a semiconvective region exists just outside a convective region mixing may occur between the two regions.

With regard to Cepheid evolution, the convective cores of the core hydrogen-burning and the core helium-burning phases are the most sensitive to any uncertainties in modeling convective overshoot and semiconvection. Convective overshoot occurs throughout the existence of a convective core while semiconvection tends to appear only outside the helium-burning convective core (see, however, Brunish & Truran, 1982a,b for a description of how massive stars behave). Both effects bring more fuel to the convective core resulting in a longer life for a given core burning phase.

Maeder & Mermilliod (1981), Matraka et al. (1982), and Huang & Weigert (1983) discuss the effect convective overshoot has during the core hydrogen-burning phase on the later evolutionary behavior of a model. Compared to models where this effect is neglected, models which include the effect of overshoot are brighter during their core helium-burning phases and their first blue loops in the H-R diagram (when they occur) tend to be somewhat smaller. Figure 3 (from Becker & Cox, 1982) shows this effect for a 9 $M_\odot$,(Y,Z) = (0.28, 0.03) model.

Robertson (1972), Robertson & Faulkner (1972) and Renzini (1977), discuss the effect of including convective overshoot and semiconvection during the core helium-burning phase. Compared to models where these effects are neglected, models which include these effects have longer core helium-burning lifetimes and their first blue loops (when they occur) may be lengthened.

FIGURE 3. The theoretical H-R diagram for a 9 $M_\odot$, (Y,Z) = (0.28,0.03) model without convective overshoot (dashed line) and with convective overshoot (solid line).

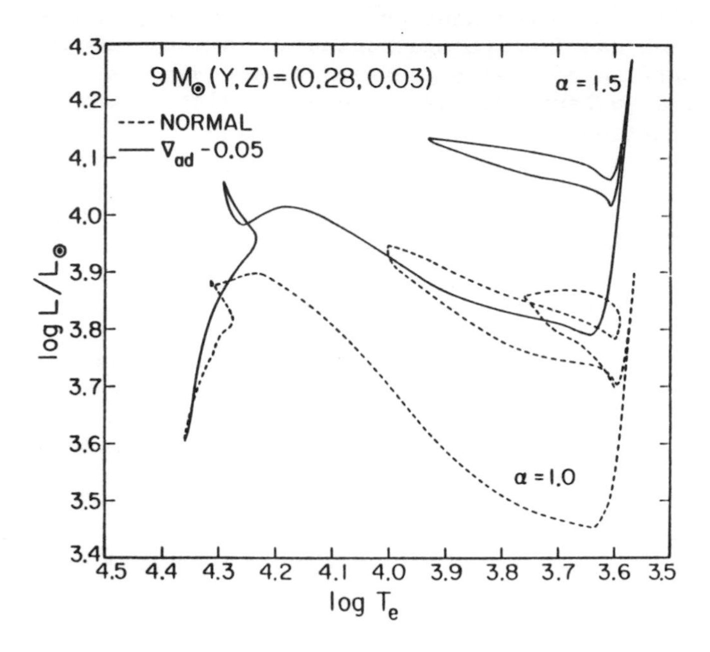

Although there is general agreement about the existence of convective overshoot and semiconvection, the degree to which these effects occur is still an open topic of discussion. One hope of determining the extent of these effects is to calibrate the models by using the observed color-magnitude diagrams of star clusters (Maeder & Mermilliod, 1981), however, observational uncertainties then enter the picture. In any case when these effects are included in stellar evolution models, the predicted evolutionary mass for a Cepheid variable of a given luminosity is smaller than the case where these effects are neglected. Matraka et al. (1982), for example, argue that this reduction may be as large as 15%, a result which is not inconsistent with predictions of other studies.

2.2 Nuclear Reaction Rates

A study of the works of Fowler et al. (1967, 1975, FCZII) and Harris et al. (1983, HFCZ III) show that many nuclear reaction rates are not known precisely and that the published rates have changed over the years. With regard to Cepheid variables, the most relevant nuclear reaction rates are those involving the burning of hydrogen and helium. Fortunately, the important hydrogen-burning rates have changed little over the years, however, such is not the case for the helium-burning reactions particularly, the $^{12}C(\alpha,\gamma)^{16}O$ rate. Iben (1972) was the first to show how uncertainty in the reduced width term of this reaction rate would affect the evolutionary behavior of stellar models that become Cepheids.

Since the $^{12}C(\alpha,\gamma)^{16}O$ rate did not change between FCZ II and HFCZ III it began to appear that this rate was on much firmer ground than it was previously. However, Kettner et al. (1982) made a new measurement of this rate and concluded that the rate should be approximately five times greater than the rate given in FCZ II! Fowler (1984) has restudied the problem from the Caltech data base and has concluded that the increase should be instead about a factor of three greater than the FCZ II rate. Settlement of this controversy will have to await new measurements of the $^{12}C(\alpha,\gamma)^{16}O$ rate.

Figure 4 shows the appearance of the core helium-burning phase in the H-R diagram for a 5 $M_\odot$,(Y,Z) = (0.28, 0.02) model as a result of using both the FCZ II rate and Fowler (1984) $^{12}C(\alpha,\gamma)^{16}O$ rate. With the new reaction rate the lifetime of the core helium-burning phase is lengthened by about 5%, and the blue loop extends to higher temperatures. The lengthening of the blue loop is only noticeable in models for which the loop was small to begin with. One important effect of this change is that the minimum mass of a star that becomes a Cepheid during core helium-burning phase would drop, going in the case of a (Y,Z) = (0.28, 0.02) composition from about 6 $M_\odot$ to 5.1 $M_\odot$ if the Fowler (1984) rate is adopted.

FIGURE 4. The theoretical H-R diagram for a 5 $M_\odot$, (Y,Z)=(0.28,0.02) model as a result of using the FCZ II $^{12}C(\alpha,\gamma)^{16}O$ rate (solid) and the Fowler (1984) $^{12}C(\alpha,\gamma)^{16}O$ rate (dash-dot). The parallel dashed lines mark the fundamental boundaries of the instability strip.

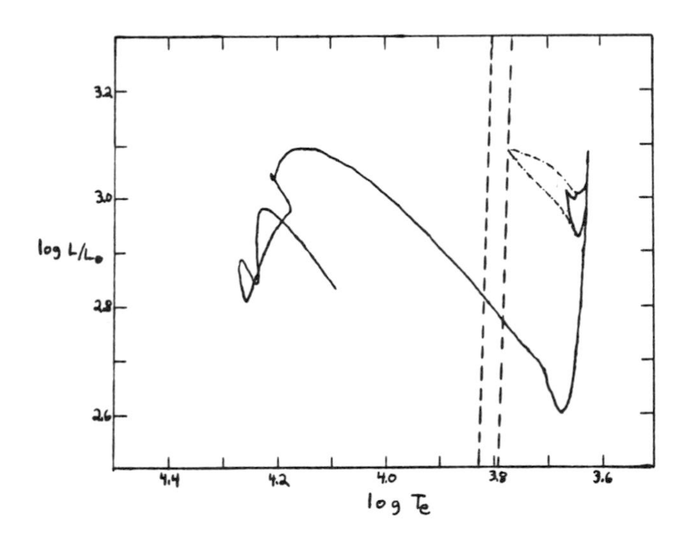

2.3 <u>Opacities and Stellar Atmospheres</u>

The studies of Fricke et al. (1971) Henyey et al. (1965), and Johnson & Whittaker (1975), to name a few, show that changing the opacity used either in the stellar interior or the stellar atmosphere can have a noticeable effect on a stellar model's evolution track in the H-R diagram. For Cepheid variables the most important question was whether the Carson opacities (see e.g. Carson & Stothers, 1976) or the Los Alamos opacities (see e.g. Cox & Tabor, 1976) were the most accurate. The primary difference in behavior between the two sets of opacities is that the Carson opacities have noticeably larger contributions from He and CNO atoms to the opacity in the temperature range of 6.5 > log T > 5.4. The question remained unresolved until last year, when Carson visited Los Alamos and a direct comparison was made between the two methods of calculation and an error was discovered in the Carson code which produced the larger opacities.

With this controversy resolved, the effect of the remaining uncertainties in the opacities or the modeling of stellar atmospheres on the evolutionary behavior of Cepheids is relatively small. For example, Simon (1982) has suggested that increasing the opacities due to heavy elements by a factor of 2 to 3 in the temperature range of 10^5 K to 2×10^6 K could help to remove the mass anomalies encountered with the double-mode and bump Cepheids. Figure 5 shows how the evolutionary track of a 6 $M_\odot$,(Y,Z) = (0.28, 0.02) model changes in going from the standard opacities to one similar to that suggested by Simon. The overall differences are small and consequently, from the stellar evolution standpoint, the changes advocated by Simon are potentially

FIGURE 5. The theoretical H-R diagram for a 6 $M_\odot$, (Y,Z)=(0.28,0.02) model using standard Los Alamos opacities (solid) and opacities in which the heavy element contribution is doubled for $0.1 < T_6 < 0.5$ (dashed). The fundamental blue edge of the instability strip is represented by the dash-dot line.

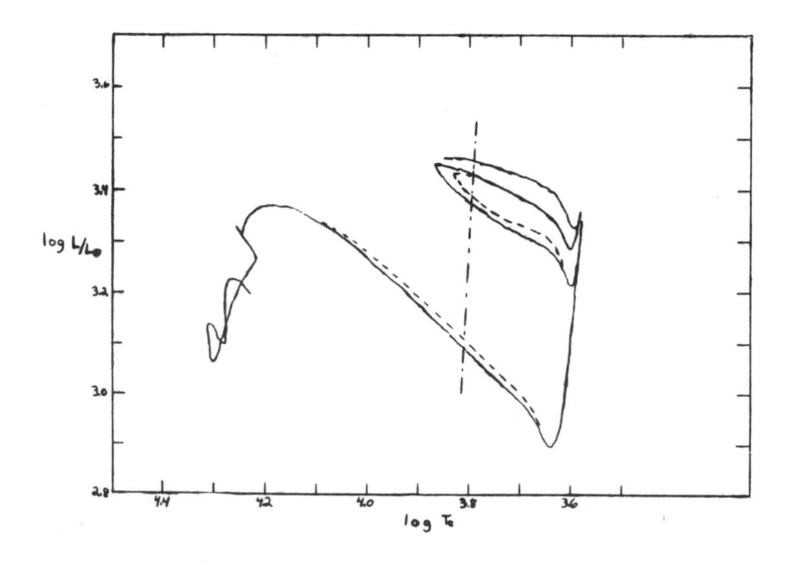

permissible. However, Magee et al. (1984) point out that there is no reason to expect an uncertainty that large in the Los Alamos opacities.

2.4 Mass Loss

Lauterborn et al. (1971) and Lauterborn & Siquig (1974) have studied the effect of mass loss during the red giant phase of intermediate-mass stars of extreme Population I compositions. They find that mass loss is able to suppress formation of blue loops in the H-R diagram if more than 10%, 13%, and 20% of the total stellar mass is lost respectively for 5 $M_\odot$, 7 $M_\odot$ and 9 $M_\odot$ models. Forbes (1968) and Lauterborn et. al. (1971) find that the blue loops remain suppressed until greater than 60% of the total stellar mass is lost. While the above results probably depend somewhat on the initial composition of the star, the implication is clear that a moderate degree of mass loss for intermediate-mass stars effectively reduces the number of crossings of the Cepheid instability strip from a maximum of five to only one. Fortunately, the observed mass loss rates (see, e.g., Lamers, 1981 and Riemers, 1975) indicate that very little mass is lost during the pre-Cepheid evolution for single intermediate-mass stars. Only low-mass stars during the red giant phase and massive stars during all their phases appear to lose a significant fraction of their mass.

For low-mass stars, mass loss plays a significant role in their evolution because without it these stars would not evolve onto the horizontal branch after core helium ignition. Without a horizontal branch it would not be possible to produce RR Lyrae stars, BL Her stars, and possibly anomalous Cepheids. For massive stars, mass loss increases with increasing luminosity so that above about 40 $M_\odot$ no star is able to evolve as red as the Cepheid strip. Mass loss when included in models of massive stars causes evolution to proceed at a lower luminosity than the case where it is not included (see e.g. Brunish & Truran, 1982a,b).

As noted in Lamers (1981) and Reimers, (1975) the mass loss rates are not precisely known and it is likely that a range of rates is possible for stars having the same mass and luminosity. For intermediate-mass stars the uncertainty in the mass loss rates is not of much consequence for Cepheid evolution. However, in case of massive stars, any underestimate of the mass loss rate will lead to an underestimate of the evolutionary mass for a Cepheid of a given luminosity.

2.5 Rotation and Stellar Evolution

Observation of the main sequence progenitor stars that evolve into Population I Cepheids shows that they rotate at speeds of 200 to 270 km/sec (see e.g. the review article of Slettebak, 1970). Observations of red giants (see e.g. Kraft, 1970) show, however, that these stars have small rotational velocities. These observations indicate that the surface angular momentum present during the main sequence is transferred to another location which most likely is the interior of the star. Clearly, since real stars have angular momentum, the neglect of rotation in most stellar evolution calculations is a deficiency. Problems arise, however, when one attempts to add rotation to a stellar model. Rotation destroys the one-dimensional symmetry of a stellar model and an accurate treatment of rotation requires two- or three-dimensional coding capability. Modern computers can accommodate codes of this complexity and efforts to develop such codes are beginning to take place. With some simplifications, however, rotating models can be studied with one-dimensional stellar evolution codes. Problems still confront any investigator using this latter approach, since the interior angular momentum distribution cannot be observed. One therefore needs to make a number of educated guesses as to what the distribution might be.

Despite the difficulties, Kippenhahn et al. (1970), Meyer-Hofmeister (1972) and Endal & Sofia (1976, 1978, 1979) have investigated from a one-dimensional standpoint the influence of rotation on the evolutionary behavior of intermediate-mass stars. Their results provide a mixed picture. For example, they all agree that rotation lengthens the core helium-burning lifetime and that this increase can result in longer lifetimes for the second and third crossings of the instability strip. However, Endal & Sofia (1976, 1978) find that rotation increases the luminosity of the first blue loop in the H-R diagram, but reduces the temperature of the blue loop tip while the work of the others finds the opposite results. Due to these disagreements, it is clear that more work needs to be done before an accurate assessment of how neglect of rotation in most stellar models is affecting theoretical predictions.

2.6 Binary Evolution

Madore (1977) notes that about 27% of the Population I Cepheids are in binaries which have companions of similar mass. Sandage & Tammam (1969) point out that one binary system, CE Cas, consists of two Cepheids. It is apparent that a significant fraction of Cepheid variables, as is the case for most Population I stars, occurs in binary systems. If the separation between the binary components is sufficiently large (on the order of a few x 100 R_o), each star will be able to evolve through its core hydrogen- and helium-burning phases in much the same way as if each star were single. If, however, the separation is such that one star fills its Roche lobe during the course of its evolution, mass transfer will occur and the evolutionary behavior of the two stars will no longer be accurately described by single star evolutionary models.

Detailed evolutionary calculations of a binary system consisting of intermediate or massive stars have not yet been calculated for a wide variety of cases. A few general statements can, however, be made. The more massive star of the system, if it fills its Roche lobe as a red giant, will probably lose enough mass to prevent it from making any further crossings of instability strip. As a result, this star can be a Cepheid only prior to its becoming a red giant. The companion star will probably accrete mass as a result of the more massive component overflowing its Roche lobe. Having accreted mass, the companion star will later evolve in a manner different from that of a single star having the companion star's original or current total mass. As a result, if the companion star later evolves into the instability strip it will do so at a luminosity that is probably inconsistent with its mass based on single star evolution.

Binary evolution is clearly an area for future research. The scenario described in the second half of the last paragraph is only a general case and even more complicated situations can be imagined as a result of binary evolution.

3 THEORY AND OBSERVATION COMPARED

Stellar evolution models can be used to make predictions which can be tested by comparison with observations. Such comparisons provide a diagnostic on the current state of the theoretical models. Areas of disagreement between theory and observation are useful because they help to provide some insight on how theoretical models can be improved. The subject of comparing theory and observations is a topic worthy of a separate paper and as a result this subject will be only briefly discussed in this section.

3.1 Cepheid Masses

Using stellar evolution calculations, a mass-luminosity relationship can be made for Cepheid variables (see e.g. Becker et. al., 1977). Using pulsational theory, a luminosity-period-temperature relationship can be derived. These two relations can be combined to relate period to mass if the exact crossing of the instability strip is known. If the exact crossing is not known, the usual procedure is to assume that the star is undergoing the second crossing. The work of van Genderen (1983), for example, shows that these theoretical relations can be used very successfully with the observed behavior of Cepheids in the Magellanic Clouds.

Cox (1980), in his review article, describes six methods of theoretically determining the mass of Cepheid variables which are: 1) the evolutionary mass, 2) the theoretical mass, 3) the pulsational mass, 4) the bump Cepheid mass, 5) the double and triple mode Cepheid mass, and 6) the Wesselink radius mass. Cox (1980) notes that the evolutionary mass is normally in agreement with all the other methods except the bump Cepheid masses and the double and triple mode Cepheid masses.

Since no Cepheid is close enough to determine distance by trigonometric parallax, the most direct method of getting Cepheid masses is from binary orbits and the results have been mixed. Evans (1980) orignally found for the case of SU Cygni that the orbital mass is consistent with masses predicted by both stellar evolution and pulsation theory, however, this interpretation is now clouded because this system has been found to be a triple (Evans, this conference). Bohm-Vitense (1984), finds, for binary systems consisting of a blue companion and a Cepheid variable, that the Cepheids are overluminous if both stars are of the same mass. The discrepancy may, however, be due to mass transfer inside the system.

The agreement in the case of Cepheid variables between evolutionary masses and pulsational masses from all but the bump and double and triple mode Cepheids shows that stellar evolution calculations are, in general, qualitatively correct (using the current distance scale). The discrepancy with the bump and double and triple mode Cepheids is a problem which is still not understood, but it is an active topic of research. The agreement between evolutionary mass and pulsation mass would end, however, if the luminosity scale of Cepheid variables is revised downward as advocated by Schmidt (1984).

3.2 Period Changes

During evolution across the instability strip, the period of a Cepheid increases as the temperature decreases. As a result of this behavior, measurements of the rate of period change can help to determine which crossing of the instability strip is taking place. For intermediate-mass stars the first, third, and fifth crossings will have a positive rate of change, while the second and fourth crossings will show a negative rate of change. The magnitude of the rate of change, $\dot{P}$, provides further information in that stars undergoing the second or third crossing will in general be evolving at a slower rate than stars undergoing the other crossings (see e.g. Becker 1981b). Parenago (1956), Payne-Gaposchkin (1974), Fernie (1979, 1984) and Szabados (1983, 1984) have studied period changes in Cepheid variables.Payne-Gaposchkin (1974), Fernie (1984) and Szabados (1983, 1984) note that measurements of the periods changes of Cepheids sometimes show nonsecular changes or noise which makes the determination of $\dot{P}$ complicated. However, a large number of sufficiently clear measurements of $\dot{P}$ have been made by Fernie (1984) and Szabados (1983,1984) for them to conclude that the observed period changes are due to effects of stellar evolution. Observation of positive and negative values of $\dot{P}$ provides direct evidence for the existence of blue loops in the H-R diagram.

3.3 Surface Changes in the Chemical Composition

Cepheid variables of intermediate mass that are in the core helium-burning phase spent part of their previous evolution as red giants. As red giants, these stars underwent the first dredge-up phase in which convection mixes the outer layers of the star with a portion of interior that in the past has undergone nuclear reactions. Stellar evolution calculations (see e.g. Becker & Iben, 1979) predict that the surface abundance of ^{14}N should increase by a factor of two to three and that the surface abundance of ^{12}C should decrease by 30 to 40% as a result of the first dredge-up phase.

Luck & Lambert (1981) were the first to measure CNO abundances in Cepheid variables and their results showed abundance changes far in excess of that predicted by standard stellar evolution theory. Furthermore, the observations of Luck & Lambert (1981) showed that mixing must take place with matter so deep in the interior that it has undergone ON processing. Becker & Cox (1982) discussed the implications of how stellar evolution models would have to be altered to account for these observations. Iben & Renzini (1983) took the opposite approach and discussed what might be wrong with the interpretation of the observations. The controversy has apparently now come full circle when

Lambert (1984) stated that the observed O depletion in Cepheids was a result of a systematic error in the values used for log g. He now finds evidence only for CN processing as predicted by standard stellar evolution calculations.

3.4 Frequency Period Distributions of Cepheids

From the observations of hundreds to thousands of Cepheid variables in the Galaxy, M31, the LMC, and the SMC, detailed but not complete frequency period distributions can be made. The distributions all show a small number of short period varibles, a rapid increase in the numbers of Cepheids at some key period and an exponential-like decline in the number of variables for larger and larger values of the period (see Figures 7 and 8 of Becker et al. 1977). The distribution for the Galaxy and M31 are very similar, while those for the LMC and the SMC show a greater spread and the peak in their distributions takes place at a noticeably smaller value for the period.

The sudden rise in the frequency period distribution provides strong evidence for the first blue loop occurring in the H-R diagram for helium burning intermediate-mass stars. The peak should arise from stars in which the tip of the blue loop is tangent to the fundamental blue edge. Becker et al. (1977) show that a galactic model based on stellar evolution calculations which involves a spread in the chemical composition and reasonable values for birthrate function can reproduce many of the features of the observed frequency-period distributions. The reason for the different peaks in the distributions for the Galaxy, the LMC, and the SMC is a result of having the average chemical composition change as one goes from the Galaxy, to the LMC, and to the SMC.

3.5 Star Clusters

Unlike the case for the Galaxy, a number of star clusters in the LMC and the SMC have a high population density and are the right age to show significant numbers of Cepheids and stars in the core helium-burning phase. Arp (1967) was the first to note this behavior in the case of NGC 1866. These clusters can be modeled with synthetic clusters constructed from stellar evolution models. For the case of NCG 1866 as shown in Figure 6, the observed color-magnitude diagram provides visual proof of the predictions of stellar evolution theory because the diagram looks like a blurred stellar evolution track with the Cepheids appearing at the blue loop tip. Becker and Mathews (1983) have analyzed the data for NGC 1866 with synthetic cluster models and have shown its age to be about 86 x 10^6 yr and its composition to be about (Y,Z) = (0.273, 0.016).

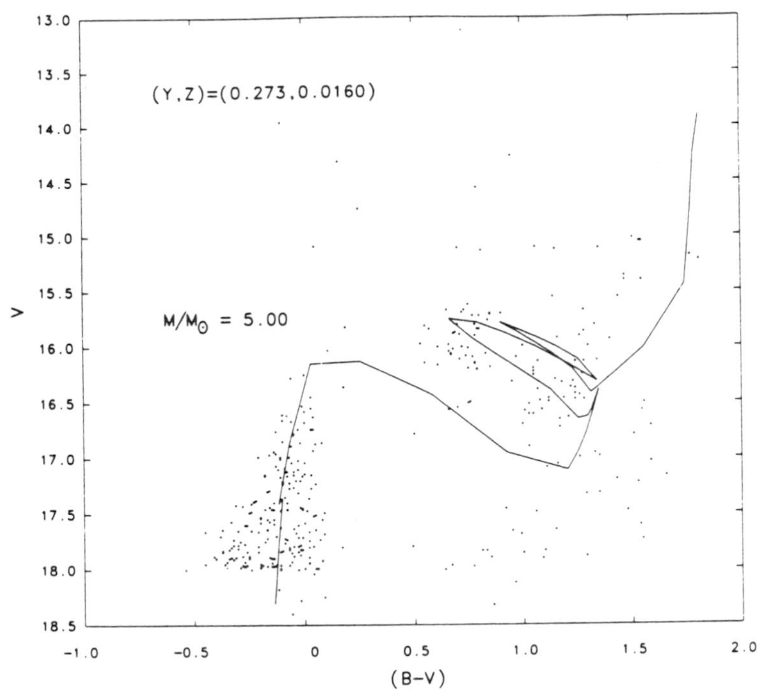

FIGURE 6. The observed color-magnitude diagram for NGC 1866 (dots) with the evolutionary track for a 5 $M_\odot$,(Y,Z) = (0.273, 0.016) model superimposed. The observed Cepheids are near the blue loop tip.

4 SUMMARY

In the review, the behavior of stellar evolution models relevant to Cepheid variables of Types I and II is discussed. Uncertainties in the theoretical models are examined, but are found not to be of such size as to seriously limit the usefulness of the theoretical predictions. Finally, observations and theoretical predictions are made for a number of cases to show areas of agreement and disagreement.

This work was performed under the auspices of the U.S. Department of Energy Contract # W-7405-ENG. 36.

Arp, H. (1967) Ap. J., 149, 91-106.
Becker, S. A. (1981a) Ap. J. Suppl. 45, 475-505.
Becker, S. A. (1981b) Ap. J., 248, 298-310.
Becker, S. A. & Cox, A. N. (1982) Ap. J., 260, 707-715.
Becker, S. A. & Iben, I. Jr. (1979) Ap. J., 232, 831-853.
Becker, S. A. , Iben, Icko Jr., & Tuggle, R. S. (1977) Ap. J., 218, 633-653.
Becker, S. A. & Mathews, S. J. (1983) Ap. J., 270, 155-168.
Bohm-Vitense (1984) In Observational Tests of the Stellar Evolution Theory, IAU Sym., 105, ed. A. Maeder, in press, Dordrecht: D. Reidel Publ. Co.
Brunish, W. M. & Truran, J. W. (1982a) Ap. J., 256, 247-258.
Brunish, W. M. & Truran, J. W. (1982b) Ap. J. Suppl., 49, 447-468.
Carson, T. R. & Stothers, R. (1976) Ap. J., 204, 461-471.
Cox, A. N. (1980) Ann. Rev. Ast. Ap., 18, 15-41.
Cox, A. N., & Tabor, J. E. (1976) Ap. J. Suppl., 31, 271-312.
de Loore, C. (1980) Spc. Sci Rev., 26, 113-155.
Despain, K. H. (1981) Ap. J. 251, 639-653.
Eggen, O. J. (1983) Aston. J., 88, 386-403.
Endal, A. S. & Sofia, S. (1976) Ap. J., 210, 184-198.
Endal, A. S. & Sofia, S. (1978) Ap. J., 220, 279-290.
Endal, A. S. & Sofia, S. (1979) Ap. J., 232, 531-540.
Evans, N. R. (1980) Bull. AAS., 12, 862.
Fernie, J. D. (1979) Ap. J., 231, 841-845.
Fernie, J. D. (1984) In Observational Tests of the Stellar Evolution Theory; IAU Sym 105, A. Maeder in press, Dordrecht: D. Reidel Publ. Co.
Forbes, J. E. (1968) Ap. J., 153, 495-510.
Fowler, W. A. (1984), private communication.
Fowler, W. A., Caughlan, G. R. & Zimmerman, B. A. (1967) Ann. Rev. Astr. Ap., 5, 525-570.
Fowler, W. A., Caughlan, G. R. & Zimmerman, B. A. (1975) Ann. Rev. Astr. Ap., 13, 69-111 (FCZII).
Fricke, K., Stobie, R. S., & Strittmatter, P. A. (1971) M.N.R.A.S., 154 ,23-46.
van Genderen, A. M. (1983) Astr. Ap., 124, 223-235.
Gingold, P. A. (1974) Ap. J., 193, 177-185.
Harris, M. J., Fowler, W. A., Caughlan, G. R., & Zimmerman, B. A. (1983) Ann. Rev. Astr Ap., 21, 165-176 (HFCZIII).
Henyey, L. G., Vardaja, M. S., & Bodenheimer, P. (1965) Ap. J., 142, 841-854.
Hirshfeld, A. W. (1980) Ap. J., 241, 111-124.
Hoppner, W., Kahler, H., Roth, M. L., & Weigert, A. (1978) Astr. Ap., 63, 391-399.
Huang, R. Q., & Weigert, A. (1983) Astr. Ap., 127, 309-312.
Iben, I. Jr. (1972) Ap. J., 178, 433-440.
Iben, I. Jr. (1974a) Ann. Rev. Ast. Ap., 12, 215-256.
Iben, I. Jr. (1974b) In Stellar Instability and Evolution, IAU Sym. 59, ed. P. Ledoux, A. Noels, & A. W. Rogers, pp.3-34. Dordrecht: D. Reidel Publ. Co.

Iben, I. Jr. (1982) Ap. J., 260, 821-837.
Iben, I. Jr. & Renzini, A. (1983) Ann. Rev. Ast. Ap., 21, 271-342.
Iben, I. Jr., & Rood, R. J. (1970) Ap. J., 161, 587-617.
Iben, I. Jr. & Tuggle, R. S. (1975) Ap. J. 197, 39-54.
Johnson, H. R. & Whitaker, R. W. (1975) M.N.R.A.S., 173, 523-526.
Kettner, K. U. et. al. (1982) Z. Phys. A - Atoms and Nuclei, 308, 73-94.
Kippenhahn, R., Meyer-Hofmeister, E., & Weigert, A. (1970), Astr. Ap., 5, 155-161.
Kraft, R. P. (1970) In Spectroscopic Astrophysics, ed. G. H. Herbig, pp. 385, Berkeley: Univ. of California Press.
Lambert, D. L. (1984) In Observational Tests of the Stellar Evolution Theory, IAU Sym., 105, ed. A. Maeder, in press, Dordrecht: D. Reidel Publ. Co.
Lamers, H. J. G. L. M. (1981) Ap. J., 245, 593-608.
Lauterborn, D., Refsdal, S., & Weigert, A. (1971) Astr. Ap., 10, 97-117.
Lauterborn, D., & Siquig, R. A. (1974) Ap. J., 191, 589-592.
Luck, R., & Lambert, D. L. (1981) Ap. J., 245, 1018-1034.
Madore, B. F. (1977) M.N.R.A.S., 178, 505-511.
Maeder, A., & Mermilliod, J. C. (1981) Astr. Ap., 93, 136-149.
Magee, N. H., Merts, A. T., & Huebner, W. T. (1984) Ap. J., in press.
Matraka, B., Wassermann, C., & Weigert, A. (1982) Astr. Ap., 107, 283-291.
Mengel, J. G., Sweigart, A, V., Demarque, P., & Gross, P. G. (1979) Ap. J. Suppl., 40, 733-791.
Meyer-Hofmeister, E. (1972) Astr. Ap., 16, 282-285.
Parenago, P. P. (1956) Perem. Zvezdy; 11, 236.
Payne-Gaposchkin, C. (1974) Smithsonian Contr. Ap., 16, 1-32.
Reimers, D. 1975, Mem. Soc. Roy. Sci. Liese 6th Ser., 8, 369.
Renzini, A. (1977), In Advanced Stages of Stellar Evolution, ed. P. Boivier and A. Maeder, pp. 151-283, Sauverny: Geneva Observatory.
Robertson, J. W. (1972) Ap. J., 177, 473-488.
Robertson, J. W. & Faulkner, D. J. (1972) Ap. J., 171, 309-315.
Sandage, A. & Tammann (1969) Ap. J., 157, 683-708.
Schlesinger, B. M. (1977) Ap. J., 212, 507-512.
Schmidt, E. G. (1984) Ap. J., in press.
Schonberner, D. (1979) Astr. Ap., 79, 108-114.
Simon, N. R. (1982) Ap. J., 260, L87.-L90.
Slettebak, A. (1970) In Stellar Rotation, IAU Coll. 4, ed. A. Slettebak, pp. 3-8, Dordrecht: D. Reidel Publ. Co.
Sreenivasan, S.R. & Wilson, W.J.F. 1978, Ap. Space Sci., 53, 193-216.
Sweigart, A. V. & Gross, P. G. (1976) Ap. J. Suppl. 32, 367-398.
Szabados, L. (1983) Astr. Sp. Sci., 96, 185-194.
Szabados, L. (1984) In "Observational Tests of the Stellar Evolution Theory"; IAU Sym 105, ed. A.Maeder, in press, Dordrecht: D. Reidel Publ. Co.
Wallerstein, G. & Cox, A. N. (1984) Publ. Ast. Soc. Pac., in press.

THEORY OF CEPHEID PULSATION: EXCITATION MECHANISMS

John P. Cox

Joint Institute for Laboratory Astrophysics, University of Colorado and National Bureau of Standards, Boulder, CO 80309

Abstract. The various excitation mechanisms (eight in all) that have been proposed to account for the vibrational instability of variable stars, are surveyed. The most widely applied one is perhaps the "envelope ionization mechanism." This can account for most of the essential characteristics of the "instability strip." A simple explanation of the period-luminosity relation of classical Cepheids is given. A few outstanding problems in pulsation theory are also listed.

1 INTRODUCTION

The main purpose of this paper is to survey the various excitation mechanisms which have been proposed down through the years to account for the observed light and radial velocity variations of Cepheids and of other types of variable stars, with particular emphasis on Cepheids themselves.

Historically, the main driving agent for the (essentially radial) pulsations of classical Cepheids was the first to be investigated, and this early investigation began about 30 years ago. Yours truly was heavily involved with this work. At the time, he was highly astonished at the paucity of research papers dealing with this seemingly fundamental problem. There were plenty of papers concerned mainly with details of light and velocity curves and the like. But almost no papers attacking the main problem of what drives the pulsations! The only ones that he recalls now were two papers by Eddington, published in the early 1940's (Eddington 1941, 1942). These papers had a large effect on the author's subsequent work. He also remembers early papers by Zhevakin and Krogdahl.

This early work culminated in two fundamental papers: one, by Baker & Kippenhahn (1962), on Delta Cephei itself; the other, by Cox (1963), on simplified Cepheid envelope models. The work of Baker and Kippenhahn was later extended and refined (Baker & Kippenhahn 1965). The term "kappa-mechanism" (see § 3) was first explicitly introduced by Baker & Kippenhahn (1962). Definitive and reliable calculations of linear, radial, nonadiabatic pulsations of stellar models first became possible with use of a computer code originated and developed by Castor (1971).

One of the important conclusions of this early work was that the pulsation amplitude in the deep stellar interior was very small. It was, in fact, far too small for the nuclear reactions, which supply the mean luminosity of the star, to have anything directly to do with the pulsations. This conclusion was based largely on an early study by Epstein (1950). It turns out that Cepheid pulsation is primarily an envelope phenomenon; the cause of such pulsations had to be sought in factors affecting these outer layers. The basic reason for the smallness of the pulsation amplitude in the deep stellar interior is ultimately that, according to current ideas, Cepheids are in an advanced stage of evolution and are consequently highly centrally concentrated.

The extension of these results into the nonlinear domain was first effected by Christy (1966) and by Cox, Cox, Olsen, King, and Eilers (1966). The latter authors introduced explicitly the term "gamma-mechanism" (see § 3).

Linear, nonradial pulsations of stars, in adiabatic theory, were first explored by Smeyers (1966) and Dziembowski (1971). (Reviews of the properties of nonradial stellar pulsations can be found in Ledoux & Walraven 1958; Ledoux 1974; Cox 1976, 1980; Unno, Osaki, Ando & Shibahashi 1979.) Extension of the theory into the nonadiabatic domain was effected by Ando & Osaki (1975) (however, they neglected the Eulerian perturbations of gravitational potential, i.e., made the Cowling approximation), Saio & Cox (1980), and perhaps others. A new technique for treating linear, nonradial, nonadiabatic oscillations of spherical stars has been developed by Pesnell (1984a). This technique uses Lagrangian, rather than Eulerian, variations, and possesses many advantages over previous schemes. It appears highly promising, and may even constitute a breakthrough in the field.

Attempts to treat convection in pulsating stars were carried out by a number of investigators. The complications introduced by convection are by no means trivial, and were ignored in all the above works. Some earlier unpublished work on this problem was carried out by Gough, which ultimately resulted in the study by Baker & Gough (1979). The problem has also been investigated by, among others, Castor (1968); Deupree (1977a,b,c,d; 1980); Saio (1980); Gonczi (1981, 1982); Gonczi & Osaki (1980); Durney (1984); and Stellingwerf (1982a,b; 1984a,b,c). The most ambitious work is that of Deupree (1977a); but the papers by Stellingwerf are also impressive. Both of the latter authors are in essential agreement. However, in contrast to Deupree (1977a), Stellingwerf (1984a) found that convection had a slight effect on the blue edge of the instability strip.

The mechanism by which convection in the envelope can result in a red edge to the instability strip, was first identified by Deupree (1977a). This mechanism is explained in § 3 below.

2 GENERAL PRINCIPLES OF EXCITATION MECHANISMS

In general, an excitation mechanism consists of a way, or mechanism, for transforming some other kind of energy into mechanical energy of stellar pulsations. The most clearcut example would be the transformation of some of the steady radiant energy flowing through the outer layers of a Cepheid variable, into pulsation energy. This latter energy may be estimated, for a typical Cepheid, to be $\sim 10^{41-43}$ ergs. We may take as the (steady) luminosity $\sim 10^{37}$ ergs s^{-1} ($\sim 10^4 L_\odot$, $L_\odot$ = solar luminosity). If only a hundred thousandth (10^{-5}) of this energy is available for such a transformation, the entire pulsation energy of such a Cepheid could be supplied in only $\sim 10^9$-10^{11} s ~ 10^{2-4} yr or $\sim 10^{3-5}$ pulsation periods.

As another example, consider the recently discovered (Kurtz 1982) "rapidly oscillating Ap stars." The total energy (kinetic) involved in the observed oscillations is perhaps $\sim 10^{34}$ ergs or larger. On the other hand, the total (kinetic) energy involved in such a star's outer convection zone may be of the same order of magnitude or, more likely, one or two orders of magnitude larger. At any rate, there seems to be enough energy residing in such a star's outer convection zone, that if only a small fraction of this energy were transformed into oscillatory energy, there would be enough to account for Kurtz's observations.

3 A SURVEY OF SPECIFIC EXCITATION MECHANISMS

In our survey of specific excitation mechanisms that have been proposed, we shall consider the following:

(1) "Envelope Ionization Mechanisms." Crucial elements: He^+, He, H, C, O. The kappa- and gamma-mechanisms. The r(radius)-mechanism.
(2) Stellingwerf "Bump" Mechanism.
(3) Epsilon (ε)-Mechanism.
(4) Kato Overstable Convection (associated with a mean molecular weight gradient).
(5) Shibahashi Overstable Magnetic Convection (associated with a magnetic field).
(6) Osaki Overstable Convection (associated with rotation).
(7) Tidally Forced Oscillations (Kato).
(8) Kelvin-Helmholtz Instability (Ando).

3.1 "Envelope ionization mechanisms"

This kind of mechanism was first suggested by Eddington (1941, 1942). However, he had the idea that hydrogen was the crucial element for classical Cepheids; this element was subsequently identified as He^+, once-ionized helium (Zhevakin 1953; Cox 1980, Chap. 10). Eddington also thought that Cepheid pulsations basically were "driven" by nuclear reactions (the "ε-mechanism," see below) (King & Cox 1968), and that the hydrogen ionization zone only served to reduce the damping.

The main purpose of such an "envelope ionization mechanism" is to furnish heat to these outer layers upon compression, or to release heat upon expansion. Nuclear energy sources can play no important role.

Therefore, the above all-important gains and losses of heat must be accomplished by a *modulation* of the radiation flowing through these layers. This is a rather bizarre means by terrestrial standards: It is the *leakage* of heat that is varied in a stellar heat engine of the kind we are describing. On the other hand, it is the *input* of heat that is varied in a conventional internal combustion engine.

The basic destabilizing effect of an "envelope ionization mechanism" can be understood as follows. Any destabilizing agent operating in the outer stellar layers will be most effective if it is located in or near the appropriate "transition region." This region is that level above which the pulsations are extremely nonadiabatic and the energy flux is effectively "frozen-in" (because the material in these outermost regions has so little heat capacity), and below which the pulsations are almost adiabatic (quasi-adiabatic) (see Fig. 1, adapted from Osaki 1982). In the "quasi-adiabatic interior" any "driving" is almost exactly cancelled by a nearly equal amount of damping, so that there is essentially no net driving in this region. In the "nonadiabatic exterior" there can be no driving because there is no modulation of the flux. In or near the "transition region" itself there can be driving immediately below this region; however, immediately above this region the damping might be eliminated by nonadiabatic effects, leaving only driving. This transition region can be roughly located in any star by the order-of-magnitude condition that the total internal energy in the layers above this region be of the same order as the energy radiated by the star (equilibrium luminosity L) in one pulsation period Π. Quantitatively, this statement may be expressed by the equation:

$$\frac{\langle c_V T_* \rangle m_*}{L\Pi} \sim 1 \qquad (1)$$

where m_* is the total mass lying above the transition region, c_V is the specific heat at constant volume, T_* is the temperature in the transition region, and angular brackets denote an appropriate average over these outer layers. Although equation (1) was derived for radial

Fig. 1. Schematic drawing of the transition region. Adapted from Osaki (1982).

SURFACE

FLUX VARIATIONS "FROZEN-IN;" VERY NON-ADIABATIC } NO DRIVING

TRANSITION REGION

QUASI-ADIABATIC } NO NET DRIVING

RADIATIVE FLUX

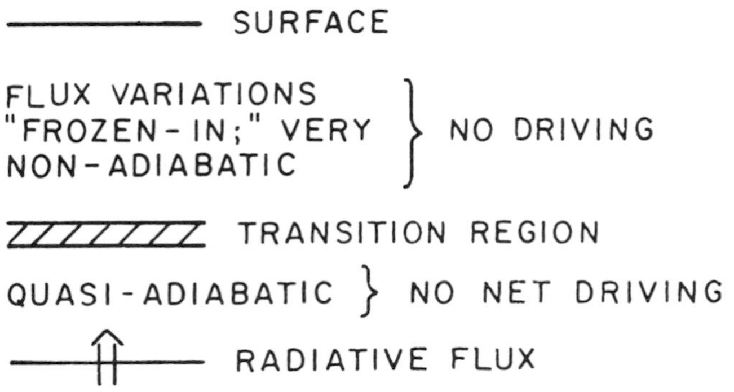

oscillations, it has recently been shown (Pesnell 1984b) that this equation also applies, to order of magnitude, to nonradial oscillations whose ℓ value is not too large, where ℓ is the latitudinal "quantum number" of a spherical harmonic. Further discussion of equation (1) may be found in Pesnell (1984b).

An "ionizing element" (a crucial element in the mid-stages of ionization) will, if this element is sufficiently abundant, produce a destabilizing effect essentially by causing the local temperature to be lower than if ionization were not occurring. This ionization represents additional internal degrees of freedom, so that any energy added to the gas (e.g., the energy supplied by adiabatic compression) goes partly into these additional degrees of freedom rather than into the kinetic energy of thermal motions. Hence, for example, such a gas will remain relatively cool upon compression. Malcolm Savedoff once expressed this relative coolness upon compression rather vividly as "putting ice cubes" into the appropriate stellar layers when they are most compressed. This relative coolness upon compression, say, will have further consequences, as follows.

The local opacity will now increase upon compression, whereas, normally, it decreases upon compression. The basic reason for this behavior is that the opacity normally varies about as the inverse cube of the local temperature, which is now remaining more-or-less constant because of the ionization. This statement is still qualitatively true even in the hydrogen ionization region of the ZZ Ceti (white dwarf) stars, where the destabilization is supposed to be occurring (see below). A figure in Winget (1981) shows that, in the regions of interest, the opacity decreases, for given density, with increasing temperature. On the other hand, this opacity is roughly proportional to the density, which is still increasing upon compression. These two behaviors explain the "kappa-mechanism" (Baker & Kippenhahn 1962), which refers to an increase of the opacity upon compression.

This increase of the local opacity upon compression will tend to "trap" the energy inside trying to get out. This "trapped" energy will instead get absorbed by the gas, and, through nonadiabatic effects, cause the pressure upon subsequent expansion to be larger than if the pulsations had been adiabatic. In this way any incipient pulsations will tend to get "pumped up." Such a region is therefore destabilizing.

Similarly, the relative coolness of layers undergoing compression will cause these layers to radiate less. This diminished radiation upon compression has been termed the "gamma-mechanism" (Cox, Cox, Olsen, King & Eilers 1966). The overall effect will be the same as for the kappa-mechanism, and will be a destabilizing tendency.

Finally, upon compression the total radiating area of the star will be reduced. This effect will also tend to "trap" radiation inside the star. Such an effect has been termed the "r (radius)-effect" (Baker 1966), and will also contribute to instability.

These considerations permit us to see rather easily how convection in the envelope can "quench" pulsations and result in a "red edge" of the instability strip (Deupree 1977a). One of Deupree's important findings was that convective transfer is most efficient when the star is most compressed. Hence, near minimum stellar radius convection causes the rate of energy loss from the star to be near maximum. This is exactly the condition for damping; this condition is just the opposite of that for driving. One may say that convection essentially "undoes" the trapping of energy effected by the kappa-, gamma-, and r-mechanisms discussed above. These last two mechanisms are still operating, even when convection in the envelope is present. However, their combined driving effect is small compared to the above damping effect of convection.

Envelope ionization mechanisms have recently been invoked to explain successfully the observed nonradial oscillations of the variable white dwarfs, the "ZZ Ceti" stars (Winget *et al.* 1982a). Only, in this case, the crucial element is apparently hydrogen (see Winget & Fontaine 1982 for a good review of the ZZ Ceti stars).

These considerations have been extended to the DB, or helium, white dwarfs (Winget *et al.* 1982b). In this case, the crucial element is apparently He (neutral helium, not once-ionized helium, as for the classical Cepheids). For the first time, a prediction as to the variability of the DB white dwarfs was made (Winget *et al.* 1982a), and subsequently a DB variable was actually detected observationally (this was GD 358) (Winget *et al.* 1982b, 1983). Three more have since been found (Winget *et al.* 1984, Winget 1984); these are PG 1654+160, PG 1116+158, and PG 1351+489.

It has recently been suggested (Starrfield, Cox & Hodson 1981, 1983; Starrfield, Cox, Kidman & Pesnell 1984) that an envelope ionization mechanism, with carbon and/or oxygen as the crucial element(s), may account for the instability of the very hot variable PG 1159-035 (see also Winget, Hansen & Van Horn 1983).

3.2 Stellingwerf "bump" mechanism

The "bump" mechanism was suggested by Stellingwerf (1978, 1979) and was applied in Stellingwerf (1978) and in Cox & Stellingwerf (1979) to the Beta Cephei stars. This may be considered a kind of "envelope ionization mechanism," although, as we shall see below, its operation does not really depend on ionization.

It is well known that the effectiveness of any destabilizing mechanism operating in regions of a star where the energy is carried predominantly or solely by radiation is very sensitive to certain features of the opacity. One of these features is the temperature dependence of the opacity law. The Stellingwerf "bump" mechanism can be effective because it introduces a "wiggle," or "bump," on the opacity versus temperature curve, for given density. Such a "bump" has a region where the opacity decreases with increasing temperature less rapidly than immediately above or below this region (i.e., the opacity vs. temperature curve has a relatively "flat" region) (see Fig. 2). This feature causes such a region to radiate less strongly than the regions immediately above or below it, when the material is compressed. Such a region can therefore impede the flow of radiation through these layers, in somewhat the same way as the kappa-, gamma-, and r-mechanisms operate.

Roughly, the bump is, essentially, a result of the coincidence of the peak of the Planck function with the ionization edge of He^+ (at 54.4 eV). The bump accordingly occurs at a temperature close to 1.5×10^5 K, and is rather insensitive to density. It turns out that the temperature of the "transition region" in Beta Cephei stars is close to this value (Cox 1967, p. 90), so it might be expected that this bump might have something to do with the instability of these stars (Cox & Stellingwerf 1979). In this study it was also shown that radiation pressure, in only modest amounts, could increase c_V quite significantly. However, in a very careful study by Lee & Osaki (1982) it was shown that this bump is not actually pronounced enough to cause instability in these stars. This "bump" mechanism is also discussed by Pesnell (1983), who reaches the same conclusion as Lee & Osaki (1982).

Fig. 2. Schematic opacity (κ) versus temperature (T) curve, at fixed density (ρ), showing the Stellingwerf "bump."

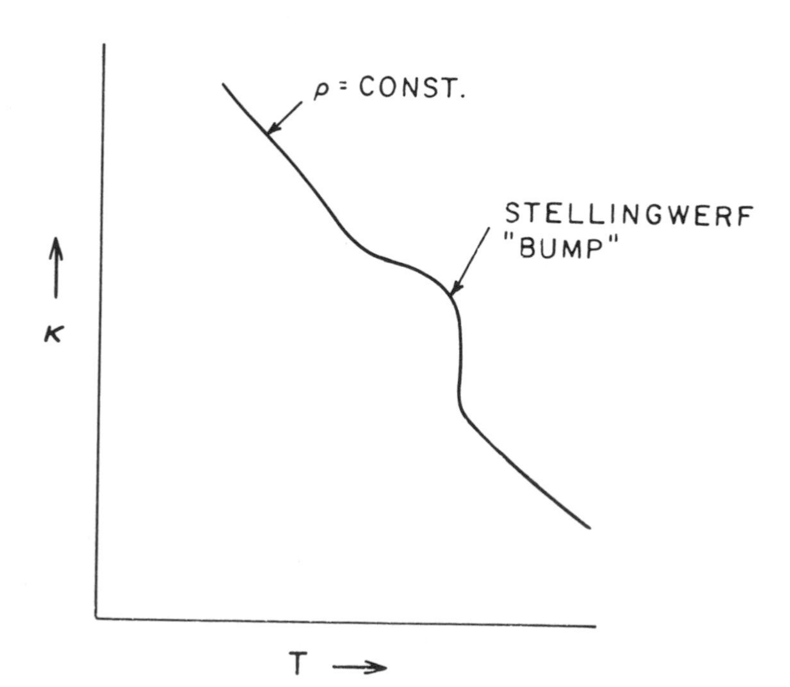

3.3 Epsilon (ε)-mechanism

The ε-mechanism is the excitation effect of thermonuclear reactions. It essentially always (at least in principle) exerts a destabilizing tendency, although the effects of this mechanism on actual variable stars are usually negligible (Cox 1967).

The destabilizing tendency arising from the ε-mechanism derives from several factors. First, the rate of energy production per unit mass ε from thermonuclear reactions is generally given to adequate accuracy by the approximate interpolation formula

$$\varepsilon \propto \rho^{\lambda} T^{\nu} \quad . \tag{2}$$

Here λ is usually 1 or 2 but ν may be quite large (e.g., $\nu \approx 15$-17 for the carbon cycle, $\nu \approx 4$ for the proton chains, and $\nu \sim 30$-40 for helium-burning reactions). Second, in regions where nuclear reactions are occurring the oscillations are usually nearly adiabatic, so that, for example,

$$\frac{\delta T}{T} = (\Gamma_3 - 1) \frac{\delta \rho}{\rho} \quad , \tag{3}$$

where a "δ" denotes the Lagrangian variation and Γ_3 is one of the adiabatic exponents. Hence, usually,

$$\frac{\delta \varepsilon}{\varepsilon} \approx \left[\lambda + \nu(\Gamma_3 - 1)\right] \frac{\delta \rho}{\rho} \quad , \tag{4}$$

where the quantity in square brackets can be rather large.

What this means is that, upon compression, a relatively large amount of thermonuclear energy may be generated. If this energy is not carried away, it may be absorbed by the local matter, enhance the local temperature increase, and so lead to overstable oscillations.

The agent that may carry this excess energy away is radiation. If we consider an "element" which has been slowly displaced inward in a region where the material is "subadiabatic" (i.e., convectively stable according to the Schwarzschild criterion), the element will be warmer than its local surroundings, and so will radiate more copiously than they (of course, an outward-displaced element will be <u>cooler</u> than its local surroundings, and so will radiate <u>less</u> copiously than they). Whether this excess energy will be carried away (or whether the deficit of energy will be eliminated by heat gains from the surroundings) depends on the size of the element under consideration. The total amount of energy lost (or gained) by radiation in a given time will be proportional to the <u>surface area</u> of the element, and hence to the square of its size. On the other hand, the total excess (or deficit) of thermonuclear energy generated in this same time will be proportional to the <u>volume</u> of the element, and hence to the cube of its size. Hence large elements are more likely, from this effect, to cause overstability than small elements.

Note that, if the temperature gradient in the relevant regions were <u>superadiabatic</u> rather than <u>subadiabatic</u> (yet such that the elements are convectively stable according to the Ledoux criterion), the roles of nuclear energy production and radiative transfer would be reversed. In this case a rising element, for example, would be hotter than its surroundings. The enhanced loss of heat from the element (as compared to its surroundings) due to radiation would now tend to lead to overstability (Kato effect, Kato 1966; also, see § 3.4), whereas the enhanced gain of heat due to an increased thermonuclear reaction rate would now tend to lead to damping.

An overstability in hydrogen-burning and helium-burning shells in evolved stars (in such stars the temperature gradient in the relevant regions was always subadiabatic) has recently been investigated by Kippenhahn (1983). He finds that, generally, hydrogen-burning shells are too thin ever to be overstable. However, such an overstability can sometimes exist in helium-burning shells. In particular, he finds that such shells can be overstable during portions of "thermal pulses," when the evolutionary track of the star "loops" for a time to much higher effective temperatures. What the observational consequences of this overstability are, are not clear (see Kippenhahn 1983).

In immediately post-main sequence stars it is known that there is a thin hydrogen-burning shell which supplies most of the energy radiated by the star. Moreover, in this shell the mean molecular weight μ increases inward from a value appropriate to predominantly hydrogen in the exterior to a value appropriate to nearly pure helium in the core. In this thin region there is, arising from the μ-gradient, an additional contribution to the restoring force acting on a displaced mass element (see § 3.4). Consequently, the Brunt-Väisälä frequency here is relatively large. One can therefore have "propagating" g modes, or spatially oscillatory g modes, in this region, with these modes "evanescent" above and below, as was first shown by Osaki (1976). He referred to this property as a "trapping" of modes. In addition, their amplitudes may be relatively large here, and their vertical wavelengths small, as was also shown by Osaki (1976). Even so, he finds that the driving resulting from ε-mechanism here was insufficient to destabilize the entire star. The fact is, the radiative dissipation in this region is very large, so large that this damping completely predominates over the nuclear driving. This large radiative damping is essentially a consequence of the very short wavelengths of these trapped modes. In a sense, short wavelengths are analogous to too-small elements in the discussion of the Kippenhahn overstability above. Thus, the conclusion was that the ε-mechanism operating in the hydrogen-burning shell of slightly post-main sequence stars could not account for the pulsations of the Beta Cephei stars, as had been suggested earlier by Chiosi (1974) and by Aizenman, Cox & Lesh (1975).

3.4 Kato overstable convection

In a region in a star where a mean molecular weight gradient is present, it is possible, as was first pointed out by Kato (1966), that overstable convection can exist. This kind of overstability might occur in a region that is convectively stable according to the "Ledoux criterion," i.e., an element of matter displaced slowly upward, for example, would have a greater density than its surroundings, and so would tend to sink. In the absence of nonadiabatic effects, such an element would simply oscillate indefinitely up and down at the local Brunt-Väisälä frequency.

What these nonadiabatic effects will do, however, depends on the local average temperature gradient. If this temperature gradient is sub-adiabatic, such an upward-displaced element will be cooler than its surroundings. It will therefore absorb heat from them during its upward swing. This absorption of heat will decrease the density of the element below the value appropriate for adiabatic motion, and so will cause its subsequent descent to be slower than for adiabatic motion. In this way it can be seen that the oscillatory motion of the element will eventually be damped out by these nonadiabatic effects.

On the other hand, if the average temperature gradient is superadiabatic, then a slowly rising element will be warmer than its surroundings, and so will lose heat to them. This loss of heat during the upward swing will lead to a further enhancement of the density of the element, and so it will subsequently sink faster than if it had lost no heat. Similarly, a sinking element will subsequently rise faster than in the adiabatic case. In this case it can be seen that the amplitude of this oscillatory motion will gradually increase, as a result of these nonadiabatic effects. This phenomenon of an oscillatory motion of a mass element with a gradually growing amplitude is known as overstable convection.

It is clear, moreover, that overstable convection can only occur when the mean molecular weight μ increases toward the center of the star. But this is, after all, the normal situation encountered in stellar evolution. We assume that the motion of the convective element is so slow that pressure equilibrium with its surroundings will have been established. Then, if a rising element is to have a net downward buoyant force, its density must be greater than that of its surroundings. Hence, the quantity (T/μ) (T = temperature) of the element must be less than that of its surroundings. The value of μ for the element is characteristic of its level of origin. If μ increases inward, it is clear that the temperature of the element could be greater than that of its surroundings (the condition for overstability), and the element could still be convectively stable (i.e., the net buoyant force on an upward-displaced element would be downward).

Note that the Kato overstable convection mechanism can also be looked at as follows. The μ-gradient is equivalent to an additional contribution to the total force acting on a displaced mass element computed as if the chemical composition were uniform. This additional contribution to the force is proportional to the factor $-(d \ln \mu / d \ln P)\delta r$, where P denotes total pressure and δr is the (radial) displacement of a mass element from its equilibrium position. Thus, if $d \ln \mu / d \ln P > 0$ (the normal situation in stellar evolution), we have: if $\delta r > 0$ (the element is displaced upward), this additional force is downward, and acts as an extra contribution to the restoring force. If $\delta r < 0$

(downward displacement), this additional force is upward, and thus in this case too it increases the restoring force. Moreover, this additional contribution to the restoring force may be considered constant with time, since this force will change only on a nuclear time.

Now, if the average temperature gradient is superadiabatic, the buoyant force on an upward-displaced mass element will be upward, in the same direction as the displacement. In the absence of a μ-gradient such a region would then be convectively unstable. However, the μ-gradient may supply enough of a downward contribution to the restoring force acting on an upward-displaced element that the motion may still be oscillatory. But a rising element in a region of the star with a superadiabatic temperature gradient will be hotter than its surroundings, and so will lose heat to them. It can thus be seen that these heat exchanges will increase the density of an upward-displaced element, and so cause the buoyant force slowly to get weaker with time. On the other hand, the force arising from the μ-gradient is remaining essentially constant with time. Thus the net restoring force (buoyant force plus the "μ-gradient force") is gradually getting stronger with time. This situation of an oscillator with a slowly strengthening spring constant clearly corresponds to overstability.

These considerations will be helpful in connection with the mechanism to be discussed next, the "overstable magnetic convection," mechanism. In the semi-convective zone of a massive star we have all the necessary ingredients for the Kato overstable convection mechanism to operate. The convective elements are likely, moreover, to be rather massive (perhaps a tenth or a hundredth of a solar mass?). The idea therefore arose that, if such an overstability occurred with the convective elements in the convective core of a massive star, some kind of instability might well arise. This instability might conceivably account for the pulsations of the Beta Cephei stars. Such a suggestion has, in fact, been made by Percy (1970) and Spiegel (1970). However, the calculations of Shibahashi and Osaki (1976) showed that, while this overstability was indeed present in the relevant regions, it was too weak to destabilize the whole star. The conclusion is that this mechanism will probably not account for the pulsations of the Beta Cephei stars nor of any other known kind of pulsating star.

3.5 Shibahashi overstable magnetic convection

The suggestion of "magnetic overstability" was made by Shibahashi (1983) to account for the observed oscillations (~6-14 min) in the recently discovered "rapidly oscillating Ap stars" (Kurtz 1982). A more physical discussion is given in Cox (1984).

It has been known for some time (e.g., Chandrasekhar 1952) that a convectively unstable region (e.g., the outer hydrogen convection zone) in the presence of only a moderately strong magnetic field can lead to overstable oscillations.

The mechanism by which overstable oscillations can arise in the present circumstances can be understood as follows. It is well known (e.g., Alfvén & Fälthammer 1963) that magnetic lines of force "embedded" in an astrophysical plasma behave in many respects as if they were material, elastic strings "glued" to the plasma. Moreover, these lines of magnetic force exert a tension along their length, and so try to remain straight. On the other hand, convective motions will tend to produce "wiggles" in these lines of force. It can easily be seen, then, that this magnetic tension will amount to an additional contribution to the restoring force on a displaced mass element. This additional contribution to the restoring force will tend to induce oscillatory motion in an otherwise convectively unstable region; in this way a magnetic field will tend to suppress convection.

It is the exchange of heat (nonadiabatic effects) that leads to overstability, just as in the Kato overstability mechanism. Consider a superadiabatic region in a star. In such a region the buoyant force is in the same direction as the displacement. Therefore, such a region would be convectively unstable in the absence of a magnetic field. A rising mass element, for example, in such a region would, however, be hotter than its surroundings and so would lose heat to them. Hence the buoyant force would gradually get weaker with time. However, the additional contribution to the restoring force, being magnetic in origin (and proportional to the factor $-(B^2/d^2)\delta r$, where B is the magnetic induction and d is the diameter of the element), will change much more slowly in time (Cox 1984). The net restoring force (buoyant plus magnetic) on the element is thus slowly getting larger in time, and this corresponds to overstability. (In this sense "magnetic overstability" and the Kato overstable convection mechanism have much in common.) This mechanism and the Kato overstable convection mechanism may be considered as examples of the thermally excited oscillator of Moore and Spiegel (1966).

The growth times of the overstability can be shown to be fairly short, say less than a year. Moreover, the oscillatory periods should be only of the order of minutes (Cox 1984).

Whether this mechanism will continue to be a viable explanation of the observed variations of the "rapidly oscillating Ap stars," remains to be seen.

3.6 Osaki overstable convection

This mechanism was proposed by Osaki (1974) in order to attempt to find an excitation mechanism for the Beta Cephei stars. It is based on the known fact that rotation of a convective region in a star will introduce, via the Coriolis forces, an oscillatory component into the otherwise dynamically unstable motions. In other words, a rising or descending element will, because of the Coriolis force, swerve sideways and hence will execute a "circular" rather than an "up-and-down" motion. The period of this oscillatory motion will be determined by the rotation period of the convective region. The amplitude of this "circular" motion will increase at a rate determined by the magnitude of the superadiabaticity of the temperature gradient in the convective region.

Thus, the convective elements in such a rotating convective region execute overstable oscillations. Note that this overstability has nothing to do with nonadiabatic effects; it would exist even if the oscillations were strictly adiabatic.

Osaki assumed that the convective region was the convective core of a massive star. If the rotation period of the core was properly chosen, this overstable convection might resonate with the nonradial f mode of the radiative envelope (this mode has a period in the proper range for these stars). Thus, the whole star might be rendered unstable, and the Beta Cephei phenomenon might be accounted for.

This idea is very ingenious, and the parameter values required for it to work properly are certainly reasonable. The trouble is, however, that in nature all the requirements may not be satisfied, and this mechanism has not been generally accepted as an explanation of the instability of the Beta Cephei stars.

3.7 Tidally forced oscillations

This mechanism was suggested by Kato (1974) in an attempt to account for the often-observed double periodicity in Beta Cephei stars. Specifically, he pointed out that if the frequency separation between two rotationally split "m-sublevels" of a rotating star is very close to the orbital frequency of a companion, a resonance might occur which could amplify one (or more) of these sublevels. However, such a resonant interaction can occur only if a mode is already excited by some other mechanism. The paper is perhaps not fundamental in the sense that it offers no suggestion as to what this other mechanism might be.

There are many interesting points raised in this paper. A number of these resonant interactions require special circumstances, which may or may not be satisfied in nature. Still, many of the points raised in this paper should be kept in mind by those working in the field.

3.8 Kelvin-Helmholtz instability

This mechanism was suggested by Ando (1981) again in an attempt to account for the instability of the Beta Cephei stars. The instability basically derives from a Kelvin-Helmholtz instability in the thin region between a supposed rapidly spinning core and a slower spinning envelope. In this thin region there is strong shear, which may be subject to a Kelvin-Helmholtz instability (Papaloizou & Pringle 1978). There is also here, incidentally, a large μ-gradient. The frequency of this Kelvin-Helmholtz instability may, moreover, be almost the same as that of one of the f or g nonradial modes of the rest of the star. If this is the case, a resonance might occur which might excite one or more of these modes. In this way the excitation of some of the observed nonradial oscillation modes of the Beta Cephei stars and of the "line profile variable B stars" ("53 Per" stars) (Smith 1978, 1980a,b; Smith & McCall 1978; Smith & Buta 1979) might be accounted for.

The author points out a number of strong points in favor of the present mechanism for the kinds of stars cited above. He also notes that this mechanism is not as restrictive as the rotating convective core mechanism of Osaki (1974) (see earlier).

The final word on this mechanism has not yet been said. The mechanism looks promising, and only time will tell how successful it is.

4 APPLICATION TO CEPHEIDS

The He^+ envelope ionization mechanism was applied to Cepheid variables by Baker & Kippenhahn (1962) and Cox (1963). One of the most important aspects of this work (verified many times both previously and subsequently) is that it showed the following: It is the approximate coincidence of the "transition region" and the region of 50 percent ionization of He^+ (both in the envelope) that defines the "Cepheid instability strip" on the Hertzsprung-Russell (H-R) diagram. This strip cuts through the classical Cepheids, the W Virginis variables, the BL Herculis stars (King, Cox & Hodson 1981; Hodson, Cox, & King 1982; Cox & Wallerstein 1984), the RR Lyrae variables, the Delta Scuti stars, the Ap stars, and may even extend down to the white dwarfs (see Fig. 3).

These works and many subsequent calculations have confirmed the somewhat ubiquitous period-mean density relation of pulsating stars:

Fig. 3. Schematic Hertzsprung-Russell diagram, showing instability strip.

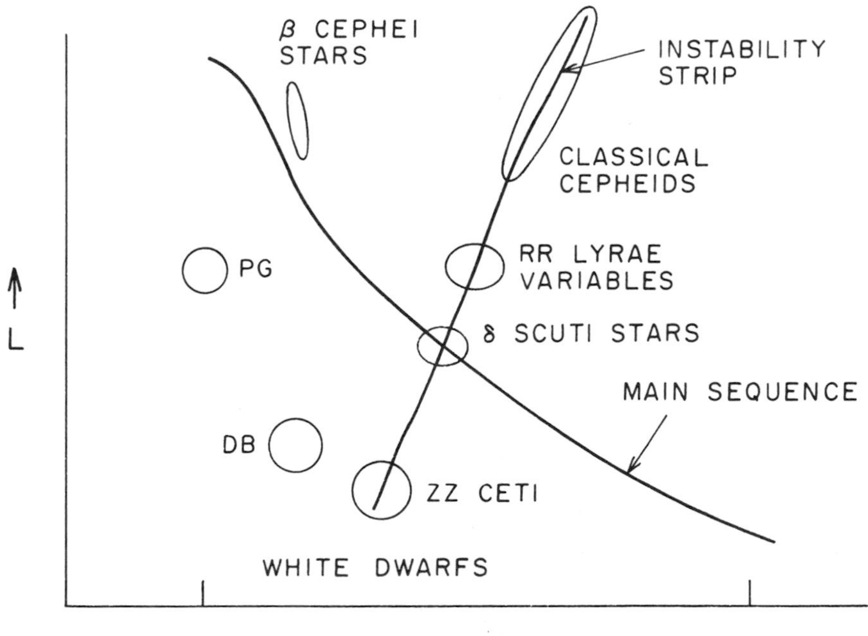

$$\Pi \sqrt{\bar{\rho}} \approx \text{const} \tag{5}$$

for a given mode, where Π denotes the period of the mode and $\bar{\rho}$ is the mean density of the star. Hence small, compact stars generally have shorter periods than do large, tenuous stars. This relation even holds, approximately, for nonradial oscillations of not-too-high orders or degrees (at least for p modes).

One feature of the above instability strip, along with the period-mean density relation, make the period-luminosity relation of Cepheids particularly easy to understand. (This relation has been confirmed many times, both observationally and theoretically.) This feature is that the strip is almost vertical on an H-R diagram. In turn, the reason for this feature is that, for a star of given mean luminosity, coincidence of the transition region and the region of 50 percent ionization of He^+ occurs for a sharply defined value of the effective temperature. Moreover, this value does not depend strongly on other stellar parameters. Thus, all stars in the strip have roughly the same effective temperatures. Hence, approximately,

$$L \propto R^2 \tag{6}$$

for such stars, where R is the mean radius of the star and L is its mean luminosity. Thus, large stars in the strip (those in its upper regions) are more luminous than smaller stars. On the other hand, according to the period-mean density relation,

$$\Pi \propto R^{3/2}/M^{1/2} \quad , \tag{7}$$

approximately, where M is the mass of the star. Eliminating R between equations (6) and (7) gives

$$\frac{L}{M^{2/3}} \propto \Pi^{4/3} \quad . \tag{8}$$

For M = constant, we have

$$L \propto \Pi^{4/3} \quad . \tag{9}$$

For $L \propto M^4$, we get

$$L \propto \Pi^{8/5} \quad . \tag{10}$$

The empirical relation is approximately

$$L \propto \Pi \quad . \tag{11}$$

The rough relations (9) and (10) are not in terribly bad agreement with the empirical relation (11), considering the crudity of the derivation. In the simplest terms, one may say that the more luminous stars in the instability strip are large, and large stars have longer periods than do small stars. Hence one might expect a positive correlation to exist between period and luminosity for stars in the instability strip.

5 SOME OUTSTANDING UNSOLVED PROBLEMS

I suppose a review of Cepheid theory and of excitation mechanisms in pulsating stars would not be complete without at least a listing of current outstanding unsolved problems. Such a list, as seen by this reviewer, is presented in this section.

5.1 Time-dependent convection

Great strides have recently been made in incorporating treatments, in some approximation or other, of time-dependent convection into pulsating star calculations. We now even have a theoretical red edge to the instability strip, thanks largely to the work of Deupree (1977a). As shown by the rather impressive set of papers by Stellingwerf (1982,a,b; 1984a,b,c), a treatment of time-dependent convection that is not too complicated can now be conveniently put into codes for the computation of radial oscillations of stars. This last work not only agrees satisfactorily with that of Deupree, but it also shows that convection may play an important role in the problem of the limiting amplitudes of pulsating stars. It would be useful if some not-too-complicated scheme for treating time-dependent convection, somewhat analogous to Stellingwerf's for radial oscillations, could be developed for use in connection with nonradial oscillations.

In spite of these successes, the general problem of the interaction of time-dependent convection with pulsations cannot be considered as solved. There is still considerable need for rather fundamental studies along these lines, such as that of Toomre and collaborators (see the review paper by Toomre 1982). As has been emphasized by Toomre (1982), the treatments referred to in the preceding paragraph still involve a number of somewhat uncomfortably arbitrary stellar parameters. There is still a great need for more fundamental work on the problem (e.g., Toomre, Hurlburt & Massaguer 1984).

5.2 Cepheid masses

Securing agreement between masses of pulsating stars as inferred, on the one hand, from evolution theory, and, on the other, from pulsation theory, is still a problem. Perhaps most severe are the "beat masses" of double-mode classical Cepheids. (These are the masses of double-mode pulsators derived from the first overtone -- fundamental period ratio and the fundamental period itself.) Despite the low values of these masses, the actual masses of these stars are probably normal, i.e., as expected from evolution theory (e.g., Cox 1982). Why is pulsation theory leading to such low mass values for these stars?

It is also puzzling that beat masses for RR Lyrae variables are not anomalously low. Both of these problems are discussed by A. N. Cox in an important review paper (Cox 1982 and references therein). Perhaps the augmented metal opacities idea of Simon (1982a,b) has some merit, despite certain rather strong objections (Huebner & Magee 1984).

5.3 Cepheid "bumps"

The physical explanation of the "bumps" which often appear on the descending portions of the light curves for Cepheids with periods in the approximate range 7-11 days (the "Hertzsprung relation," Hertzsprung 1926) is not yet available. These "bumps," when they occur, usually also show up in the appropriate velocity curves. The "bumps" were attributed by Christy (1967) to an "echo" phenomenon. In this process a shock wave was generated in the stellar envelope, travelled to the stellar center, was reflected, and subsequently travelled again to the surface, and produced the "bump." On the other hand, it was suggested by Simon & Schmidt (1976) that the "bumps" were the result of a resonance condition, in which the second overtone (radial) period was almost exactly one-half of the fundamental period.

At first sight, these two explanations appear quite different and independent of each other. However, as Simon has said (quoted from Whitney 1982), "...the pulse and resonance models may merely be opposite sides of a single coin." This possibility has been explored in a preliminary way by Whitney (1982), and further elaborated on by Aikawa & Whitney (1984). The results of Buchler (1984) are also consistent with this idea.

Petersen (1981, 1984) has noted that Population II Cepheids also exhibit "bumps" whose phase of occurrence depends on period, analogously to the Hertzsprung relation in classical Cepheids.

5.4 Double mode cepheids

As yet, initial-value type stellar pulsation calculations have not succeeded in producing true double-mode behavior (i.e., a long-lasting admixture of both fundamental and first overtone pulsation modes). The physical cause of double-mode behavior continues to elude us. An excellent and very comprehensive review (with lots of references!) has been published by A. Cox (1982). I think I can add nothing to what has been said here. As has been emphasized in this last reference, the stars which exhibit double-mode behavior probably have normal masses for their luminosities (i.e., as given by evolution theory). One would like to know not only what causes this kind of behavior, but also why the ratio of first overtone to fundamental periods is so low for the classical Cepheids.

5.5 The excitation mechanism for Beta Cephei stars

The cause of the pulsations of the Beta Cephei stars remains as one of the outstanding problems of pulsation theory. This problem has persisted for more than three-quarters of a century, despite many attempts at solution.

The most recent review article on possible excitation mechanisms for these stars is the excellent one by Osaki (1982). Earlier reviews are by Kato (1976) and Cox (1976).

The most recent mechanism is that suggested by Ando (1981), based on a Kelvin-Helmholtz instability in a thin shear layer surrounding a rapidly spinning core. While not as restrictive as the rapidly spinning convective core model of Osaki (1974), still special assumptions are required with this model.

5.6 Miscellaneous problems

Besides the above problems, there are the following rather general ones: (1) the R Coronae Borealis stars specifically, and very nonadiabatic stellar pulsations generally. (2) The Mira variables specifically, and long period variables generally; here time-dependent convection might have important applications; also, the relation between pulsations and mass loss should be investigated. (3) The very hot star PG 1159-035 and others of its kind; here the excitation mechanism should be clarified; temporal changes due to stellar evolution are probably important. (4) The general problem of the relation between magnetic fields and stellar pulsations should be further investigated. (5) The complicated nature of neutron star oscillations should be studied further. (6) The nature and cause of solar oscillations, and what they can tell us about the solar interior (helioseismology), should be studied more. (7) Here I put all other problems which I have not thought of and which are not listed above.

6 SUMMARY

In this paper we have surveyed the main destabilizing mechanisms (eight in all) that have been proposed to account for the instability of pulsating stars of known types. Perhaps the most widely applied type of mechanism is the "envelope ionization mechanism." This mechanism can account for most of the essential characteristics of the "instability strip" (e.g., Cox 1980, p.146). A simple interpretation of the period-luminosity relation of classical Cepheids is given. Finally, we list a few outstanding problems in pulsation theory.

We are grateful for helpful conversations with Art Cox, Carl J. Hansen, W. Dean Pesnell, Bernhard Durney, and Don Winget. Many thanks for preprints, etc., to H. Ando, Robert Buchler, Art Cox, Bernhard Durney, George Gonczi, Carl Hansen, Icko Iben, Phil Marcus, Barbara & Dimitri Mihalas, Chander Mohan, W. Dean Pesnell, Jorgen Petersen, H. Saio, H. Shibahashi, Norman Simon, Joe Smak, Paul Smeyers, Myron Smith, Sumner Starrfield, Bob Stellingwerf, Mine Takeuti, Juri Toomre, Craig Wheeler, Chuck Whitney, Lee Anne Willson, Don Winget and many others. We also thank Dean Pesnell and Carl Hansen for reading and commenting on the manuscript. This work was supported in part by National Science Foundation grant No. AST83-15698 through the University of Colorado.

REFERENCES

Aikawa, T. & Whitney, C. A. (1984). Astrophys. J., in press.

Aizenman, M. L., Cox, J. P. & Lesh, J. R. (1975). Astrophys. J., 197, 399.

Alfvén, H. & Fälthammer, C. G. (1963). Cosmical Electrodynamics. Oxford: Clarendon Press.

Ando, H. (1981). M.N.R.A.S., 197, 1139.

Ando, H. & Osaki, Y. (1975). Publ. Astron. Soc. Japan, 27, 581.
Baker, N. (1966). In Stellar Evolution, eds. R. F. Stein & A. G. W. Cameron, p. 333. New York: Plenum Press.
Baker, N. & Gough, D. O. (1979). Astrophys. J., 234, 232.
Baker, N. & Kippenhahn, R. (1962). Z. Astrophys., 54, 114.
Baker, N. & Kippenhahn, R. (1965). Astrophys. J., 142, 868.
Buchler, R. (1984). Private communication.
Castor, J. I. (1968). Unpublished manuscript.
Castor, J. I. (1971). Astrophys. J., 166, 109.
Chandrasekhar, S. (1952). Phil. Mag., Ser. 7, 43, 501.
Chiosi, C. (1974). Astron. Astrophys., 37, 281.
Christy, R. F. (1966). Astrophys. J., 144, 108.
Christy, R. F. (1967). In Aerodynamic Phenomena in Stellar Atmospheres, ed. R. N. Thomas, p. 105. New York: Academic.
Cox, A. N. (1982). In Pulsations in Classical and Cataclysmic Variable Stars, eds. J. P. Cox & C. J. Hansen, p. 157. Boulder: JILA.
Cox, A. N. & Wallerstein, G. (1984). Preprint.
Cox, J. P. (1963). Astrophys. J., 138, 487.
Cox, J. P. (1967). In Aerodynamic Phenomena in Stellar Atmospheres, ed. R. N. Thomas, p. 3. New York: Academic Press.
Cox, J. P. (1976). Ann. Rev. Astron. Astrophys., 14, 247.
Cox, J. P. (1976). In Proc. Solar and Stellar Pulsation Conf., eds. A. N. Cox and R. G. Deupree, p. 127. Los Alamos: LANL.
Cox, J. P. (1980). Theory of Stellar Pulsation. Princeton: Princeton Univ. Press.
Cox, J. P. (1984). Astrophys. J., in press.
Cox, J. P., Cox, A. N., Olsen, K. H., King, D. S. & Eilers, D. D. (1966). Astrophys. J., 144, 1038.
Cox, J. P. & Stellingwerf, R. F. (1979). Pub. A.S.P., 91, 319.
Deupree, R. G. (1977a). Astrophys. J., 211, 509.
Deupree, R. G. (1977b). Astrophys. J., 214, 502.
Deupree, R. G. (1977c). Astrophys. J., 215, 232.
Deupree, R. G. (1977d). Astrophys. J., 215, 620.
Deupree, R. G. (1980). Astrophys. J., 236, 225.
Durney, B. (1984). Preprint.
Dziembowski, W. (1971). Acta Astron., 21, 289.
Eddington, A. S. (1941). M.N.R.A.S., 101, 182.
Eddington, A. S. (1942). M.N.R.A.S., 102, 154.
Epstein, I. (1950). Astrophys. J., 112, 6.
Gonczi, G. (1981). Astron. Astrophys., 96, 138.
Gonczi, G. (1982). In Pulsations in Classical and Cataclysmic Variable Stars, eds. J. P. Cox & C. J. Hansen, p. 206. Boulder: JILA.
Gonczi, G. & Osaki, Y. (1980). Astron. Astrophys., 84, 304.
Hertzsprung, E. (1926). B.A.N., 3, 115.
Hodson, S. W., Cox, A. N. & King, D. S. (1982). Astrophys. J., 253, 260.
Huebner, W. & Magee, N. (1984). Preprint.
Kato, S. (1966). Publ. Astron. Soc. Japan, 18, 374.
Kato, S. (1974). Publ. Astron. Soc. Japan, 26, 341.
Kato, S. (1976). In Multiple Periodic Variable Stars, ed. W. S. Fitch, p. 3. Dordrecht: Reidel.

King, D. S. & Cox, J. P. (1968). Publ. A.S.P., 80, 365.
King, D. S., Cox, A. N. & Hodson, S. W. (1981). Astrophys. J., 244, 242.
Kippenhahn, R. (1983). Preprint.
Kurtz, D. W. (1982). M.N.R.A.S., 200, 807.
Ledoux, P. (1974). in Stellar Instability and Evolution, eds. P. Ledoux, A. Noels & A. W. Rodgers, p. 135. Dordrecht: Reidel.
Ledoux, P. & Walraven, Th. (1958). Handbuch der Physik, ed. S. Flügge, 51, p. 353. Berlin: Springer-Verlag.
Lee, U. & Osaki, Y. (1982). Publ. Astron. Soc. Japan, 34, 39.
Moore, D. W. & Spiegel, E. A. (1966). Astrophys. J., 143, 871.
Osaki, Y. (1974). Astrophys. J., 189, 469.
Osaki, Y. (1976). Publ. Astron. Soc. Japan, 28, 105.
Osaki, Y. (1982). In Pulsations in Classical and Cataclysmic Variable Stars, eds. J. P. Cox & C. J. Hansen, p. 303. Boulder: JILA.
Papaloizou, J. & Pringle, J. E. (1978). M.N.R.A.S., 182, 423.
Percy, J. R. (1970). Astrophys. J., 159, 177.
Pesnell, W. D. (1983). Ph.D. Dissertation, University of Florida.
Pesnell, W. D. (1984a). Preprint.
Pesnell, W. D. (1984b). Astrophys. J., in press.
Petersen, J. O. (1981). Astron. Astrophys., 96, 146.
Petersen, J. O. (1984). Preprint.
Saio, H. (1980). Astrophys. J., 240, 685.
Saio, H. & Cox, J. P. (1980). Astrophys. J., 236, 549.
Shibahashi, H. (1983). Astrophys. J. (Lett.), 275, L5.
Shibahashi, H. & Osaki, Y. (1976). Publ. Astron. Soc. Japan, 28, 199.
Simon, N. R. (1982a). In Pulsations in Classical and Cataclysmic Variable Stars, eds. J. P. Cox & C. J. Hansen, p. 221. Boulder: JILA.
Simon, N. R. (1982b). Astrophys. J. (Lett.), 260, L87.
Simon, N. R. & Schmidt, E. G. (1976). Astrophys. J., 205, 162.
Smeyers, P. (1966). Acad. Roy. des Sci., des Lett. et des Beaux.-Arts d. Belgique, 5e Ser., 52, 1126.
Smith, M. A. (1978). Astrophys. J., 224, 927.
Smith, M. A. (1980a). In Nonlinear and Nonradial Oscillations of the Sun and Stars, eds. H. A. Hill & W. Dziembowski, p. 60. New York: Springer-Verlag.
Smith, M. A. (1980b). In Highlights of Astronomy, ed. P. A. Wayman, 5, p. 457. Dordrecht: Reidel.
Smith, M. A. & Buta, R. J. (1979). Astrophys. J. (Lett.), 232, L193.
Smith, M. A. & McCall, M. L. (1978). Astrophys. J., 223, 221.
Spiegel, E. A. (1970). Comments on Astrophys. and Space Sci., 1, 57.
Starrfield, S. G., Cox, A. N. & Hodson, S. W. (1981). Space Sci. Rev., 27, 621.
Starrfield, S. G., Cox, A. N. & Hodson, S. W. (1983). Astrophys. J. (Lett.), 268, L227.
Starrfield, S. G., Cox, A. N., Kidman, R. B. & Pesnell, W. D. (1984). Astrophys. J., in press.
Stellingwerf, R. F. (1978). Astron. J., 83, 1184.
Stellingwerf, R. F. (1979). Astrophys. J., 227, 935.

Stellingwerf, R. F. (1982a). Astrophys. J., 262, 330.
Stellingwerf, R. F. (1982b). Astrophys. J., 262, 339.
Stellingwerf, R. F. (1984a). Astrophys. J., 277, 322.
Stellingwerf, R. F. (1984b). Astrophys. J., 277, 327.
Stellingwerf, R. F. (1984c). Preprint.
Toomre, J. (1982). In Pulsations in Classical and Cataclysmic Variable Stars, eds. J. P. Cox & C. J. Hansen, p. 170. Boulder: JILA.
Toomre, J., Hurlburt, N. E. & Massaguer, J. M. (1984). Preprint (to appear in Proc. Small-Scale Dynamical Processes in Quiet Stellar Atmospheres, ed. S. Keil, National Solar Observatory.
Unno, W., Osaki, Y., Ando, H. & Shibahashi, H. (1979). Nonradial Oscillations of Stars. Tokyo: Tokyo Univ. Press.
Whitney, C. A. (1982). In Pulsations in Classical and Cataclysmic Variable Stars, eds. J. P. Cox & C. J. Hansen, p. 226. Boulder: JILA.
Winget, D. E. (1981). Ph.D. dissertation, Univ. of Rochester.
Winget, D. E. (1984). Private communication.
Winget, D. E. & Fontaine, G. (1982). In Pulsations in Classical and Cataclysmic Variable Stars, eds. J. P. Cox & C. J. Hansen, p. 46. Boulder: JILA.
Winget, D. E., Hansen, C. J. & Van Horn, H. M. (1983). Nature, 303, 781.
Winget, D. E., Robinson, E. L., Nather, R. E. & Balachandran, S. (1984). Astrophys. J. (Lett.), 279, L15.
Winget, D. E., Robinson, E. L., Nather, R. E. & Fontaine, G. (1982b). Astrophys. J. (Lett.), 262, L11.
Winget, D. E., Van Horn, H. M., Tassoul, M., Hansen, C. J. & Fontaine, G. (1983). Astrophys. J. (Lett.), 268, L33.
Winget, D. E., Van Horn, H. M., Tassoul, M., Hansen, C. J., Fontaine, G. & Carroll, B. W. (1982a). Astrophys. J. (Lett.), 252, L65.
Zhevakin, S. A. (1953), Russian A.J., 30, 161.

NON-ADIABATIC EFFECTS ON PULSATION PERIODS

TOSHIKI AIKAWA
Faculty of Engineering, Tohoku-Gakuin
University, Tagajo, 985 Japan

INTRODUCTION

It is well known that the difference between the adiabatic pulsation periods and the corresponding non-adiabatic periods is small in classical Cepheids. However, the difference becomes significantly large in low surface-gravity models (e.g. Aikawa 1984a). In this paper we shall discuss the origins of the difference between the two pulsation periods. The weight functions for non-adiabatic pulsation periods are introduced in analogy to those of Epstein (1950) for adiabatic periods.

ACOUSTIC WAVES IN RADIATION FIELDS

Non-adiabatic pulsations are influenced by the interaction between acoustic waves and radiation fields. Two important parameters characterize the interaction (e.g. Vincenti and Kruger 1965). They are:

$$N_{Bo} = \rho C_{\rho} a/\sigma T^3, \tag{1}$$

$$\tau_a = \kappa/\omega , \tag{2}$$

where $\underline{a}$ is the adiabatic sound speed, κ is the specific opacity, and ω is the frequency of the acoustic motions. We examined the behavior of these quantities in the envelope of classical Cepheids and the low surface-gravity stars and we find the star is divided into three regions according to the coupling between the the acoustic waves and the radiation (Mihalas and Mihalas 1983; Zhugzhda 1983):

I: adiabatic region,
II: interaction region,
III: isothermal region.

We also plotted the peak of the weight functions of the fundamental mode. The peak is in Region I for the classical Cepheid while it is in Region II for the model with low surface gravity. This fact shows that the pulsation periods of classical Cepheids are determined essentially in the adiabatic region and, on the contrary, those of the low surface-gravity are strongly affected by the region where the acoustic waves are coupled tightly to the radiation field. This is the first conclusion of this paper.

PROPAGATION SPEED OF ACOUSTIC WAVES IN RADIATION FIELDS

The travel time of the adiabatic sound wave from the center to the surface of a star is nearly equal to the adiabatic pulsation period (Lamb 1945). We shall discuss the difference between the adiabatic and the corresponding non-adiabatic pulsation periods in terms of the propagation speed of acoustic waves.

In Region I, the acoustic waves propagate with the adiabatic sound speed. Thus non-adiabatic effects do not significantly influence pulsation periods. In Region III, the acoustic waves propagate with the isothermal sound speed, so acoustic waves have longer travel times than adiabatic sound waves. It is expected that the non-adiabatic effects in this region make the pulsation periods longer.

Region II is very complicated. The acoustic mode changes its speed from the adiabatic to the isothermal sound speed gradually. On the other hand, the thermal-diffusion mode has a phase speed of 1 to 2 times the adiabatic sound speed in this region. It is noted that the acoustic waves are still tightly coupled to the radiation in the latter mode. Thus the effective propagation speed of acoustic waves must be considerably higher than the adiabatic sound speed in this region. Hence, the non-adiabatic effects will make the pulsation period shorter.

We conclude that the characteristics of the non-adiabatic weight functions are explained qualitatively by these arguments. This is the second conclusion of this paper.

Finally, we conclude that the non-adiabatic effects are quite important in low surface-gravity pulsating stars. The details will appear in Aikawa (1984b).

REFERENCES

Aikawa, T. (1984a). Astrophys. & Space Science, submitted.
Aikawa, T. (1984b). Astrophys. & Space Science, submitted.
Epstein, I. (1950). Astrophys. J., 112, 6.
Lamb, H. (1945). Hydrodynamics. Dover. New York.
Mihalas, D. & Mihalas, S.W. (1983). Astrophys. J., 273, 355.
Vincenti, W.D. & Kruger, C.H. (1965). Introduction to Physical Gas Dynamics. Wiley. New York.
Zhugzhda, Yu.D. (1983). Astrophys. & Space Science, 95, 255.

NONADIABATIC, NONLINEAR PULSATIONS OF BUMP CEPHEIDS - A NEW APPROACH

J. R. Buchler, M. J. Goupil & J. Klapp
Physics Department
University of Florida, Gainesville, FL 32611

Abstract

Finite amplitude pulsations of realistic stellar models are analyzed within the framework of a promising asymptotic perturbation approach and the results are compared with those of numerical hydrodynamic studies.

The pulsations of Cepheid variables are characterized by the fact that the growthrates of the excited modes are fairly small compared to their oscillation frequencies. While this is a distinct drawback for the numerical hydrodynamic modelling of the nonlinear behavior of such stars, it can be used to one's advantage in a nonadiabatic nonlinear perturbation formalism (**NNPF**) (Buchler & Goupil 1984; **BG84**). Previous perturbation approaches (e.g. Buchler 1978, Regev & Buchler 1981, Dziembowski & Kovacz 1984, Takeuti & Aikawa 1982) have assumed the pulsations to be quasiadiabatic throughout the whole star, that is also in the obviously nonadiabatic outer layers, and thus conveniently used the real adiabatic eigenvectors as a perturbation basis. While quite useful for analytical considerations, the quasiadiabatic approximation unfortunately does not lead to an unambiguous practical computational scheme (Pesnell & Buchler 1984). In contrast, the NNPF includes the nonadiabatic effects already in lowest order through the use of the linear nonadiabatic (**LNA**) eigenvectors. The price one has to pay is that these eigenvectors are not only complex, but, in addition, necessitate the introduction of the left (adjoint) LNA eigenvectors, which form a dual set orthogonal to the right eigenvectors. One of the small parameters of problem is then the ratio r of the growthrate, κ, to the oscillation frequency, Ω. Since r is a global parameter, nonadiabatic effects can be arbitrarily large as long as r remains small. The basic assumption of the NNPF is that the LNA modes can be split into two groups, one containing the strongly damped (slave) modes and one containing the marginally unstable, excited modes together with those marginally stable modes which may get entrained either because of a low order internal resonance or because of a nonlinear loss of stability. When these conditions are satisfied the center manifold theorem and the theory of normal forms guarantee us that amplitude equations (**AE**) can be found, which involve only the amplitudes of the second group of modes.

So far, we have applied (Klapp, Goupil and Buchler 1984; **KGB84**) the NNPF to the case of Cepheid models, of population I (Bump Cepheids) and of population II (paradigmatized by BL Her), in the regime where they have an internal resonance of the type $2\Omega_0 \simeq \Omega_2$. The subscripts 0 and 2 refer to the fundamental and second overtone, respectively. The apposite AE are given by (BG84)

$$\frac{da_0}{dt} = \sigma_0 a_0 + N_0 a_0^* a_2, \quad \text{and} \quad \frac{da_2}{dt} = \sigma_2 a_2 + N_2 a_0^2 , \qquad (1)$$

where $\sigma_\nu = i\Omega_\nu + \kappa_\nu$ are the LNA eigenvalues, a_0 and a_2 are the complex amplitudes for the two resonant modes and the coefficients N_0 and N_2 involve the quadratic nonlinearities of the momentum and heat flow equations, expanded around hydrostatic and thermal equilibrium. These quadratic operators are then sandwiched between appropriate right and left LNA eigenvectors (BG84). In lowest order the displacement can be written as

$$\delta R(m) = 0.5 \left[a_0(t)\ \xi_0(m) + a_2(t)\ \xi_2(m) + \text{c.c.}\right] + \ldots, \qquad (2)$$

where the $\xi_\nu(m)$ are the radial parts of the right LNA vectors.

The solution of eq. 1 provides the time dependence of the amplitudes in eq. 2. By introducing moduli and phases for the amplitudes, $a_\nu(t) = A_\nu(t)\exp(i\Omega_\nu t)\exp(i\theta_\nu(t))$, it is possible to reduce the two complex AE to a set of three real AE involving A_0, A_2 and $\Gamma = \theta_2 - 2\theta_1$ only. Of particular interest are the solutions with constant A_0, A_2 and Γ (stable fixed points of the AE), corresponding to oscillations in which the amplitudes of the two modes are constant. Interesting, more complicated behavior is also possible (KGB84), namely periodic energy transfer between the two modes or irregular (chaotic) transfer.

The physical content of our AE (1) is very simple. The fundamental mode is unstable, saps thermal energy from the star and converts it initially exponentially fast into mechanical (pulsational) energy. As the oscillation amplitude gets sufficiently large energy is shared with the resonant overtone through the nonlinear terms. The latter mode, bring vibrationally stable, restores energy back to the thermal reservoir. Stable fixed points (and more general bounded solutions as well) can result from a balance between these effects.

Figure 1 shows an H-R diagram for a specific set of models of 4 solar masses. Our AE assume a near resonance to hold and we have indicated lines of constant period ratio, P_2/P_0 marked 0.48, 0.50 and 0.52. Also exhibited is the line (F) below which the fundamental mode is unstable. Finally to the right of the line marked (FPB) our AE have no fixed points.

Fig. 1 H-R Diagram

Fig. 2 Sequence of models

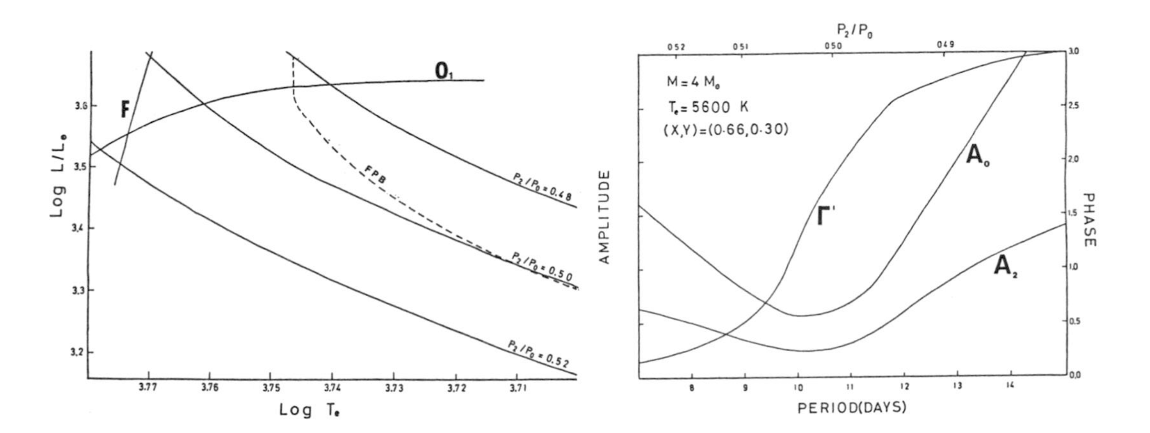

Fig. 2 shows the behavior of the amplitudes and phases for a sequence of models of constant effective temperature. This behavior, namely the variation of $\Gamma' \equiv \Gamma + \pi$ through $\pi/2$ and the dip of the amplitudes near the resonance is typical and can readily be derived from a mere inspection of the AE (KGB84). Similarly it can be shown that these features do not occur exactly at the resonance, the deviation depending on the magnitude of the ratios ($\mathrm{Re}\, N_\nu / \mathrm{Im}\, N_\nu$). This, while having been held against a resonance origin of the bumps (Vemury & Stothers 1978), in fact corroborates it. If the quadratic terms were evaluated with adiabatic eigenvectors, the N_ν would be purely imaginary. The size of these ratios and the shift of the above features from exact resonance are then an overall measure of the nonadiabaticity of the pulsations.

In fig. 3 we show the (scaled) surface velocities for the sequence of models of fig. 2, clearly exhibiting the so-called Herzsprung progression of the bump. The calculated surface velocities agree to within a factor of two with those obtained with comparable numerical hydrodynamic models.

We have studied a variety of resonant Cepheid models (KGB84). In all the models in which our AE predict a steady bump, hydrodynamic models indeed also find one.

To be fair we should also mention some of the limitations of the AE. The AE cannot claim to reproduce all the physics of the envelope, for example, shock waves, which involve high nonlinearities. Similarly they cannot pretend to reproduce the skewed observed velocity profiles in the atmosphere which occur in the dilute atmosphere. However, as long as the shock waves or the

Fig. 3 Surface velocity for the sequence of models of fig. 2.

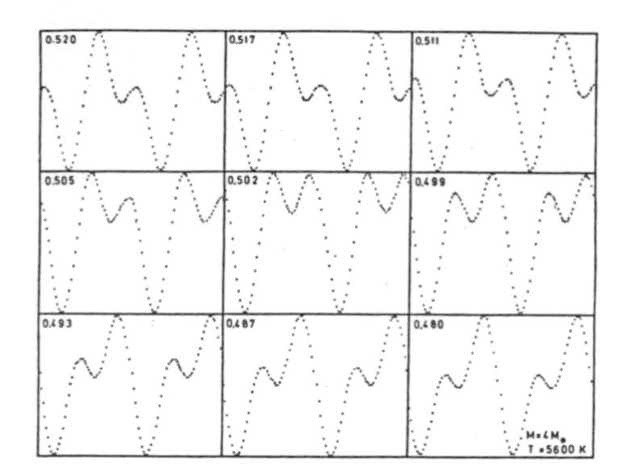

outermost dilute layers are not instrumental in the saturation mechanism, our AE are a useful and fast computational tool for determining the gross features of the nonlinear behavior of a given stellar model, i.e., the type of finite amplitude behavior (stable fixed point, limit cycle, irregular attractor), the magnitude of the velocities and the position (phase) of the bump. The NNPF is thus seen to be complementary to the much more (hundredfold) expensive, but more detailed numerical hydrodynamic approach.

We conclude that the good agreement with the hydrocode suggests that our formalism captures the basic saturation mechanism and can be used with some confidence to study the nonadiabatic nonlinear behavior of large classes of stellar models. To finish we want to emphasize that the NNPF is not limited to the resonant models considered in this contribution, but applies to other models as well.

This work has been supported in part by NSF.

References

Buchler, J. R., 1978, Ap. J., **220**, 629.
Buchler, J. R. and Goupil, M. J., 1984, Ap. J. **279**, 394.
Dziembowski, W. and Kovacz, G., 1984, M.N.R.A.S. **206**, 497.
Klapp, J., Goupil, M. J. and Buchler, J. R., 1984, Ap. J. (submitted).
Pesnell, W. D. and Buchler, J. R., 1984, Ap. J. (submitted)
Regev, O. and Buchler, J. R., 1981, Ap. J., **250**, 769.
Takeuti, M. and Aikawa, T., 1981, Sci. Rep. Tokohu Univ., 8th Ser. **2**, No. 3.
Vemury, S. K. and Stothers, R., 1978, Ap. J. **225**, 939.

THEORETICAL STUDY OF CEPHEID LIGHT CURVES

C.G. Davis
Los Alamos National Laboratory, Los Alamos, New Mexico

INTRODUCTION

Starting with the initial understanding that pulsation in variable stars is caused by the heat engine of Hydrogen and Helium ionization in their atmospheres (A.S. Eddington in Cox 1980) it was soon realized that non-linear effects were responsible for the detailed features on their light and velocity curves. With the advent of the computer we were able to solve the coupled set of hydrodynamics and radiation diffusion equations to model these non-linear features (Christy 1968, Cox et. al. 1966). Calculations including the effects of multi-frequency radiative transfer (Davis 1975) showed that grey diffusion was adequate for modeling Cepheids but not for the RR Lyrae or W Virginis type variables. In 1977, in collaboration with J. Castor and D. Davidson (1977), we developed a non-Lagrangian method to resolve the region of the ionization front and remove the zoning effects previously found in theoretical light curves (Keller and Mutschlecner 1972). The new code (DYN) has been used in recent studies by Takeuti (1983), with Simon (1983), in an attempt to understand modal coupling, and with Moffett and Barnes (1981) in a detailed study of X Cygni. A minor success for DYN was obtained by comparing "phase lags", as determined from the first order modal expansions between maximum light and maximum outward velocity, with some low period Cepheid observations (Simon 1984). Another support for the "dips" as observed by Moffett and Barnes (1984) is the observation by Schmidt and Parsons (1982) of a "dip" in the IUE spectra of δ Cep near the phase of 0.85.

In this paper we want to describe some recent model results for long period (LP) Cepheids in an attempt to understand these "dips" and possible get another handle on Cepheid masses. In section II we discuss these results and in section III we consider the implications of these model results on the problem of the Cepheid mass discrepancy.

LONG PERIOD CEPHEID MODELS

It appears that a classification of Cepheids could be: low amplitude low period sinusoidal Cepheids, "bump" Cepheids with periods from 7 to 10 days, and long period Cepheids with periods longer than 10 days. In this paper we are considering this latter class of variables. The models we have selected have effective temperatures near the middle of the instability strip and nearly constant at 5300 or 5500 K. We are mainly interested in seeing the effects of mass on the light curves so we selected near evolutionary masses and 60% or so of the evolutionary masses. The period range was picked in the range from 12 to 22 days with selected stars from the Moffett and Barnes list for comparisons (Table I).

TABLE I

LONG PERIOD CEPHEID MODELS

PERIOD (PROTOTYPE)	LOW MASS LUM	LOW MASS TEFF	LOW MASS P2/PO	HIGH MASS LUM	HIGH MASS TEFF	HIGH MASS P2/PO
12.9 (Z SCT)	1.5	4.5 M_o .54	.499	2.5	7.5 M_o .55	.536
14.8 (RW CAS)	2.1	5.0 M_o .551	.497	3.0	8.5 M_o .54	.527
16.4 (X CYG)	2.45	5.7 M_o .54	.486	3.34	9.0 M_o .53	.520
22.0 (WZ SGR)	3.5	6.0 M_o .54	.475	5.5	10.0 M_o .54	.507

LUM ($\times 10^{37}$ erg/cm^2-S) TEFF($\times 10^4$ ^{o}K)

First we note, from Table I, that the low and high mass models stradle the linear theory resonance boundary parameter of $P_2/P_0=0.5$. Near 0.5 the resonance interaction is the strongest and the Christy "bump" should appear near maximum light. A reasonable explanation of the connection between resonance theory and the "echo" is given by Whitney (1983). In Fig. 1 we illustrate the DYN results of this phenomena for model of a 10 day Cepheid with evolutionary mass and 60% of evolutionary mass. In the low mass model the "echo", which causes the "bump" at the surface, can clearly be seen near the inner boundary.

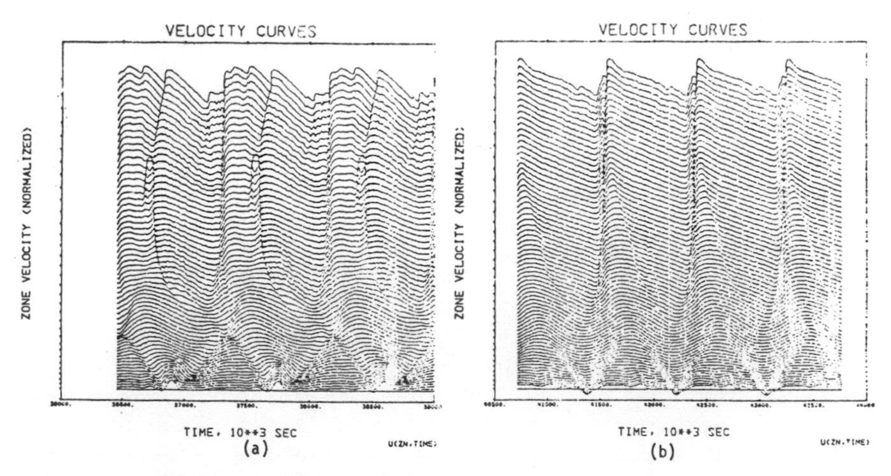

Figure 1 - Velocity distributions in models of 10 day Cepheids a) low mass (4.0 M_{θ}) and b) high mass (7.4 M_0) using the DYN code and KingIVA opacities.

Now it is clear that simply eyeballing the resulting light curves for comparison to observations is highly subjective (Simon and Davis 1983) but for these long period Cepheids, the spiked peak makes Fourier analysis difficult. It is also possible that the so called "bumps" or shoulders on the rising branch of LP Cepheid light curves could be the "dips" now resolved by Moffett and Barnes. These "dips" appear to be due to a surface phenomena that is more easily described by the methods of dynamic zoning used in DYN. Even though the shock is treated by the use of "Pseudo-viscosity" it is well resolved by the fine zones included in the shock forming region. In the atmosphere of the star the strongest shock that develops is the one caused by the stopping of the infall of the envelope near the phase of maximum inward velocity. This shock gives rise to the "artificial viscosity dip" discussed by Davis (1975) and others. In most models another shock develops at the time of rapid expansion. This shock is associated with the brief transition of the ionization front from a "D" to a "R" type (Kahn's notation see Castor and Adams 1974) as it moves rapidly inward in mass. This inward shock remains below optical depth unity and therefore should not effect the light curves. Without the capability to follow line transport in a moving media we must rely on the photospheric continuum results for our comparisons to the observations.

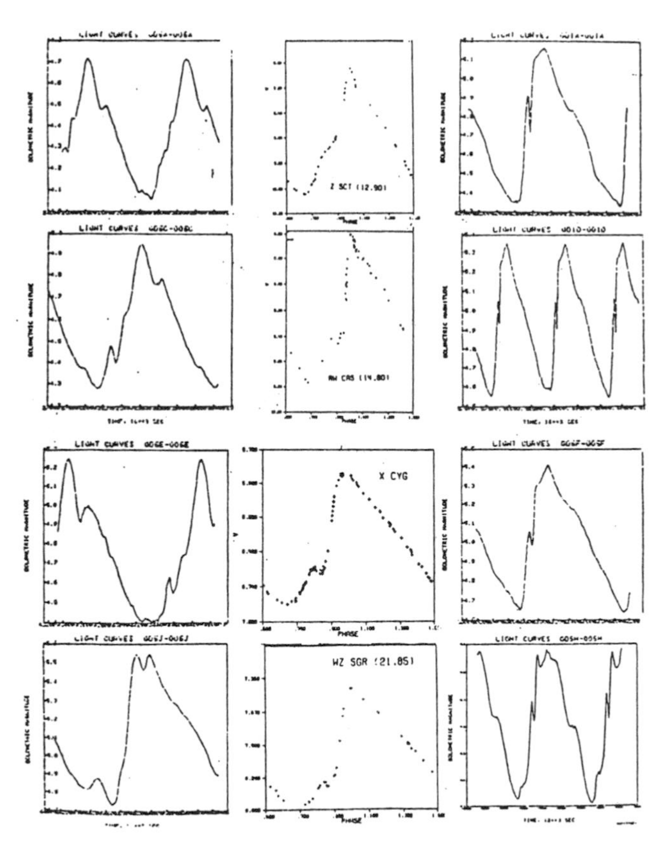

Fig. 2 - The Dyn non-linear long period Cepheid model results with 60% evolutionary mass (left panel) and near evolutionary mass (right panel) compared to the observations of Moffett and Barnes (center panel).

THE RESULTS AND CEPHEID MASSES

In modeling the non-linear hydrodynamics that occurs in pulsating stars we have considered the effects of radiative transfer but not the effect of convection. It is believed that convection causes the limit to the red edge of the instability strip but is not very important to the light output otherwise. For long period Cepheids (observe Fig. 2) we generally find a "dip" in the rising part of the light curve which signals "shock dissipation" and possibly the limit on the amplitude of the pulsation. These "dips" may have previously been mistaken for "bumps" or shoulders and therefore the resonance effect. As discussed in our paper on X Cygni to remove the bumps on the light curves we need to use masses near the evolutionary masses. More and detailed velocity measurements, in synchronism with the light measurements, may help answer the question, "Are there "bumps" as well as "dips" in the long period Cepheid light curves?"

REFERENCES

Adams, T.F. and Castor, J.I., Ap. J. 230, 826.
Castor, J.I., Davidson, D.K., and Davis, C.G., LA-6664.
Christy, R.F. (1968). Quart. J.R. A.S. 171, 593.
Connolly, L.P., (1980). Pub. A.S.P. 92, 165.
Cox, A.N., Brownlee, R.R. and Eilers, D.D. (1966). Ap. J. 144, 1024.
Cox, J.P. (1980) Theory of Stellar Pulsation, Princeton, Princeton University Press).
Davis, C.G. (1975) Cepheid Modeling, NASA SP-383.
Davis, C.G. Moffett, T.J. and Barnes, T.G. (1981) Ap. J. 246, 914.
Keller, C. and Mutschlecner, J.P. (1972) Ap. J. 171, 593.
Moffett, T.J. and Barnes, T.G., 11984 Ap. J. Suppl. in press.
Schmidt, E.G. and Parsons, S.B. (1982) Ap. J. 48, 165.
Simon, N.R. and Davis, C.G. (1983) Ap. J. 266, 787.
Simon, N.R. (1984) to be published.
Takeuti, M., UJI-IVE, K. and Aikawa, T. (1983) Sei. Rept. Tohuku University Vol. 14. No. 1.
Whitney, C. (1983) Ap. J. 274, 830.

MAGELLANIC CLOUD AND OTHER EXTRAGALACTIC CEPHEIDS: SOME CURRENT TOPICS

M.W. Feast
South African Astronomical Observatory

INTRODUCTION

Table 1 lists galaxies in which Cepheids are known. The early work on the detection and period determination for these stars forms the basis of subsequent studies. These later studies include the infrared photometry of Cepheids pioneered by the Toronto group and the efforts by various workers to improve the optical photometry. An example of the importance of this latter work is the recent study of M33 by Sandage (1983) in which a revised magnitude scale leads to a distance modulus 0.67 greater than that previously adopted.

Table 1.

M31	Baade and Swope 1963, 1965
M33	Sandage 1983
IC1613	Sandage 1971
NGC 6822	Kayser 1967
LMC	many writers
SMC	many writers
NGC 2403 and the M81 group	Tammann and Sandage 1968
NGC 300	Graham (proceeding)
Sextans A	Sandage and Carlson 1982
WLM)	
Sextans B)	
NGC 3109)	Sandage and others
IC5152)	(cf. Sandage and Carlson 1982)
Leo A)	
Pegasus)	

So as to avoid too much overlap with other reviews this paper concentrates on a limited number of topics, mainly related to Magellanic Cloud Cepheids. The aim is to see what these Cepheids tell us about the physical and chemical properties of Cepheids and to assess their potential as distance indicators. Work on Magellanic Cloud Cepheids has been recently reviewed (Feast 1984) and the present paper emphasises work since that review was written. Useful references to various aspects of work on extragalactic Cepheids are: Sandage (1972), van den Bergh (1968, 1977), Fernie (1969), Sandage and Tammann (1981).

CHEMICAL ABUNDANCES OF EXTRAGALACTIC CEPHEIDS

So far the Magellanic Clouds are the only galaxies other than our own for which attempts have been made to determine the chemical abundances in Cepheids. Three methods have been used: (1) Washington four colour photometry (Harris 1981), 1983); (2) Walraven five colour photometry (Pel, van Genderen and Lub 1981, Pel 1981, Pel 1984); (3) Rough curves of growth (Laney 1983). The results are given in Table 2.

Table 2. Abundances [A/H] in Magellanic Cloud Cepheids.

	LMC	SMC
Harris	-0.09	-0.65
Pel et al.	-0.2	-0.6
Laney	-0.06	-0.50
Adopted	-0.15	-0.60
Depletion Factor (D)	1.4	4

The Washington four colour results are calibrated partly from galactic stars of known metallicity and partly from the models of Kurucz (1979) and Bohm-Vitense (1972). The Walraven results depend on Kurucz (1979) models. The adopted values (Table 2) are close to those derived from HII regions (cf. Laney's tabulation).

The Washington data shows an apparent scatter in abundance in both Clouds. The most likely cause is either a real spread in abundance in each Cloud or else the effect of a range of temperatures at a given period (i.e. the effect of finite width of the instability strip), or of course a combination of both. Harris (1983) tentatively suggests that there may be a real spread in [A/H] of ±0.10 in the LMC and ±0.15 in the SMC. The work discussed below on the reddenings of Magellanic Cloud Cepheids and on the P-L-C relation leaves rather little room for abundance variations. Evidently this matter needs further study. It may be possible to combine the Washington and the BVI data for individual Cepheids to provide self consistent reddenings, intrinsic colours and abundances.

THE SIGNIFICANCE OF ACCURATE INTERSTELLAR REDDENINGS FOR MAGELLANIC CLOUD CEPHEIDS

Caldwell and Coulson (1984a) have now thoroughly investigated the use of BVI colours to derive accurate individual reddenings of Cepheids and have derived reddenings for both LMC and SMC Cepheids. In the SMC these results depend to a large extent on 1060 BVRI observations of 53 Cepheids (Caldwell and Coulson 1984b). The reddenings are based on the B-V/V-I intrinsic line of Dean et al. (1978) corrected for abundance effects using Bell-Gustafsson (1978) models. We are of course dependent on the accuracy of the models for the reddening zero points, but within this framework the reddenings are not very sensitive to the abundances. A change in [A/H] of 0.1 (D = 1.26) only changes E_{B-V} by 0.011. For the sample of Cepheids observed the mean reddenings are 0.074 (LMC) and 0.054 (SMC). Mean values close to this are found for a range of other types of object in the Clouds.

Both Harris (1981) and Isserstedt (1976, 1980) have used indirect arguments based on Cepheids to suggest that the ratio $R = Av/E_{B-V}$ in the SMC is different to that in the Galaxy. Harris finds R to be three times larger in the SMC than in the Galaxy. If this were really so it would seriously complicate the discussion of Magellanic Cloud Cepheids. Fortunately it is possible to derive R rather directly from infrared photometry of some highly reddened A type supergiants in the SMC and a value close to the galactic value (3.1) is found (Feast and Whitelock 1984).

For the sample of Cepheids studied the dispersion in reddenings is remarkably small in each Cloud. The observed dispersions are only 0.037 (LMC) and 0.026 (SMC). Some of the scatter will be due to observational errors in the colours. Making allowance for this reduces the dispersions to 0.032 (LMC) and 0.019 (SMC). This result has several important consequences.

(1) It is not reasonable to assume there is no real scatter in the reddenings. This means that there is very little leeway left for any significant spread in the intrinsic line. We conclude that the BVI locus is narrow and that reddenings of high precision can be obtained from it.

(2) The fact that the relative values of the derived reddenings are at least as accurate as the observed scatter is of relevance to the discussion of the P-L-C relations. Recent suggestions that the P-L-C relation is not firmly based depend essentially on adopting a much greater spread in Cepheid reddenings than is consistent with the BVI data (cf. Feast 1984 and references there).

(3) Although the derived reddenings are relatively insensitive to abundance the observed scatter in E_{B-V} values places limits on the scatter in abundance. For instance Harris (1983) suggest a possible scatter of ±0.15 in [A/H] in the SMC. This would produce an apparent scatter of ±0.016 in E_{B-V}. The observed scatter corrected for observational uncertainty is only 0.019 so that such abundance effects would make the real reddening scatter negligible which appears somewhat unlikely.

The ability to derive reddenings of high accuracy from BVI photometry will, one hopes, be exploited to the full in our own and nearby galaxies.

THE P-L-C RELATION AND THE STRUCTURE OF THE MAGELLANIC CLOUDS

It was shown by Martin *et al.* (1979) that, for LMC Cepheids with individually determined reddenings a P-L-C relation was a distinctly better fit than a P-L relation. Caldwell has been re-investigating this problem for both Clouds allowing for a possible distribution of Cepheids along the line of sight. This is a particularly illuminating study as it shows the power of Cepheids to determine accurate relative distances within the Clouds. The potential of Cepheids in this regard was shown some years ago by Gascoigne and Shobbrook (1978). They observed Cepheids on the far eastern and western edges of the LMC and confirmed de Vaucouleurs'(1960) model of a flattened system tilted with respect to the line of sight.

It would be valuable to obtain BVI reddenings for these Cepheids to improve the accuracy of the discussion. However even the available sample of LMC Cepheids with individual reddenings shows clear evidence of a tilt. Taking this tilt into account the scatter about a P-L-C relation is very small (0.10 ± 0.01). Purely internal errors in magnitudes, colours and reddenings lead to an estimated scatter of 0.08 ± 0.01). Within the accuracy of the observations therefore the P-L-C relation may be considered an exact relationship. Furthermore the thickness of the LMC disc must be very small.

Most Cepheids are expected to be on their second crossing of the instability strip. If first crossing Cepheids were numerous they would be expected to introduce a scatter in the P-L-C relation which is not found. There is also no room for significant scatter due to abundance variations, binarity or other effects. Caldwell (cf. Feast 1984) has estimated the effect of abundance on the P-L-C zero point. Using this we can estimate roughly that a scatter in [A/H] of ±0.10 as proposed by Harris for the LMC when added to the internal scatter given above would lead to a predicted total scatter of ±0.13. This is becoming uncomfortably large compared with the observational scatter of 0.10 ± 0.01 and suggests at least tentatively that the scatter in abundance is less than Harris estimates.

The case of the SMC is particularly interesting. Here again a P-L-C relation and a tilt can be determined simultaneously. The tilt found is remarkably large (70°). There have been indications from work on early type stars (e.g. Ardeberg and Maurice 1979, Florsch <u>et al</u>. 1981) and from a preliminary report on infrared observations of Cepheids (Welch and Madore 1984) that there is a considerable depth to the SMC. Mathewson and Ford (1984) have interpreted 21cm work to mean that the SMC is divided into two sub-clouds. It is possible that the new results could be represented by groups of Cepheids at distinctly different distances rather than by a single plane. The distances involved are large Δ (Modulus) $\sim$0.4 (i.e. $\sim$13 kpc). This in fact is just about the separation of fragments expected if the SMC were disrupted $\sim 2 \times 10^8$ years ago by a close passage of the LMC (cf. Mathewson 1984).

The model of the SMC as a tilted plane may thus be too simple a picture. Nevertheless even on this fairly crude model the fit to the data is very good. The scatter is only 0.11 ± 0.01, comparable to the value in the LMC. Despite the small value, Caldwell's error analysis suggest that there is some small extra scatter in addition to the internal photometric errors. This is likely to be due to the complex structure of the SMC but it remains possible, at least in principle, that the presence of first crossing Cepheids or other causes contribute.

The superiority of a P-L-C relation to a P-L relation is shown by the fact that the scatter about the latter is 0.23 (LMC) and 0.28 (SMC) compared with 0.10 and 0.11 respectively for the P-L-C relation.

The ability to determine relative distance moduli of Cepheids to an accuracy of 0.10 (including photometric and reddening errors) is very impressive. It is rather striking that this power of the Cepheids has not yet been exploited to the full in our own Galaxy.

For our understanding of Cepheid variables it is important to have an accurate value for the coefficient of the colour term (β) in the P-L-C relation. Initially it was difficult to disentangle the colour term from interstellar reddening. Martin et al. (1979) were able to do this and obtained $\beta = 2.70$ from a maximum likelihood solution which is the appropriate method for such a multivariant problem. A least squares solution is inappropriate but the value found ($\beta = 2.18$) at least gives some lower limit to β. Feast and Balona (1980) and Balona (1984) have investigated the effects of errors in the reddenings on the value of β. These turn out to be small, perhaps reducing β from 2.7 to $\sim$2.6. Caldwell (private communication) has now studied this problem extensively for both the LMC and the SMC taking into account effects due to the structure of the Clouds discussed above. It is fairly obvious for instance, that if we assume that all the scatter (above the internal photometric errors) is due to real differences in distance moduli of the Cepheids, unconstrained by any overall structure of the Clouds, this will give a solution which is mathematically similar to one in which the bulk of the errors are in the V magnitudes. This is the limit in which a least squares solution applies and it is not surprising therefore that in each Cloud a low value of β ($\sim$2.1) is found. This is an extreme case and the most physically realistic models yield values of β near 2.5 in both Clouds. Because there is some sensitivity to the models, to the period range selected and to the ratio of total to selective absorption used, the standard error is estimated at 0.3.

Within the errors the values of β are the same in the LMC and the SMC. Considerably higher precision is required to determine whether or not there is a dependence of β on metallicity as suggested theoretically (e.g. Gascoigne 1974).

PERIOD-COLOUR RELATION

Figure 1 shows the $(\langle B\rangle - \langle V\rangle)_0$ against log P plot for Cepheids with BVI reddenings in the LMC and SMC together with galactic Cepheids with BVI reddenings (or equivalent) in Dean et al. (1978) and in the same period range. Dean et al. determined a slope for the galactic relation (over a wider period range than plotted here) of 0.46. Lines of this slope are used in Figure 1 to define the limits for the three galaxies. Each galaxy populates a strip of considerable width. There is no evidence that the slope varies from galaxy to galaxy. The widths are also not significantly different (Galaxy = 0.32, LMC = 0.29, SMC = 0.37). On Johnson's temperature scale for supergiants this corresponds to a width of about 750K. With such a wide instability strip and the possibility that this will be non-uniformily filled, it is difficult to determine precisely the mean colour displacement from galaxy to galaxy. The LMC strip shown is 0.08 to the blue of the one for the Galaxy and the SMC strip 0.07 bluer still. Some of this is due to the effects of abundances on the (B-V) colours and it is not yet clear whether there are any significant variations of the temperature limits to the strip with metallicity.

Figure 1. The Period-Colour strips for the Galaxy (G), LMC (L) and SMC (S).

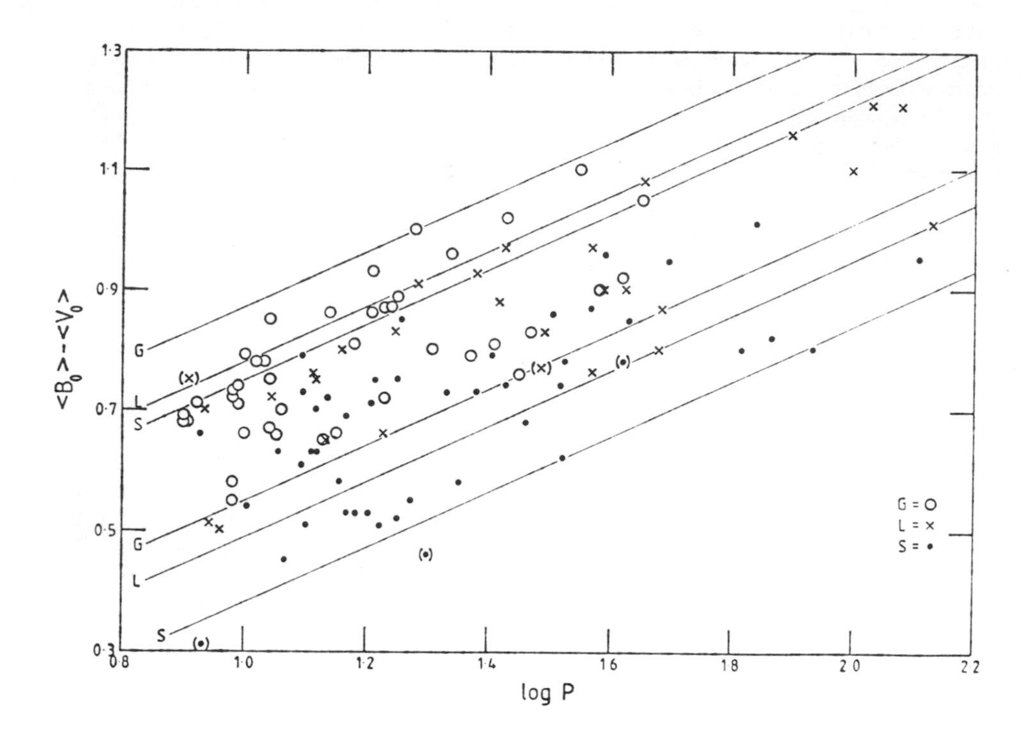

The widths of the P-C and P-L relations are of course related. There is evidence that both these strips become narrower at shorter periods (e.g. the P-L relation of Martin et al. (1979) in the LMC and the P-C relation in the Clouds and in the Galaxy of Butler (1976, 1978), Martin (1980) and Dean et al. (1978) see also Pel and Lub 1978). The theoretical work of Deupree (1980) in fact suggests that the instability strip will decrease in width at lower luminosities. This problem needs further careful examination since there is always the possibility of a spuriously narrow strip resulting from observational bias against fainter, redder variables.

The recent work by Sandage (1983) in M33 shows a P-L strip which fits that of the LMC well. There is no evidence that any galaxy violates the P-L relation of the Magellanic Clouds although there is the well known problem of the small slope (and small scatter) in IC1613. Sandage (1971) has attributed this to poor filling of the instability strip.

RADII AND SURFACE BRIGHTNESS

It is important to compare Cepheids in our Galaxy with those in other galaxies in as many ways as possible to check for any fundamental differences. The period-radius relation provides one way of doing this. Radii can be determined by combining light, colour and radial velocity curves and until recently no radial velocity curves had been determined for extragalactic Cepheids. A programme has been in progress for some while at ESO using CORAVEL to determine velocity curves for long period Cepheids in the LMC and SMC (Imbert and Prévot 1981). Photon counting systems on spectrographs can also be used to determine velocity curves for Magellanic Cloud Cepheids and a few Cepheids have been studied in this way at SAAO. 3 Cepheids in the LMC and one in the SMC with $\log P \sim 1$ give radii which fit the mean period-radius relation for galactic Cepheids (Balona 1977). They also show that the slope of the surface brightness-colour relation (A) is the same (2.30 ± 0.15) in the Clouds as in the Galaxy (2.15 ± 0.02).

INFRARED WORK ON MAGELLANIC CLOUD CEPHEIDS

The Toronto workers have pioneered the use of JHK infrared magnitudes in the study of extragalactic Cepheids. They have shown that even random phase observations define a P-L relation of acceptable accuracy ($\sigma = 0.25$). The application of this method to Cepheids beyond the Magellanic Clouds has been summarized by Madore (1984). Laney and Stobie (1984) have obtained high accuracy JHK photometry of a sample of LMC and SMC Cepheids. One aim has been to obtain good light and colour curves so that the infrared behaviour of Cepheids can be studied in detail. Some Cloud Cepheids have quite large light amplitudes ($\Delta H \gtrsim 0.5$). A P-L relation

$$H_o = 3.37 \log P + \gamma \qquad \sigma = 0.14$$

is found for the LMC and the main body of the SMC. However as with the BVI data a significant reduction in errors is obtained (especially when *all* the SMC observations are used) by introducing tilts to the Clouds and a colour term. The tilts strongly support the parallel BVI results. In particular the infrared work confirms the great range of distances in the SMC found from the BVI data.

The observations also strongly suggest that in a P-L-C relation of the form

$$H_o = \alpha_1 \log P + \beta_1 (J-K)_o + \gamma_1$$

there is a significant colour term with $\beta_1 \approx 3 \pm 1$. An infrared colour coefficient of about this amount is in fact anticipated from the observed optical value of β and the present results therefore provide additional confirmation of the optical value. For many applications of course the infrared P-L relation quoted earlier is quite satisfactory.

It has long been known that the very long period Cepheids ($P > 100$ days) are anomalously faint with respect to linear P-L and P-L-C relations in BV wavelengths in both Clouds. The results of Laney and Stobie show that the effect persists into the infrared. It cannot therefore be due to the effects of circumstellar reddening (as has sometimes been suggested) unless the extinction is essentially neutral. An explanation of these Cepheids is much to be desired.

ZERO POINT OF THE P-L-C RELATION

Work on the zero point of both the P-L-C and the (infrared) P-L relations was summarized recently (Feast 1984 see also Schmidt 1984) and distance moduli of 18.7 (LMC) and 19.2 (SMC) were adopted. Support for these distances comes from the OB stars in the Magellanic Clouds

(Crampton 1979, Crampton and Greasley 1982). On the other hand some recent work suggests that these distance moduli may be too large by about 0.2 (e.g. Hβ photometry, Schmidt 1984, Balona and Shobbrook, 1984, Balona 1984). There are welcome signs that new observational data will help materially to solve this problem. For instance, besides the Hβ work, there is an extensive re-investigation by Pel (1984) of the cluster NGC 6087 which contains the Cepheid S Nor. Another example is the cluster NGC 6067 which probably contains two Cepheids (Coulson and Caldwell, 1984). Walker and Coulson (1984) have obtained a greatly improved c-m diagram of this cluster using a CCD camera. Their new result tends to support the Shobbrook-Balona distance scale.

The present calibration of the P-L-C zero point is probably good to 0.2. Hopefully the accuracy will be considerably improved in the near future.

ACKNOWLEDGEMENTS

I am very grateful to my colleagues at SAAO and elsewhere for their extensive help in preparing this review.

REFERENCES

Ardeberg, A. & Maurice, E. (1979). Ast. Astrophys. 77, 277-85.
Baade, W. & Swope, H.H. (1963). Astron. J. 68, 435-69.
Baade, W. & Swope, H.H. (1965). Astron. J. 70, 212-68.
Balona, L.A. (1977). Mon. Not. R. astr. Soc. 178, 231-43.
Balona, L.A. (1983). In Statistical Methods in Astronomy, Proceedings of an International Colloquium, Strasbourg, 1983 (ESA SP-201), pp. 187-9. Paris, ESA.
Balona, L.A. & Shobbrook, R.R. (1984). Mon. Not. R. astr. Soc. In press.
Balona, L.A. (1984). This volume.
Bell, R.A. & Gustafsson, B. (1978). Astr. Astrophys. Suppl. 34, 229-40.
Böhm-Vitense, E. (1972). Astr. Astrophys. 17, 335-53.
Butler, C.J. (1976). Astr. Astrophys. Suppl. 24, 299-356.
Butler, C.J. (1978). Astr. Astrophys. Suppl. 32, 83-126.
Caldwell, J.A.R. & Coulson, I.M. (1984a). Mon. Not. R. astr. Soc. Submitted.
Caldwell, J.A.R. & Coulson, I.M. (1984b). SAAO Circ., no. 8. In press.
Coulson, I.M. & Caldwell, J.A.R. (1984). Mon. Not. R. astr. Soc. In press.
Crampton, D. (1979). Astrophys. J. 230, 717-23.
Crampton, D. & Greasley, J. (1982). Publs. astr. Soc. Pacif. 94, 31-5.
Dean, J.F. et al. (1978). Mon. Not. R. astr. Soc. 183, 569-83.
Deupree, R.G. (1980). Astrophys. J. 236, 225-9.
de Vaucouleurs, G. (1960). Astrophys. J. 131, 265-81.
Feast, M.W. & Balona, L.A. (1980). Mon. Not. R. astr. Soc. 192, 439-43.
Feast, M.W. (1984). In Structure and Evolution of the Magellanic Clouds, I.A.U. Symp. 108, eds S. van den Bergh & K.S. de Boer, pp. 157-170. Dordrecht, Reidel.
Feast, M.W. & Whitelock, P.A. (1984). Observatory. In press.
Fernie, J.D. (1969). Publs. astr. Soc. Pacif. 81, 707-31.

Florsch, A. et al. (1981). Astr. Astrophys. 96, 158-63.
Gascoigne, S.C.B. (1974). Mon. Not. R. astr. Soc. 166, 25P-27P.
Gascoigne, S.C.B. & Shobbrook, R.R. (1978). Proc. astr. Soc. Aust. 3, 285-6.
Harris, H.C. (1981). Astron. J. 86, 1192-9.
Harris, H.C. (1983). Astron. J. 88, 507-17.
Imbert, M. & Prévot, L. (1981). Messenger (ESO), no. 25, p. 6-7.
Isserstedt, J. (1976). Astr. Astrophys. 47, 463-6.
Isserstedt, J. (1980). Astr. Astrophys. 83, 322-7.
Kayser, S.E. (1967). Astron. J. 72, 134-48.
Kurucz, R.L. (1979). Astrophys. J. Suppl. 40, 1-340.
Laney, C.D. (1983). Publs. astr. Soc. Pacif. Submitted.
Laney, D.L. & Stobie, R.S. (1984). In preparation.
Madore, B.F. (1984). This volume.
Martin, W.L. et al. (1979). Mon. Not. R. astr. Soc. 188, 139-57.
Martin, W.L. (1980). Ph. D. Thesis, University of Cape Town.
Mathewson, D. (1984). Mercury 13, 57-9.
Mathewson, D. & Ford, V.L. (1984). In Structure and Evolution of the Magellanic Clouds, I.A.U. Symp. 108, eds. S. van den Bergh & K.S. de Boer, pp. 125-36. Dordrecht, Reidel.
Pel, J.W. & Lub, J. (1978). In The HR Diagram, I.A.U. Symp. 80, eds A.G. Davis Philip and D.S. Hayes, pp. 229-36. Dordrecht, Reidel.
Pel, J.W. et al. (1981). Astr. Astrophys. 99 L1-L4.
Pel, J.W. (1981). 2nd Asian-Pacific Regional Meeting. In press!
Pel, J.W. (1984). In Structure and Evolution of the Magellanic Clouds, I.A.U. Symp. 108, eds. S. van den Bergh and K.S. de Boer, p. 170. Dordrecht, Reidel.
Sandage, A. (1971). Astrophys. J. 166, 13-35.
Sandage, A. (1972). Q. Jl. R. astr. Soc. 13, 202-21.
Sandage, A. & Carlson, G. (1982). Astrophys. J. 258, 439-56.
Sandage, A. & Tammann, G.A. (1982). In Astrophysical Cosmology: Proc. of the Study Week on Cosmology and Fundamental Physics, 1981, eds H.A. Brück et al., pp. 23-83. Specola Vaticana.
Sandage, A. (1983). Astron. J. 88, 1108-25.
Schmidt, E.G. (1984). In press.
Tammann, G.A. & Sandage, A. (1968). Astrophys. J. 151, 825-60.
van den Bergh, S. (1969). Comm. David Dunlap Observatory, no. 195.
van den Bergh, S. (1977). In Décalages vers le rouge et Expansion de l'Univers, I.A.U. Coll. 37, eds C. Balkowski & B.W. Westerlund, pp. 13-41. Paris, CNRS.
Walker and Coulson, I. (1984). preprint.

INVITED REVIEW

Cepheid Variables as Extra-Galactic Distance Indicators

Barry F. Madore
David Dunlap Observatory
University of Toronto

Summary

The role of cepheid variables in establishing the inner distance scale to nearby galaxies is discussed. Emphasis is placed on the necessity for broad wavelength coverage in attempting to account for metallicity differences and reddening internal to the parent galaxies. In addition linear detectors are essential in minimizing the effects of any unresolved background contibution to the photometry. Recent infrared observations of Cepheids in Local Group galaxies are surveyed and all published data on extra-galactic Cepheids are presented for convenient access.

I. Introduction

Cepheid variables are no strangers to the extragalactic distance scale. On the contrary it can be argued that they are the fundamental basis for it. From Leavitt and Shapley, to Hubble, Baade and Sandage, and now to the host of other workers in the field, Cepheids have acted as the cornerstone in establishing the distances to nearby galaxies.

It is not without some trepidation that such reliance has been placed on a single primary distance indicator, but the alternatives are limited and at best statistical in nature. Cepheids are relatively bright, they are plentiful, their properties are stable with time. We have every reason to believe that we understand the basic physics governing the pulsation and we now have the strongest indicators ever that we have a reliable calibration and systematically correct data available to us. With such a firm basis for proceeding it is not unlikely that we have missed something vital. What that is, we of course don't know, but workers on the extragalactic distance scale are notoriously optimistic. This review will carry on with that tradition.

II. Observing in the Blue

So often the available technology has defined the ways in which we can practice our science of observing. Initially the photographic plate provided a unprecedented opportunity for astronomers. It immediately produced semi-quantitative, permanent and panoramic surveys of the sky. But its non-linear response demanded the utmost in care in calibration and unending external checks and verification. As a wide-field detector, the photographic plate in combination with the appropriate telescope, was and still is superb; as a means of quantifying data to the systematic precision needed for the distance scale, numerous difficulties persisted.

Perhaps just as unfortunate as the non-linear intensity response of the photographic plate, is the fact that the spectral response of most emulsions was dictated again by technology rather than astronomical imperatives. Astronomers had no option but to observe at wavelengths accessible to the available detectors. Certainly the choice provided originally (i.e., that of a photographic blue magnitude or a photovisual green magnitude) seemed innocuous enough. And for many purposes this "forced choise" was fortuitously good; the blue/green combination of magnitudes acted as excellent luminosity and temperature discriminants for the majority of visible stars. This was evidenced by the successful construction of colour-magnitude diagrams for dozens of star clusters, and by the discovery of hundreds of galactic and extragalactic Cepheid variables which undergo temperature-induced light variations of a magnitude or more at the blue wavelengths. Had the first panoramic detectors been sensitive to the infrared, the discovery process would have been spectacular; but as we shall see it might not have been quite so uncertain.

With the advent of photoelectric devices the situation changed fundamentally on one front, but stood rigidly still on the other. The linear intensity response of the photoelectric photometers, available after World War II, eliminated the fear of continued scale errors in stellar magnitudes; but all of the existing photographic studies needed to be redone if old errors were to be eliminated. Because the photoelectric photometers were single channel devices, sample size had to be sacrificed for systematic accuracy. It should nevertheless be emphasized that despite their limited photometric capabilities, photographic surveys were, and continued for many decades to be the principal means by which new Cepheids could be discovered and their periods determined. The panoramic surveys were an absolutely necessary first step in the course of accurately calibrating and applying the Period-Luminosity relation beyond the Galaxy.

III. Testing the Calibration

A fine review of the contemporary studies of the Cepheids in the Magellanic Clouds is given by Feast (1984) in these Proceedings so the specifics of that review need not be repeated here. However, because of their proximity and their abundant supply of Cepheids, the Magellanic Clouds, cannot be over-looked because, as always, they provide the testing ground for all fundamental work on the distances to nearby galaxies. In the context of this review, the importance of the Magellanic Clouds cannot be over-emphasized. If we cannot successfully intercompare the Cepheids in the two Magellanic Clouds, we certainly cannot go further afield with any confidence. For instance, knowing that the Large and Small Magellanic Clouds differ in mean metallicity by a factor of three or so, is this idfference in composition reflected in any of the observed properties of the Cepheids which must be taken into account when using these stars as distance indicators? Specifically, is the slope, the zero point or the width of the Period-Luminosity relation systematically shifted in the SMC with respect to the LMC? If the answer is yes, can we then understand the shift, can we correct for it, or can we avoid it by undertaking new and specific observations?

In some of the first modern photographic studies of the Cepheids in the Magellanic Clouds, claims were made that comparisons of the Period-Colour distributions with that of the Cepheids in the Galaxy, indicated that differences existed between the two populations. The SMC Cepheids were claimed to be redder than their galactic counterparts and this difference was presumably because of the different metallicities of the two systems. Unfortunately, photographic colours are notoriously susceptible to slight scale errors and therefore required confirmation.

It was not until the photoelectric study of the modest sample of Cepheids in each of the Magellanic Clouds, undertaken by Gascoigne and Kron (1965) and Gascoigne (1969), that the question of colour differences could be reliably discussed. The photoelectric studies indicated for the shortest-period Cepheids in the Small Magellanic Cloud that they might be systematically bluer than Cepheids of similar period, observed in the Large Cloud and the Galaxy. The colour difference was now in the opposite sense from the photographic survey, but nonetheless it was concluded that differences in metallicity between the two galaxies were the cause of the colour difference.

It was later argued that the colour difference of the ensemble might be only a reflection of the different degree of filling of the universal instability strip. Given a large instability strip which is defined to be independent of metallicity variations itself, there is room to postulate that low metallicity systems preferentially populate the blue side of the short-period portion of the strip, while high-metallicity systems, such as our galaxy, preferentially have Cepheids which reside toward the red edge of the instability strip. This population effect will manifest itself in different Period-Colour distributions and in different mean Period-Luminosity relations; but if the strip itself is invariant to metallicity then a Period-Luminosity-Colour relationship should be sufficient to define a unique luminosity for each star, compensating for this class of postulated, but optimistically harmless, effect of metallicity.

Some support for this suggestion was obtained from stellar evolution theory which indicated that stars in the cepheid-phase of evolution (core He burning) make deeper excursions to the blue when their interior metallicity is low. Thus stars of low metallicity might be expected to preferentially populate blue parts of the instability strip. But the logic is not complete. If one allows metallicity to effect the theoretical interior solution, one must also consider the effects on the atmosphere, (i.e., the observed colours at fixed effective temperature and bolometric luminosity) and ultimately one must consider the effects on the pulsational characteristic of the Cepheids (i.e., the effects on the position of the instability strip itself).

The first effect was estimated by Gascoigne (1974) who used the synthetic spectra of Bell and Parsons (1972) to estimate the differential colour effect of metallicity variations on the atmospheres of Magellanic Cloud Cepheids. He concluded that a correction of -0.4 to the V magnitude was probably in order for a star with a metallicity typical of an SMC Cepheid as compared to one with a solar abundance. The latter effect, that of the calculating the position of the instability strip as a function of metallicity was undertaken by Iben and Tuggle (1970). They parameterized the position of the blue edge of the instability strip in terms of the input metallicity as well as the helium abundance; and gave explicit formulae for the Period-Luminosity-Colour relation also as a function of those parameters of the models.

But as secure as the theory might appear, the necessary corrections were becoming rather appreciable. Even for workers in the extragalactic distance scale these uncertainties in the cepheid calibration started to sound unacceptably large, and some looked afield for other distance indicators or at very least cast a suspicious glance at distances which relied heavily on cepheid variables. But the worst fears however were not to be realized, and so now with a better understanding of the problems and no obvious contenders for the better distance indicator, Cepheids are seeing a renewed interest in their precise calibration.

In summary, there is good reason to expect that in the blue and visual regions of the spectrum, atmospheric metallicity effects can have significant observational effects on the magnitudes and colours of Cepheids, through line blanketing in the ultraviolet and blue, and energy redistribution towards the visual and red. In addition, the evolutionary tracks will indipendently respond to differences in the interior metallicities of the stars, and finally, the instability strip itself is probably a function of metallicity as well. Attempts to map some of these three effects onto the observational data have been made, most recently by the South African workers, but much work remains to be done.

Still, perhaps the biggest single concern in attempting to make serious corrections for metallicity variations, especially for the blue wavelengths, is the additional corrections for individual reddening which must be applied before any other analysis. Extinction and reddening is suffered by the Cepheid's light due to dust internal to the parent galaxy and due to a foreground galactic component. Without prior knowledge of the intrinsic colours of the individual Cepheids concerned, the reddening and metallicity effects are difficult to disentangle. Explicit (Martin, Warren, and Feast 1979) and implicit (Madore 1982) attempts have been made but the results are equivocal. If it is our intention to study the intrinsic properties of Cepheids at all wavelengths then many effects await our considered attention; if we need to use Cepheids as a tool and a probe then perhaps we can judiciously choose the best properties now that a broad spectrum of observations are relatively accessible.

IV. Moving to the Red

Rather than confront the problems of reddening and atmospheric metallicity variations head on, it is now possible to avoid those problems by obtaining new observations at wavelengths considerably less sensitive to obscuration and line blanketing. This side-stepping of the problem has come about with the availability of new, and sensitive detectors which allow the observations to be made in the near-infrared. Detailed justifications for the expected advantages were given by McGonegal et al. (1982), and the empirical justification is still gaining momentum with each new galaxy that is observed. It appears that from a purely empirical and pragmatic point of view, observations of Cepheids made in the red and near infrared are insensitive to most of the above problems which plague the optical observations of the past. Furthermore, the reduced response of the infrared portion of the stellar spectrum to temperature variations means that both the width of the observed instability strip and the periodic light variations of individual Cepheids are so reduced in the infrared that single observations, made at randomly chosen times, give unexpectedly accurate distances.

V. A Survey of Modern Studies of Extra-Galactic Cepheids

a) The Magellanic Clouds

The published UBVRI modern photometric data are compiled for the LMC and SMC, respectively in Tables 1 and 3. The final columns in those tables also present the near-infrared data avilable to date; the first row contains the random-phase observation which results in the predicted time-averaged magnitude (based on the methods outlined in Welch et al. 1984) given in the second row. A compilation of references to modern observations in all photometric systems is given in Tables 2 and 4.

TABLE 1

Large Magellanic Cloud Cepheid Photometry

No.	P(days)	log P	V	U-B	B-V	V-R	V-I	J	H	K
S 65-48	250.	2.398	09.64		0.66	0.31	0.60	08.71	08.48	08.37
S 65-08	250.	2.398	10.00		0.98	0.43	0.81	08.72	08.46	08.35
HV 883	134.0	2.127	12.12	0.83	1.18	0.59	1.09	10.29	09.88	09.73
								10.23	09.88	09.73
HV 2447	119.3	2.077	11.99	1.12	1.26	0.60	1.12	09.97	09.57	09.42
								10.07	09.58	09.44
HV 2883	108.0	2.033	12.41	1.25	1.23	0.57	1.07	10.90	10.52	10.40
								10.54	09.97	09.81
HV 5497	98.86	1.995	11.92	1.07	1.20	0.58	1.11	09.99	09.64	09.52
								10.11	09.62	09.54
HV 2827	78.89	1.897	12.30		1.24		1.08	10.25	09.87	09.79
								10.37	09.93	09.83
HV 2369	48.31	1.684	12.62	0.84	0.96	0.49	0.97	10.70	10.35	10.26
								10.90	10.48	10.41
HV 953	47.89	1.680	12.28	0.68	0.87	0.45	0.89	10.65	10.28	10.18
								10.73	10.38	10.37
HV 900	47.53	1.677	12.77	0.79	0.92		0.92	11.21	10.91	10.82
								10.95	10.68	10.66
HV 877	45.16	1.655	13.41	0.98	1.20		1.19			
								11.25	10.83	10.73
HV 2338	42.15	1.625	12.79	0.85	0.94		0.95	11.12	10.88	10.80
								11.26	10.84	10.77
HV 2257	39.29	1.592	13.03	0.82	0.96		1.04	11.60	11.16	11.04
								11.76	11.30	11.17
HV 909	37.57	1.575	12.76	0.86	0.80		1.01	11.44	11.03	10.93
								11.50	11.10	11.01
HV 879	36.81	1.566	13.35		1.03		1.06			
								11.90	11.47	11.32
HV 2294	36.52	1.563	12.64	0.68	0.83	0.44	0.87	11.23	10.82	10.74
								11.25	11.35	10.77

TABLE 1 (cont.)

Large Magellanic Cloud Cepheid Photometry

No.		P(days)	log P	V	U-B	B-V	V-R	V-I	J	H	K
HV	881	35.76	1.553	13.14	0.80	0.88					
									11.48	11.00	10.88
HV	873	34.35	1.536	13.52	0.81	1.09			11.77	11.40	11.29
									11.65	11.18	11.07
HV	882	31.81	1.503	13.42	0.74	0.93					
HV	899	31.03	1.492	13.43	0.81	0.94		1.10			
HV	1002	30.47	1.484	12.95	0.68	0.77	0.41	0.79	11.39	11.20	11.14
									11.48	11.11	11.04
HV	875	30.36	1.482	13.04	0.54	0.80					
HV	8036	28.38	1.453	13.60		0.90					
HV	2251	27.99	1.447	13.10		0.75			11.78	11.37	11.29
									11.66	11.23	11.12
HV	1023	26.61	1.425	13.74		1.04		1.06			
HV	902	26.36	1.421	13.22		0.70			12.09	11.74	11.66
									11.77	11.47	11.17
HV	12815	26.18	1.418	13.46		0.95		1.03			
HV	889	25.85	1.412	13.71	0.73	1.00					
HV	1003	24.43	1.388	13.25		0.71			11.66	11.32	11.27
									11.82	11.45	11.36
HV	1013	24.10	1.382	13.83		1.04		0.96			
HV	886	23.97	1.380	13.34	0.73	0.88			12.09	11.79	11.72
									11.94	11.63	11.57
HV	878	23.30	1.367	13.57	0.67	0.90			12.13	11.72	11.61
									12.07	11.73	11.68

TABLE 1 (cont.)

Large Magellanic Cloud Cepheid Photometry

No.	P(days)	log P	V	U-B	B-V	V-R	V-I	J	H	K
HV 2749	23.12	1.364	14.73		1.24			12.43	11.92	11.78
								12.42	11.89	11.76
U 1	22.54	1.353	14.10		0.98					
U 11	20.07	1.303	13.99	0.88	0.86					
HV 2793	19.19	1.283	14.09		1.01		1.08	12.55	12.16	12.04
								12.45	12.12	12.02
HV 1005	18.71	1.272	13.96		0.82					
HV 2836	17.62	1.246	14.62		1.01		1.13			
HV 2580	16.94	1.229	13.96		0.75		0.86			
								12.49	12.08	12.06
HV 2549	16.18	1.209	13.70		0.72					
HV 2262	15.85	1.200					0.92			
HV 12471	15.85	1.200	14.68		0.99					
HV 2324	14.45	1.160	14.35		0.85		0.92			
HV 5655	14.22	1.153					0.95			
HV 955	13.74	1.138	14.07		0.73			12.87	12.52	12.41
								12.70	12.30	12.20
HV 2352	13.61	1.134	14.17		0.75		0.86			
HV 2579	13.43	1.128	13.95		0.66					
HV 997	13.15	1.119	14.52		0.85		0.95			

TABLE 1 (cont.)

Large Magellanic Cloud Cepheid Photometry

No.	P(days)	log P	V	U-B	B-V	V-R	V-I	J	H	K
HV 2260	13.00	1.114	14.86		0.89		1.00			
HV 2527	12.94	1.112	14.59		0.80			12.92	12.65	12.55
								11.76	11.30	12.61
HV 12253	12.56	1.099	14.41		0.76					
HV 12716	11.24	1.051	14.70		0.76					
HV 2864	10.99	1.041	14.70		0.79		0.89			
HV 12248	10.91	1.038	14.47		0.75					
HV 2432	10.91	1.038	14.23		0.52			13.10	12.79	12.70
								12.94	12.60	12.57
HV 6105	10.45	1.019	14.91		0.79					
HV 12474	9.84	0.993	14.65		0.69					
HV 2301	9.51	0.978	13.94		0.84					
HV 971	9.29	0.968	14.43		0.68					
HV 12816	9.21	0.960	14.51		0.57		0.72			
HV 12717	8.85	0.947	14.69		0.70					
HV 12452	8.73	0.941	14.79		0.71					
HV 2733	8.73	0.941	14.69		0.62		0.76			
HV 2854	8.63	0.936	14.64		0.72		0.78			

TABLE 1 (cont.)

Large Magellanic Cloud Cepheid Photometry

No.	P(days)	log P	V	U-B	B-V	V-R	V-I	J	H	K
HV 12823	8.30	0.919	14.57		0.55					
HV 12700	8.15	0.911	14.85		0.73		0.75	13.44	13.09	13.04
								13.44	13.14	13.08
HV 12976	7.85	0.895	14.99		0.74					
HV 12079	6.93	0.841	14.94		0.66					
HV 2694	6.93	0.841						13.54	13.26	13.19
								13.58	13.31	13.24
HV 2405	6.92	0.840						13.91	13.56	13.46
								13.82	13.47	13.54
HV 2279	6.90	0.839						13.12	12.75	12.65
HV 914	6.89	0.838						13.49	13.22	13.15
								13.49	13.24	13.23
HV 2337	6.87	0.837						13.80	13.54	13.41
								13.62	13.35	13.23
HV 13048	6.86	0.836						13.58	13.29	13.23
								13.66	13.40	13.32
HV 6065	6.84	0.835						13.79	13.46	13.38
								13.87	13.51	13.44
HV 12797	6.82	0.834						13.42	13.16	13.11
								13.52	13.17	13.13
HV 2523	6.78	0.831						13.16	12.71	12.63
HV 12077	5.05	0.703	15.30		0.61					
HV 6093	4.79	0.680	15.35		0.64			14.26	13.96	13.81
								14.16	13.86	13.72
HV 12231	4.75	0.677	15.43		0.64					

TABLE 1 (cont.)

Large Magellanic Cloud Cepheid Photometry

No.	P(days)	log P	V	U-B	B-V	V-R	V-I	J	H	K
ROB 22	4.67	0.669	15.55		0.70					
HV 12869	4.50	0.653	15.06		0.62					
HV 12720	4.32	0.635	15.74		0.63					
ROB 9	3.71	0.569	15.80		0.58			14.55	14.22	14.11
HV 6022	3.65	0.562						13.78	13.54	13.52
HV 12747	3.60	0.556	15.76		0.56			14.59	14.27	14.17
								14.56	14.27	14.18
HV 12765	3.43	0.535	15.29		0.57			14.11	13.81	13.79
ROB 25	3.38	0.529	16.03		0.60					
HV 6089	3.21	0.507						13.60	13.29	13.25
HV 2353	3.11	0.492	15.37		0.53					
HV 5672	3.06	0.486						14.11	13.85	13.85
HV 12225	3.01	0.478						14.89	14.60	14.48
								14.96	14.64	14.52
ROB 24	2.69	0.429	16.22		0.51					
HV 5541	2.60	0.415						14.93	14.60	14.45
								14.88	14.62	14.48
ROB 44	2.56	0.408	16.36		0.58					

Table 2

Modern Observations of LMC Cepheids

	Description of Data	Reference
(1)	Photographic (BV) photometry, periods and identifications for 41 Cepheids	Wooley et al. (1962)
(2)	Photoelectric (BV) photometry and periods for 13 Cepheids.	Gascoigne and Kron (1965)
(3)	Identification charts and periods for ∿ 600 Cepheids.	Hodge and Wright (1967)
(4)	Photoelectric (BV) photometry and periods for 9 Cepheids.	Gascoigne (1969)
(5)	Random-phase (BVRI) photoelectric photometry of 2 Cepheids: HV 821 and HV 829.	Mendoza (1970)
(6)	Near-infrared (JHK) photometry of HV 883	Glass (1974)
(7)	Photoelectric (UBV) photometry of 22 Cepheids.	Madore (1975)
(8)	Photographic (BV) photometry of 119 Cepheids.	Connolly (1975)
(9)	Photoelectric (UBVRI) photometry of 11 Cepheids.	Eggen (1977)
(10)	Photographic (BV) photometry of 83 Cepheids.	Butler (1978)
(11)	Photoelectric (BV) averaged photometry for 17 Cepheids for a tilt measurement.	Gascoigne and Shobbrook (1978)
(12)	Photoelectric (UBVI) photometry of 45 Cepheids.	Martin and Warren (1979)
(13)	Photoelectric (BV) photometry of 7 sinusoidal Cepheids.	Connolly (1980)
(14)	Photoelectric (BVI) photometry of 10 cepheids.	Martin (1980)
(15)	Photoelectric (BVI + DDO) photoelectric photometry of 16 Cepheids.	Dean (1981)
(16)	Photographic (BV) photometry for 213 Cepheids.	Martin et al. (1981)
(17)	Photoelectric (BV) photometry for 10 short-period Cepheids.	Martin (1981)
(18)	Photoelectric (Washington) Photometry of 40 Cepheids.	Harris (1983)
(19)	Near-infrared (H) photometry for 40 Cepheids.	McGonegal et al. (1982)
(20)	Photographic (UBVR) photometry for 165 Cepheids.	Wayman, Stift and Butler (1984)
(21)	Photoelectric (BVRI) photometry of 8 Cepheids.	Madore et al. (1985a)
(22)	Photoelectric (BVRI) and near-infrared (JHK) photometry of two Leavitt variables.	Grieve, Madore and Welch (1985)

TABLE 3

Small Magellanic Cloud Cepheid Photometry

No.	P(days)	log P	V	U-B	B-V	V-R	V-I	J	H	K
HV 1956	207.5	2.322	12.19		1.30	0.65	1.20	10.54	10.02	09.83
HV 821	127.6	2.106	11.94	0.79	1.05	0.54	1.04	10.56	10.13	10.00
								10.28	09.90	09.76
HV 829	85.92	1.934	11.91	0.60	0.85	0.46	0.89	10.54	10.07	09.97
								10.40	10.06	09.96
HV 834	73.41	1.866	12.21	0.55	0.87	0.46	0.89	10.89	10.54	10.45
								10.68	10.33	10.19
HV 11157	69.06	1.839	12.94	0.80	1.10	0.56	1.07	11.35	10.86	10.73
								11.12	10.69	10.58
HV 824	65.84	1.818	12.37		0.88	0.46	0.91	11.05	10.58	10.51
								10.85	10.41	10.31
HV 1877	49.80	1.697	13.18	0.47	1.02	0.51	0.99			
HV 837	42.71	1.631	13.23	0.62	0.93	0.49	0.95	11.55	11.17	11.11
								11.72	11.25	11.18
HV 2195	41.80	1.621	13.00	0.61	0.81	0.44	0.85	11.49	11.16	11.11
								11.61	11.23	11.14
HV 11182	39.19	1.593	13.70	0.93	1.04	0.52	1.02	11.88	11.42	11.34
								12.05	11.61	11.55
HV 2231	36.68	1.564	13.52		0.98	0.53	1.00			
								11.83	11.40	11.30
HV 2064	33.68	1.527	13.73	0.63	0.91	0.49	0.96			
								12.37	11.87	11.82
HV 865	33.33	1.523	13.21		0.73	0.42	0.83			
								11.85	11.52	11.41
HV 1636	32.68	1.514	13.84		0.70	0.45	0.91			
HV 10357	32.03	1.505	14.02		0.88					

TABLE 3 (cont.)

Small Magellanic Cloud Cepheid Photometry

No.	P(days)	log P	V	U-B	B-V	V-R	V-I	J	H	K
HV 840	33.04	1.519	13.58		0.87	0.49	0.94			
HV 1636	32.68	1.514	13.81		0.70	0.45	0.90			
HV 823	31.93	1.504	13.77	0.67	0.96	0.51	0.97	12.32	11.86	11.76
								12.15	11.71	11.67
HV 1369	30.93	1.490	14.42		0.66					
HV 1451	30.08	1.478	14.01		1.05					
HV 863	28.96	1.462	13.34		0.77	0.43	0.84			
HV 1967	28.94	1.461	13.63		0.78					
HV 819	28.44	1.454	14.05	0.60	0.86					
HV 1501	27.41	1.438	14.04		0.83					
HV 847	27.07	1.432	13.92		0.86	0.48	0.93			
HV 2205	25.43	1.405	14.07		0.87	0.47	0.92			
HV 1430	23.98	1.380	14.30		0.86	0.50	0.95			
HV 2209	22.64	1.355	13.55	0.34	0.64	0.38	0.74	12.15	11.85	11.80
								12.27	11.91	11.85
HV 1522	22.15	1.345	14.36		0.93					
HV 11211	21.39	1.330	13.84		0.83	0.46	0.90			

TABLE 3 (cont.)

Small Magellanic Cloud Cepheid Photometry

No.	P(days)	log P	V	U-B	B-V	V-R	V-I	J	H	K
HV 1543	20.48	1.311	14.78		0.90					
HV 817	18.90	1.276	13.85		0.65	0.38	0.77	12.76	12.37	12.28
								12.60	12.24	12.18
HV 1884	18.11	1.258	14.47		0.92	0.48	0.92			
HV 1342	17.94	1.254	14.21	0.36	0.59	0.35	0.72	13.06	12.71	12.63
								12.97	12.64	12.57
HV 10386	17.74	1.249	14.23		0.86		0.93			
HV 1925	17.20	1.236	13.98	0.29	0.62			12.83	12.42	12.35
								12.68	12.29	12.23
HV 822	16.74	1.224	14.52	0.56	0.83			13.04	12.56	12.51
								13.22	12.71	12.64
HV 1954	16.69	1.223	13.84	0.38	0.62	0.38	0.74	12.73	12.31	12.24
								12.82	12.41	12.34
HV 11210	16.43	1.216	14.42		0.87	0.49	0.95			
HV 1787	16.20	1.210	14.27		0.78		0.85			
HV 854	15.97	1.203	14.25		0.65	0.38	0.77			
HV 1328	15.85	1.200	14.13		0.61			12.89	12.54	12.52
								13.07	12.66	12.62
HV 1482	15.82	1.199	14.88		0.76					
HV 1560	15.51	1.191	14.79		0.98					
HV 1442	15.29	1.184	14.81		0.80					

TABLE 3 (cont.)

Small Magellanic Cloud Cepheid Photometry

No.	P(days)	log P	V	U-B	B-V	V-R	V-I	J	H	K
HV 2233	15.17	1.181	13.94		0.61	0.38	0.72			
HV 843	14.71	1.168	15.02		0.93					
HV 1695	14.60	1.164	14.71		0.75	0.48	0.92			
HV 2088	14.58	1.164	14.66		0.79	0.44	0.86			
HV 1579	14.57	1.164	14.43		0.53					
HV 1335	14.38	1.158	14.77		0.69	0.40	0.81			
HV 1933	13.78	1.139	14.24		0.57					
HV 1326	13.72	1.137	14.88		0.76					
HV 1438	13.65	1.135	15.42		0.71					
HV 827	13.46	1.129	14.49		0.60			13.04 12.96	12.79 12.70	12.76 12.68
HV 2189	13.46	1.129	14.50		0.68					
HV 2202	13.19	1.120	14.41		0.70	0.40	0.79			
HV 2225	13.15	1.118	14.78		0.83	0.46	0.92			
HV 1873	12.94	1.112	14.88		0.73	0.42	0.84			
HV 1744	12.62	1.101	14.56		0.63	0.39	0.76			

TABLE 3 (cont.)

Small Magellanic Cloud Cepheid Photometry

No.	P(days)	log P	V	U-B	B-V	V-R	V-I	J	H	K
HV 1484	9.00	0.954	14.41		0.96					
HV 2103	8.98	0.951	15.12		0.65					
HV 1338	8.50	0.929	15.16		0.49	0.31	0.63			
HV 1437	8.38	0.923	15.50		0.50					
HV 11112	6.70	0.826						14.46	14.14	14.05
								14.54	14.16	14.06
HV 1400	6.65	0.823						14.08	13.84	13.67
								14.11	13.83	13.66
HV 1492	6.30	0.799						13.99	13.61	13.60
								14.06	13.69	13.68
HV 1425	4.55	0.658						14.47	14.38	14.24
								14.63	14.50	14.37
HV 214	4.21	0.624						14.52	14.32	14.29
								14.62	14.42	14.37
HV 11113	3.21	0.507						15.19	14.90	14.84
								15.31	14.95	14.88
HV 11216	3.12	0.494						15.56	15.13	15.23
								15.56	15.11	15.20
HV 1906	3.06	0.486						15.32	15.05	14.93
								15.27	15.06	14.95
HV 1779	1.78	0.251						15.47	15.23	15.00
								15.50	15.24	15.01
HV 1907	1.64	0.216						16.00	15.69	15.75
								16.04	15.75	15.82
HV 1897	1.24	0.094						16.05	15.84	16.64

TABLE 3 (cont.)

Small Magellanic Cloud Cepheid Photometry

No.	P(days)	log P	V	U-B	B-V	V-R	V-I	J	H	K
HV 2052	12.58	1.100	14.26		0.61					
HV 2230	12.53	1.098	14.67		0.81	0.44	0.83			
HV 2227	12.47	1.096	14.79		0.79	0.45	0.87			
HV 1365	12.41	1.094	15.02		0.70	0.41	0.81			
HV 856	12.16	1.085	14.91		0.72					
HV 1682	12.15	1.085	14.71		0.59					
HV 857	11.98	1.078	14.46		0.59					
HV 1610	11.64	1.066	14.67		0.63	0.37	0.78			
HV 2017	11.41	1.057	14.71		0.71	0.39	0.80			
HV 1630	11.40	1.057	14.99		0.65					
HV 2063	11.17	1.048	14.76		0.72					
HV 6320	10.09	1.004	14.83		0.59	0.37	0.70			
HV 1334	9.45	0.975	14.87		0.59					
HV 836	9.40	0.973	14.79		0.52					
HV 2087	9.16	0.962	15.24		0.71					

Table 4

Modern Observations of SMC Cepheids

	Description of the Data	Reference
(1)	Photographic (BV) photometry of 64 Cepheids (known scale errors).	Arp (1960)
(2)	Photoelectric (BV) photometry of 14 Cepheids.	Gascoigne and Kron (1965)
(3)	Photographic <m>pg photometry and periods for 1151 Cepheids.	Payne-Gaposchkin (1966)
(4)	Photographic (BV) photometry, periods and identifications for 105 Cepheids.	van Genderen (1969a)
(5)	Photoelectric (Walraven) photometry of 12 Cepheids.	van Genderen (1969b)
(6)	Photoelectric (BV) photometry of 12 Cepheids.	Gascoigne (1969)
(7)	Photographic mpg photometry, periods and light curves.	Gaposchkin (1970)
(8)	Photoelectric (UBV) photometry of 17 Cepheids.	Madore (1975)
(9)	Photographic (BV) photometry and periods for 72 Cepheids.	Butler (1976)
(10)	Identification charts for $\sim$1100 Cepheids.	Hodge and Wright (1977)
(11)	Photoelectric (UBVRI) photometry for 11 Cepheids.	Eggen (1977)
(12)	Photoelectric (UBVI) photometry for 25 Cepheids.	Martin and Warren (1979)
(13)	Photographic (BV) photometry for 181 Cepheids.	Martin et al.
(14)	Photoelectric (BV) photometry for 10 short-period Cepheids.	Martin (1981)
(15)	Photoelectric (Walraven) photometry for ** Cepheids.	Pel, van Genderen, and Lub (1981)
(16)	Photoelectric (Washington) photometry of 45 Cepheids.	Harris (1981)
(17)	Photoelectric (Walraven) photometry for HV 1369.	van Genderen (1981)
(18)	Photoelectric (BVRI) photometry for 4 Cepheids.	Madore et al. (1985)

b) NGC 6822

Beyond the Small Magellanic Cloud, NGC 6822 is the closest gas-rich dwarf galaxy so far discovered to contain Cepheid variables. Unfortunately, NGC 6822 is located in a line of sight that is relatively close to the galactic plane (l = -18). Accordingly, there is considerable contamination of this galaxy's colour-magnitude diagram by galactic foreground stars, and in comparison to other galaxies in the sample, NGC 6822 has a higher and certainly more uncertain obscuration correction. The only systematic search for Cepheids in this galaxy was undertaken by Kayser (1967) using photographic plate material obtained at the prime focus of the Palomar 5m by Chip Arp and Walter Baade. She found periods and light curves in B and V for 13 Cepheids. Her data are given in Table 5. Other than for a few single-phase photoelectric observations of the brightest Cepheids published by Hodge (1977) and by van den Bergh and Humphreys (1979), no modern study of the Cepheids in the optical, using CCD's, for instance, has been undertaken. The large angular size of NGC 6822 is probably the main disincentive for such a study at the present time. However, McAlary et al. (1983) have observed many of the Cepheids in NGC 6822 in the near infrared and they derive a distance modulus of 23.5.

As alluded to above, the most controversial aspect of the study of the distance to NGC 6822 is the reddening and extinction corrections to be adopted for the brightest stars and the Cepheids. The galactic foreground component is probably in excess of E(B-V) = 0.2, and the component internal to the gas-rich dwarf irregular is likely to be of a similar amount and probably patchy in its distribution across the galaxy itself. Multi-colour photometry in addition to the infrared data will be needed to sort out the reddening problem if it is to be solved for explicitly.

TABLE 5

NGC 6822 Cepheid Photometry

No.	P(days)	Log P	<V>	B-V	H
V 13	90.682	1.957	17.75	1.08	14.79
V 7	65.45	1.816	17.56	0.90	15.18
V 25	37.4432	1.573	18.52	0.89	16.24
V 28	34.6672	1.540	18.85	1.02	15.91
V 29	31.835	1.503	18.80	0.75	16.39
V 1	30.4994	1.484	19.02	0.88	
V 3	29.2111	1.466	19.32	0.86	
V 6	21.1450	1.325	19.87	0.50	
V 17	19.2968	1.285	19.22	1.18	16.39
V 21	17.457	1.242	19.76	0.59	17.98
V 4	17.3471	1.239	19.67	0.50	
V 5	13.3550	1.126	19.98	0.68	17.31
V 30	10.9039	1.038	19.73	0.18	17.58

Table 6

Modern Observations of Cepheids in NGC 6822

	Description of the Data	Reference
(1)	Photographic (BV) photometry, periods and identifications for 13 Cepheids.	Kayser (1967)
(2)	Scattered photoelectric (BV) observations of V7.	Hodge (1977)
(3)	Scattered photoelectric (BV) observations of V7.	van den Bergh and Humphreys (1979)
(4)	Near-infrared (H) photometry of 9 Cepheids.	McAlary, et al. (1984)

c) IC 1613

Like the previous galaxy, IC 1613 is a dwarf irregular member of Local Group. Its Cepheids were marked by Walter Baade and eventually photometered and analyzed by Sandage (1971). The data were obtained again from prime focus plates taken at the Palomar 5m. The disconcerting result of the photographic analysis was that the resulting PL relation can be ambiguously interpreted. As pointed out by Sandage (1971), the slope of the blue PL relation for the Cepheids in IC 1613, taken at face value, is significantly different from the slope of the relation found for the Magellanic Clouds, say. Baade saw this as indicating trouble for the universality hypothesis; however Sandage quite rightly pointed out that the observed PL relation is sufficiently wide in magnitude at any given period (i.e., $B \sim 1.5$ mag at fixed log P), that small number statistics, systematic population effects across the strip, or the preferential selection of only the brightest Cepheids at short period could result in an apparently different slope to the PL relation despite the stars being drawn from identical instability strips. That is, the Cepheids may still obey the same Period-Luminosity-Colour relation while not necessarily being representative of the full dimensions of the instability strip. Unfortunately no colour data for these Cepheids was available to test this sample.

An alternative interpretation of the photographic data was offered by McAlary, Madore and Davis (1984) after they obtained near-infrared data for many of the same Cepheids in IC 1613. The infrared data is entirely consistent with the PL relations derived for other Local Group galaxies. Therefore, either the effects that lead to a discrepancy in the blue are diminished in the infrared, or possibly the optical data is suffering from a slight magnitude scale error in the photometry, in the sense that the photographic magnitudes are systematically too bright at the faint end. A scale error in the photographic photometry is also suggested by an analysis of the stellar luminosity function for IC 1613 which is apparently too steep compared to all other galaxies studied to date (Freedman 1984a).

The published data for the Cepheids are given in Table 7. The distance to IC 1613 based on the near-infrared data is 24.3.

TABLE 7

IC 1613 Cepheid Photometry

No.	P(days)	Log P	<B>	H
V 22	146.35	2.166	19.07	15.47
V 20	41.953	1.622	19.86	16.66
V 39	28.720	1.458	19:3	16.19
V 11	25.7719	1.411	20.47	17.04
V 2	23.4611	1.370	20.30	17.69
V 18	16.4353	1.216	20.89	
V 37	12.4140	1.084	20.87	
V 16	10.43584	1.019	20.66	
V 6	9.43048	0.974	21.04	18.77
V 25	9.2112	0.960	20.67	18.62
V 34	8.47833	0.928	21.12	
V 24	6.74350	0.829	21.08	
V 27	6.66043	0.820	21.16	18.50
V 26	5.81614	0.765	21.49	
V 17	5.73687	0.759	21.21	19.21
V 1	5.59210	0.748	21.12	
V 9	5.57738	0.746	21.37	
V 14	5.14450	0.711	21.15	
V 13	4.84448	0.685	21.37	
V 12	4.28604	0.632	21.63	
V 30	4.26963	0.630	21.64	19.22
V 15	4.22744	0.626	21.33	
V 10	4.06529	0.609	21.31	
V 3	3.96789	0.599	21.61	
V 29	2.869059	0.458	21.75	

TABLE 8

Modern Observations of Cepheids in IC 1613

	Description	Reference
(1)	Photographic (B) photometry, periods and identifications for 25 Cepheids.	Sandage (1971)
(2)	Near-infrared (H) photometry of 10 Cepheids.	McAlary, Madore and Davis (1984)

d) M33

This galaxy was first studied for its variable star content by Hubble (1926) who discovered 35 Cepheids in the nuclear regions of this galaxy. The remarkable history of the finder charts for Hubble's comparison stars is given by Sandage (1983) who recently made a first transformation of the old photographic data to the modern B system. As can be seen in Sandage's Figure 6 the corrections found by him are enormous, amounting to 2.5 mag at $B \sim 22$. Considering the difficulty of doing eye photometry on variable background plates it should not be surprising that subsequent studies, using linear detectors find that convergence has not yet been reached on the final magnitudes for these stars (see Freedman 1984b).

TABLE 9

M33 Cepheid Photometry

No.	P(days)	Log P	<B>	H
H 10	69.5	1.842	20.8	16.32
H 19	54.706	1.733	19.9	15.81
H 30	46.03	1.663	20.6	17.28
H 3	41.68	1.620	20.7	16.79
B 1	37.6179	1.575	20.2	
H 31	37.33	1.572	20.7	
H 29	36.31	1.560	21.0	
H 20	35.95	1.556	20.	
H 36	35.80	1.554	20.6	
H 18	34.00	1.531	20.7	
H 35	30.5094	1.484	20.7	16.88
H 42	30.34	1.482	20.7	
H 44	30.123	1.479	21.3	16.92
H 4	27.366	1.437	20.8	
H 7	26.556	1.424	21.0	16.59
G 6	26.3218	1.420	20.8	
H 38	25.04	1.399	20.4	17.03
H 11	23.43	1.370	21.1	
H 17	23.297	1.367	21.4	17.44
H 26	23.258	1.367	21.8	
H 27	22.448	1.351	21.2	
H 34	22.16	1.346	21.1	18.34
H 22	21.916	1.341	21.4	
H 12	21.681	1.336	21.1	
F 8	21.2764	1.328	21.3	17.92
H 33	20.50	1.312	21.7	
H 43	20.166	1.305	21.7	
H 9	18.90	1.276	21.1	
H 28	18.58	1.269	21.8	
F 12	18.4823	1.267	20.6	18.35
F 9	18.0955	1.258	21.4	
H 41	17.98	1.255	21.1	
H 37	17.602	1.246	21.4	
G 14	17.5576	1.244	21.8	18.12
H 16	17.50	1.243	21.2	
H 39	16.166	1.209	21.2	17.18
H 5	14.60	1.164	21.2	
H 23	13.564	1.132	21.3	
H 25	13.438	1.128	21.7	
G 4	13.0022	1.114	20.8	17.52
H 40	12.922	1.111	21.4	
H 24	12.894	1.110	21.4	
E 14	12.8247	1.108	21.4	
G 5	8.7849	0.944	21.4	
D 54	8.7415	0.942	21.3	
A 1	8.5406	0.931	21.6	
B 8	3.2325	0.510	22.2	

As a modern photographic check on the distance to M33, Sandage and Carlson (1983) obtained new plate material and discovered new Cepheids in the less crowded, relatively clearer regions in the outer spiral arms of M33. They found 35 new Cepheids. The corrected Hubble data yielded an apparent blue distance modulus to M33 of 25.48, while the Sandage-Carlson data gave a slightly lower value of 25.28. This difference could be indicating systematic shifts in the zero points of the two independent data sets, differences in internal reddening (uncorrected for in the Sandage analysis) between the inner and outer samples of Cepheids, or simply a reflection of the uncertainty inherent in working with small data sets.

Two other studies of the Cepheids in M33 indicate that reddening internal to the parent galaxy itself is important. The first is the CCD multi-colour PL relations for the M33 Cepheids discussed by Freedman (1984b) and the second is the near-infrared data analysed by Madore et al. (1985). Both sets of linear-detector data indicate considerable reddening along the line of sight, amounting to $A_V \sim 0.6$ mag. with a derived true distance modulus of 24.1 as compared to Sandage's preferred modulus of 25.3. The latter modulus, it must be remarked is an apparent modulus, dependent upon a weighted average of the two photographic studies and a much different adopted distance modulus for the Magellanic Clouds, compared to the linear detector studies.

Table 10

Modern Observations of Cepheids in M33

	Description of the Data	Reference
(1)	Identifications and periods for 35 Cepheids (known scale errors in the Photometry).	Hubble (1926)
(2)	Identifications and very approximate colours for 5 Cepheids.	van den Bergh, Herbst and Kowal (1975).
(3)	Photoelectrically calibrated photographic B photometry and periods for 13 Cepheids.	Sandage and Carlson (1983).
(4)	Recalibration of Hubble's Argelander step-scale photometry of 35 Cepheids.	Sandage (1983)
(5)	Near-infrared (H) photometry for 15 Cepheids.	Madore et al. (1985).

e) M31

Next in the progression outwards is the Great Nebula in Andromeda, M31. Several exhaustive photographic studies of the stellar content of selected fields along the major axis of M31 have been published by Baade and Swope (1963, 1965) and by Gaposchkin (1962). In Field I 31 Cepheids are identified and have published periods and B light curves. In Field II 31 Cepheids have been found, while in Field III 232 Cepheids are known. Field IV has the distinction of having 20 Cepheids with measured periods and both B and V light curves measured. Field IV is also the region most distant from the nucleus of M31 and its photometry is therefore probably least affected by the unresolved background of the galaxy and possibly it is the least reddened of the four regions also, although considering the scales over which obscuration changes rapidly, this is at best a statistical statement.

No infrared data on the Cepheids in M31 are yet available because of the severe crowding of the candidate objects partially due to the unfavourably high inclination of the plane of the galaxy to the line of sight. Programmes are underway, both using near-infrared aperture techniques and CCD detectors.

The BV photographic data for the Cepheids in Field IV are collected together in Table 11.

TABLE 11

M31 Cepheid Photometry (Field IV)

No.	P(days)	Log P	<V>	B-V
V 15	21.263	1.328	19.80	0.95
V 31	13.336	1.125	19.81	0.66
V 30	12.878	1.110	20.38	1.05
V 5	12.840	1.109	20.35	1.01
V 3	12.714	1.104	20.38	0.91
V 8	9.643	0.984	20.37	0.74
V 9	8.508	0.930	20.55	0.76
V 17	6.732	0.828	20.83	1.01
V 2	4.368	0.640	21.24	0.59
V 13	3.803	0.580	21.72	0.86
V 46	3.711	0.580	22.15	0.93
V 36	3.593	0.555	21.74	0.78
V 48	3.403	0.532	22.02	0.73
V 21	3.348	0.525	21.64	0.63
V 10	3.043	0.483	21.70	0.60
V 11	2.978	0.474	21.09	0.59
V 27	2.593	0.414	21.68	0.71

Table 12

Modern Observations of Cepheids in M31

	Description of the Data		Reference
(1)	Field I:	Photographic (B) photometry, periods and identifications for 31 Cepheids.	Baade and Swope (1965)
(2)	Field II:	Photographic (B) photometry, periods and identifications (?) for 131 Cepheids.	Gaposchkin (1962)
(3)	Field III:	Photographic (B) photometry, periods and identifications for 232 Cepheids.	Baade and Swope (1965)
(4)	Field IV:	Photographic (BV) photometry, periods and identifications for 20 Cepheids.	Baade and Swope (1963)

f) Sextans A

Located somewhat beyond the canonical limits of the Local Group is the dwarf irregular galaxy Sextans A. This galaxy has been the object of two recent studies: The first was a photographic study by Sandage and Carlson (1982) who discovered 8 variables and derived light curves and periods for 5 of them that appear to be Cepheids. The second study was conducted by Hoessel, Schommer, and Danielson (1983) using CCD detectors, and was primarily interested in the general stellar content of the system, although some of the brighter Cepheids were also measured.

TABLE 13

Sextans A Cepheid Photometry

No.	P(days)	Log P	B(med)
V 28	25.4370	1.405	21.12
V 3	21.2115	1.327	20.88
V 25	18.5590	1.269	21.43
V 1	15.5522	1.192	21.62
V 24	10.1791	1.008	21.57

TABLE 14

NGC 3109 Cepheid Photometry

No.	P(days)	log P	V	B-V
V 1	65.	1.81	20.54	0.94
V 2	46.	1.66	20.74	0.84
V 3			20.94	0.66

Table 15

Modern Observations of Cepheids in Nearby Dwarf Galaxies

	Description of the Data	Reference
	(a) WLM	
(1)	22 variables announced	Sandage (1979)
	(b) Sextans A	
(1)	Photographic (B) photometry of 5 Cepheids: 3 other variables identified.	Sandage and Carlson (1982)
	(c) Sextans B	
(1)	8 variables announced.	Sandage (1979)
	(d) NGC 3109	
(1)	15 certain variables announced.	Sandage (1979)
(2)	Photographic (BV) photometry of 3 Cepheids. periods for 2.	Demers (1984)

g) NGC 300

Recently, Graham (1984) has reported the results of a photographic study of the Cepheids in NGC 300 begun by him and Malcolm Smith some ten years ago. There are now 18 Cepheids with periods known in NGC 300, one of the closest members of the South Polar (Sculptor) Group of galaxies. The plates, gathered by a number of observers using the CTIO 4m, were obtained in B and V but the photometry is only accurate to 0.1 mag. so the colours are at best indicative. Graham makes his comparison directly with the Magellanic Cloud PL relation and derives two distance moduli $(m-M)_B = 26.14$ and $(m-M)_V = 26.04$, which he averages to give a ture distance modulus of 26.09. The distance so derived assumes that the average internal extinctions suffered by Cepheids in NGC 300 and the LMC cancel. Furthermore the fit assumed by Graham is largely determined by the shortest-period (faintest) Cepheids which have the least certain calibration; slightly larger apparent distance moduli would result from fits to the brightest Cepheids alone.

The photographic data are gathered together in Table 16. To date no near-infrared or CCD photometry of these Cepheids is published but observing programmes are underway to remedy both of these shortcomings.

TABLE 16

NGC 300 Cepheid Photometry

No.	P(days)	Log P	V	B
V 24	126.9	2.103	20.9	20.0
V 18	94.35	1.975	20.9	19.9
V 12	89.40	1.951	20.9	19.9
V 3	56.52	1.752	21.1	20.3
V 32	52.80	1.723	21.2	20.3
V 8	43.35	1.637	21.8	20.6
V 27	34.90	1.543	21.6	20.7
V 13	34.02	1.532	21.4	20.4
V 10	25.00	1.398	21.9	21.0
V 33	24.26	1.385	21.7	21.0
V 28	23.59	1.373	22.1	21.3
V 29	23.43	1.370	21.8	20.9
V 22	20.64	1.315	22.3	21.3
V 9	18.21	1.260	21.6	20.9
V 2	17.84	1.251	21.9	21.2
V 25	16.19	1.209	22.0	20.9
V 26	15.60	1.193	21.9	21.0
V 21	9.67	0.985	22.4	21.4

h) NGC 2403

The late-type spiral NGC 2403 has long been one of the most important galaxies to have its stellar content investigated in detail. Tammann and Sandage (1968) published the results of their extensive study of the resolved population of stars, including photometry of some 17 Cepheids observed primarily near maximum light. The study, done photographically, was at the leading edge of what the best detectors and the largest telescope in the world could produce at that time. Because the Cepheids in NGC 2403 could only be detected near to maximum light (i.e., close to the plate limit) the photometry, (especially in the visual band pass, where few plates, of brighter limiting magnitude were obtained) has been the subject of much discussion and re-interpretation (Madore 1976, Hanes 1982, McCall 1984).

Because the colours of many of the Cepheids in the Tammann and Sandage (1968) study are found to be significantly redder that their galactic counterparts there has been cause for concern, ranging from suggestions that the Cepheids are intrinsically different, to the possibility that they are reddened and obscured, to the final and reluctantly offered possibility that the photometry is in error. The photometry is being checked with CCD detectors by the Toronto Group using the facilities of KPNO and CFHT. However, a complementary study has already been completed which indicates that the original blue photographic distance modulus is a slight underestimate of the true distance modulus. For the Cepheids in NGC 2403 the blue data and the modified J-band near-infrared photometry (McAlary and Madore 1984) are reasonably in accord. Since the near-infrared photometry will provide a close approximation to the true modulus it now appears unlikely that the red colour of the Cepheids is due to absorption. The anomalous colours are probably due to errors in the visual photographic photometry, but until such time as the results of the CCD survey are in, the remote possibility of intrinsic differences still exists, although now with diminishing probability.

The importance of NGC 2403 to the distance scale cannot be over-stated. As an assumed member of the M81 Group its distance ties in a large number of other 'calibrating galaxies', by association. Of course if the distance to NGC 2403 is in error, or if it is not a member of the M81 Group, or if the Group has a large back-to-front depth along the line of sight then the calibration based on its association will be systematically in error at that level.

The published BV data for the Cepheids in NGC 2403 as observed at maximum light, are collected in Table 17.

TABLE 17

NGC 2403 Cepheid Photometry

No.	P(days)	Log P	B(max)	V(max)	J
V 3	87.48	1.942	21.97	20.68	19.43
V 19	81.493	1.911	21.91	20.64	20.08
V 46	58.156	1.765	22.31	21.25	20.09
V 33	56.246	1.750	22.19	21.11	20.95
V 25	47.110	1.673	21.24	20.7	
V 54	47.058	1.673	22.87	21:60	
V 5	46.460	1.667	21.55	21.00	
V 21	39.506	1.597	22.67	21:50	21.90
V 1	39.374	1.595	22.02	21.36	
V 4	38.306	1.583	22.38	21.20	
V 34	37.742	1.577	22.62	21:52	
V 8	34.354	1.536	22.01	21.45	
V 15	33.558	1.526	21.98	21.15	
V 29	32.992	1.518	21.41	21.20	
V 6	20.260	1.481	22.12	21.49	
V 40	20.840	1.319	22:70	21:80	
V 42	20.230	1.306	22:70	22:70	

Table 18

Modern Observations of Cepheids in NGC 2403

	Description of the Data	Reference
(1)	Phtographic (BV) maximum-light photometry for 17 Cepheids.	Tammann and Sandage (1968)
(2)	Near-infrared (J) photometry of 5 Cepheids.	McAlary and Madore (1984)

i) M81

The first indication that the M81 Group might be more widely distributed than was first hoped, came from Sandage (1984) who discusses the resolved stellar population of M81 itself and concludes that NGC 2403 and M81 are separated in distance modulus by about 0.9 mag, with M81 being the more distant. The original reason for not attributing the difference in the resolution of the brightest stars in the two systems as being due to a distance effect, was that the two galaxies are of different morphological type; M81 being of earlier type than NGC 2403 and certainly less active in star formation at the present epoch.

Cepheids are known to exist in M81. Baade (1963) mentions knowing of at least one, and Sandage (1984) has found several with periods he estimates to be around thirty days. More details than that are not yet published although it is certain that ground-based photometry of these stars will not be easy due to the intense crowding problems in the filamentary arms of M81, the less favourable inclination of the galaxy to our line of sight and the promise that the Cepheids are fainter than those in NGC 2403. Several independent CCD surveys of M81 are underway by Cohen, Freedman and Madore.

j) M101

The distance to M101 is at best controversial. This uncertainty will probably persist until Cepheids are found in this galaxy. Sandage and Tammann (1974) have studied the stellar content of M101 in detail and find no cepheid candidates brighter than $B \sim 22.5$ mag. From this and other lines of evidence they conclude that the distance to M101 is in excess of 29.3. On the other hand Humphreys and Strom (1983) claim a distance modulus of 28.9 for M101 based on their discovery and photometry of red supergiants in M101. In either case the Cepheids should be accessible to modern detectors on large ground-based reflectors. Both Illingworth and Aaronson at KPNO as well as Freedman and Madore at the CFHT are undertaking to find the Cepheids in selected regions of this galaxy. The results are eagerly awaited.

Table 19

Modern Observations of Cepheids in the Most Distant Systems

	Description of the Data	Reference
	(a) M81	
(1)	One Cepheid reported.	Baade (1963)
(2)	37 variables announced, periods for 3 Cepheids.	Sandage (1984)
	(b) M101	
(1)	No Cepheids confirmed to Bmax = 22.5 V7 is possibly a Cepheid.	Sandage and Tammann (1974)

VI. The Next Step

All of the known samples of Cepheids in external galaxies are being re-observed in the near infrared in order to put the Cepheid distance scale on a homogeneous basis. The data available to date and the published results for each of the galaxies surveyed are now discussed in tern. It is anticipated that within the new year or two, and certainly before Space Telescope is launched, that all of the galaxies will be surveyed in the near-infrared and re-worked at the shorter wavelengths using ground-based CCD detectors. The combination of new linear-detector data and a broad spectral range of coverage should allow a refreshed discussion of the precision with which Cepheids can continue to be the principal tool in establishing the distances to nearby galaxies. From all accounts the future role for Cepheids in the extragalactic distance scale looks bright.

Acknowledgements:

This review written completed during the Aspen Summer Workshop in the Physical Basis for the Extragalactic Distance Scale. Grants from the National Aeronautics and Sapce Administration, the Natural Sciences and Engineering Research Council of Canada, and the Henri Chrétian Fund administered by the American Astronomical Society are gratefully acknowledged.

References

Arp, H. C. 1960, A. J., 65, 404.
Baade, W. 1963, in Evolution of Stars and Galaxies, ed. C. Payne-Gaposchkin, (Harvard: MIT Press).
Baade, W. and Swope, H. H. 1963, A. J., 68, 435.
Baade, W. and Swope, H. H. 1965, A. J., 70, 212.

Bell, R. A., and Parsons, S. B. 1972, Ap. Letters, 12, 8.
Butler, C. J. 1976, Astron. Ap. Suppl., 24, 299.
Butler, C. J. 1978, Astron. Ap. Suppl., 32, 83.
Connolly, L. 1975, Doctoral Thesis, Univ. of Arizona.
Connolly, L. 1980, P.A.S.P., 92, 165.
Dean, J. F. 1981, South African Ap. Obs. 1, No.6, 10.
Demers, S. 1984, paper given at Canadian Astr. Soc. General Assembly, Ottawa, Canada.
Eggen, O. J. 1977, Ap. J. Suppl., 31, 1.
Feast, M. W. 1984, in I.A.U. Colloquium No. 82, Cepheids: Theory and Observations, ed B. F. Madore, (Cambridge: Cambridge Univ. Press), p. 157.
Freedman, W. L. 1984a, Doctoral thesis, Univ. of Toronto.
Freedman, W. L. 1984b, in I.A.U. Colloquium No. 82, Cepheids: Theory and Observations, ed B. F. Madore, (Cambridge: Cambridge Univ. Press),p.225.
Gaposchkin, S. 1962, A. J., 67, 334.
Gaposchkin, S. 1970, Smithsonian Ap. Obs. Spec. Report, 310, 1.
Gascoigne, S. C. B. 1969, M.N.R.A.S.,146, 1.
Gascoigne, S. C. B. 1974, M.N.R.A.S.,166, 25P.
Gascoigne, S. C. B., and Kron, G. 1965, M.N.R.A.S., 130, 333.
Gascoigne, S. C. B., and Shobbrook, R. R. 1978, Proc. Astr. Soc. Australia, 3, 285.
Glass, I. S. 1974, M.N.R.A.S., 168, 249.
Graham, J. A. 1984, A. J., 89, in press.
Grieve, G., Madore, B. F., and Welch, D.L. 1985, Ap. J., (submitted).
Hanes, D. A. 1982, M.N.R.A.S., 201, 145.
Harris, H. C. 1981, A. J., 86, 1192.
Harris, H. C. 1983, A. J., 88, 507.
Hodge, P. W. 1977, Ap. J. Suppl., 33, 69.
Hodge, P. W., and Wright, F. W., 1977, The Small Magellanic Cloud, (Seattle: Univ. Washington Press).
Hodge, P. W., and Wright, F. W., 1967, The Large Magellanic Cloud, (Washington, D.C.: Smithsonian Press).
Hoessel, J. G., Schommer, R. A., and Danielson, G. E. 1983, Ap. J., 274, 577.
Hubble, E. 1926, Ap. J., 63, 236.
Humphreys, R. M., and Strom, S. E. 1983, Ap. J., 264, 458.
Iben, I. and Tuggle, R. S. 1970, Ap. J., 173, 135.
Kayser, S. E. 1967, A. J., 72, 134.
Madore, B. F. 1976, M.N.R.A.S., 177, 157.
Madore, B. F. 1975, Ap. J. Suppl., 29, 219.
Madore, B. F. 1982, Ap. J., 253, 575.
Madore, B. F., Freedman, W. L., Grieve, G., McAlary, C. W., and Davis, L. E. 1985a, Ap. J. Suppl., (submitted).
Madore, B. F., McAlary, C. W., McLaren, R. A., Welch, D. L., Neugebauer, G., Soifer, B. T., and Matthews, K. 1985b, Ap. J., (submitted).
Martin, W. L. 1980, South African Ap. Obs. Circ., 1, No. 5, 172.
Martin, W. L. 1981, South African Ap. Obs. Circ., 1, No. 6, 96.
Martin, W. L., Thomas, Y., Thomas, Y., Carter, B. S., and Davies, H. E., 1981, South African Ap. Obs. Circ., 1, No. 6, 31.

Martin, W. L., and Warren, P. R. 1979a, South African Ap. Obs. Circ., 1, No. 4, 98.
Martin, W. L., and Warren, P. R., and Feast, M. W., 1979, M.N.R.A.S., 188, 139.
McAlary, C. W., and Madore, B. F. 1984, Ap. J., 282, 101.
McAlary, C. W., Madore, B. F., and Davis, L. E. 1984, Ap. J., 276, 487.
McAlary, C. W., Madore, B. F., McGonegal, R., McLaren, R.A., and Welch, D. L. 1983, Ap. J., 273, 539.
McCall, M. L. 1984, M.N.R.A.S., 207, 801.
McGonegal, R., McLaren, R. A., McAlary, C. W., and Madore, B.F. 1982, Ap. J. (Letters), 257, L33.
Mendoza, E. 1970, Bol. Obs. Tonanzintla Tacubaya, 5, 269.
Payne-Gaposchkin, C. 1971, Smithsonian Contr. Ap., 16, 1.
Payne-Gaposchkin, C., and Gaposchkin, S. 1966, Smithsonian Contr. Ap., 9, 1.
Pel, J. W., van Genderen, A. M., and Lub, J. 1981, Aptr. Ap., 99, L1.
Sandage, A. R. 1971, Ap. J., 166, 13.
Sandage, A. R. 1979, Carnegie Inst. Wash., Annual Report.
Sandage, A. R. 1983, A. J., 88, 1108.
Sandage, A. R. 1984, A. J., 89, 621.
Sandage, A. R., and Carlson, G. 1982, Ap. J., 258, 439.
Sandage, A. R., and Carlson, G. 1983, Ap. J. (Letters), 267, L25.
Sandage, A. R., and Tammann, G. A. 1974, Ap. J., 194, 223.
Tammann, G. A., and Sandage, A. R. 1968, Ap. J., 151, 825.
Wayman, P. A., Stift, M. J., and Butler, C. J., Astr. Ap. Suppl., 56, 169.
Welch, D. L., Weiland, F., McLaren, R., McAlary, C. W., Madore, B.F. and Neugebauer, G. 1984, Ap. J. Suppl., 54, 647.
Wooley, R. v. d. R., Sandage, A. R., Eggen, O. J., Alexander, J. B., Mather, L., Epps, E., and Jones, S. 1962, Royal Obs. Bull., 58, 1.
van den Bergh, S., Herbst, E., and Kowal, C. T. 1975, Ap. J. Suppl., 29, 303.
van den Bergh, S., and Humphreys, R. M. 1979, A. J., 84, 604.
van Genderen, A. M. 1969a, Bull. Astr. Inst. Netherl. Suppl., 3, 221.
van Genderen, A. M. 1969b, Bull, Astr. Inst. Netherl., 20, 317.
van Genderen, A. M. 1981, Astr. Ap., 101, 289.

THE CEPHEID LUMINOSITY SCALE

Edward G. Schmidt
Behlen Observatory, The University of Nebraska
Lincoln, NE 68588-0111

The distances of clusters containing classical Cepheids are central to the calibration of the period-luminosity relation. As a step in improving the reliability of the calibration, uvbyβ photometry was obtained for B stars in eight of these clusters. These data were then used to determine the distances of the clusters and, thus, the absolute magnitudes of the Cepheids and the zero point of the PLC relation (See Schmidt 1984 for details).

The chief difference among various Cepheid luminosity calibrations is in the zero point. Following de Vaucouleurs (1978), we can use the quantity $< M_V(0.8) >$, the absolute magnitude implied by a PLC relation at log P = 0.8 and a color of $<B>_0 - <V>_0 = 0.65$, to intercompare them. In Table 1 we summarize values of $< M_V(0.8) >$ for several types of calibrations. It can be seen that the zero point for the Cepheid luminosity scale covers a range of more than 0.6 magnitudes. The present calibration and those based on main-sequence fitting of UBV photometry are at nearly opposite ends of this range. The reasons for the disagreement are discussed in detail by Schmidt (1984) where it was argued there that the values based on four-color and Hβ photometry are the more reliable.

Table 1
Zero Points of PLC Relations

Method	$< M_V (0.8) >$	References
uvbyβ	-3.50 to -3.55	1
Radii	-3.71 to -3.81	2
Main-sequence fitting	-3.75 to -3.92	3
Theory	-3.86 to -4.14	4

References for Table (listed in order of the derived value of $<M_V(0.8)>$)
1. Schmidt (1984), Balona & Shobbrook (1984).
2. Martin et al. (1979), Stothers (1983), Barnes (1979).
3. Caldwell (1983), van den Bergh (1976), deVaucouleurs (1978), Sandage & Tammann (1969), Fernie & McGonegal (1983).
4. Cox (1979), Stothers (1983).

With the luminosities of the cluster Cepheids from the uvbyβ photometry and temperatures based on Pel's (1978) scale, pulsational masses were calculated from the theoretical expressions of Cox (1979). The pulsational masses are 50% of the evolutionary masses for periods near 4 days and increase to about 75% of the evolutionary mass for periods around 10 days. Although this is a relatively small range in period, the difference in slope between the evolutionary masses and the pulsational masses is obvious in a plot of log M vs. log P (see Figure 5 of Schmidt 1984). An inspection of masses based on the larger distance scales (see for example the masses given by Cox (1979) shows the same slope difference for stars in common.

The small pulsational masses are a result of the small distance scale proposed here and, as many authors have discussed, the pulsational masses and the evolutionary masses are in agreement for the larger luminosity scales. However, the present pulsational masses are in good agreement with the masses inferred from the beat periods of double-mode Cepheids and with the masses inferred from the secondary bumps on velocity curves of Cepheids with periods near 10 days if standard models are used for both. Thus, the present luminosity scale produces consistency among the various masses from pulsation theory while the evolutionary masses are discrepant. This suggests that the attempts to reconcile the various masses individually with the evolutionary masses may have been misdirected. The disagreement is possibly more fundamental and may need to be rectified by improvements to either the theory of stellar evolution or to the theory of stellar pulsation.

Balona, L.A. and Shobbrook, R.R. 1984, preprint.
Barnes, T.G. 1979 In Highlights of Astronomy Vol 5 (Dordrecht: Reidel) p. 479
Caldwell, J.A.R. 1983, Observatory 103, 244.
Cox, A.N. 1979, Ap. J. 229, 212.
de Vaucouleurs, 1978, Ap. J. 223, 351.
Fernie, J.D. and McGonegal, R. 1983, Ap. J. 275, 732.
Martin, W.L., Warren, P.R. and Feast, M.W. 1979, M.N.R.A.S. 188, 139.
Pel, J.W. 1978, Astron. and Ap. 62, 75.
Sandage, A. and Tammann, G.A. 1969, Ap. J. 157, 683.
Schmidt, 1984, Ap. J. (in press).
Stothers, R.B. 1983, Ap. J. 274, 20.
van den Bergh, S. 1976, in IAU Colloquium No. 37, p. 13.

THE ZERO-POINT OF THE CEPHEID LUMINOSITY SCALE
FROM A CALIBRATION OF THE LUMINOSITIES OF EARLY-TYPE STARS

L.A. Balona
South African Astronomical Observatory

R.R. Shobbrook
University of Sydney

Abstract. A new calibration of the absolute magnitudes of early-type stars in terms of the (β, c_0) photometric system is used to establish the distance moduli of clusters containing Cepheids. The zero points of the period - luminosity and period - luminosity - colour relations are calculated and compared to previous determinations.

We have used published uvbyβ photometry of young clusters to calibrate the absolute magnitudes of early-type stars in terms of the β and c_0 indices (Balona & Shobbrook 1984). A total of 421 stars in 13 open clusters is used. All the clusters independently tie up with the Pleiades sequence, which eliminates a major source of uncertainty in previous calibrations. Our calibration is:

$$M_V = 3.499 + 7.203 \log(\beta - 2.515) - 2.319[g] + 2.938[g]^3$$

where $[g] = \log(\beta - 2.515) - 1.60 \log(c_0 + 0.322)$. This is valid for all luminosity classes.

This calibration was applied to eight galactic clusters containing Cepheids. Table 1 shows the resulting distance moduli based on uvbyβ photometry by Schmidt (1980ab, 1981, 1982ab, 1983). Also shown are distance moduli obtained by Schmidt using Crawford's (1978) calibration and distance moduli by Caldwell (1983) using zero-age main sequence fitting with the revised Hyades distance.

Table 1. Distance moduli of clusters containing Cepheids. N is the number of stars used in Schmidt's and our determinations.

Cluster	V_0-M_V	N	Schmidt	Caldwell
NGC129	11.07 ± 0.04	11	10.9 ± 0.1	11.12
NGC6087	9.64 ± 0.09	11	9.6 ± 0.1	9.81
NGC6664	10.90 ± 0.12	11	10.7 ± 0.1	10.88
NGC7790	12.15 ± 0.10	13	12.0 ± 0.1	12.69
IC4725	8.91 ± 0.08	27	8.8 ± 0.1	9.14
CV Mon	11.35 ± 0.24	5	10.9 ± 0.3	11.35
Ru 79	11.83 ± 0.21	5	11.4 ± 0.3	12.33
Ly 6	10.55 ± 0.10	3	10.6 ± 0.3	10.84

The distance moduli of Ru 79 and Ly 6 differ from the distance moduli of the corresponding Cepheids by about one magnitude, no matter which Cepheid calibration is used. Omitting these two stars, we can use the above moduli to fix the zero points of the period - luminosity (PL) and period - luminosity - colour (PLC) relationships as defined by Caldwell (1983). We then obtain the following calibrations:

$$\langle M_V \rangle = -2.79 \log P - 1.14 \ (\pm 0.11)$$
$$\langle M_V \rangle = -3.80 \log P + 2.70(\langle B \rangle_0 - \langle V \rangle_0) - 2.19 \ (\pm 0.06)$$

The PLC relation leads to absolute magnitudes which are fainter by 0.27 ± 0.08 mag than Caldwell's. Schmidt's distance moduli lead to absolute magnitudes 0.42 ± 0.10 fainter than Caldwell's.

The greatest uncertainty arising from ZAMS fitting is the poorly defined sequences in most of the calibrating clusters. Differences of as much as one magnitude are known to occur in the estimation of the distance moduli using this method. For this reason, the absolute magnitude estimates using uvbyβ photometry are expected to be more reliable, even though fewer stars are used. The standard deviation for one star in the absolute magnitude calibration is 0.4 mag, so that measurements of just a few stars should enable distance moduli to be obtained accurate to better than 0.2 mag.

A serious source of uncertainty which could affect the above calibration for Cepheids is the possibility of a systematic error in Schmidt's β index arising from systematic errors in the standard stars. There is no reason to suspect that this is the case, but differences of as much as 0.01 mag in β between different workers are not unknown. This could lead to zero point errors of as much as 0.3 mag in the PL and PLC relationships. For this reason it is important to obtain independent β observations for stars in the clusters containing Cepheids.

In conclusion, we find that a slight downward revision of the luminosities of Cepheids is indicated. The zero points determined by Caldwell (1983) are, however, within the uncertainties at present. The much lower luminosities found by Schmidt reflects the difference between Crawford's calibration and our calibration for the mid- to late-B stars.

REFERENCES

Balona, L.A. & Shobbrook, R.R., 1984. Mon. Not. R. astr. Soc., in press.
Caldwell, J.A., 1983. Observatory, 103, 244.
Crawford, D.L., 1978. Astr. J., 83, 48.
Schmidt, E.G., 1980a. Astr. J., 85, 158.
Schmidt, E.G., 1980b. Astr. J., 85, 695.
Schmidt, E.G., 1981. Astr. J., 86, 242.
Schmidt, E.G., 1982a. Astr. J., 87, 650.
Schmidt, E.G., 1982b. Astr. J., 87, 1197.
Schmidt, E.G., 1983. Astr. J., 88, 104.

ON THE DISTANCE TO THE OPEN CLUSTER LYNGÅ 6

E. R. Anderson, B.F. Madore and M.H. Pedreros
Dept. of Astronomy, Univ. of Toronto, Toronto, Ontario

The determination of the membership of the Cepheid TW Normae to the galactic cluster Lyngå 6 is important because TW Nor would then be the intrinsically brightest Cepheid associated with a cluster.

A series of nine photographic exposures (three each in U,B and V) were obtained during open time at the Anglo-Australian Telescope. The plates were measured with an iris photometer and the data calibrated using the photoelectric standards of van den Bergh and Harris (1976). The data for 300 stars were then analysed using a new computerized method of main sequence fitting and cluster membership determination developed by Pedreros (1984). We found 41 member stars with a mean reddening $E(B-V)=1.33 \pm 0.10$ and a distance modulus of 11.1 ± 0.4 magnitudes.

Figure 1 is a photo of the cluster with the member stars labelled. Figure 2 shows the colour-magnitude diagram for the cluster. The solid line is the zero age main sequence drawn for an assumed distance modulus of 11.1. The Cepheid TW Nor is represented by a solid triangle. Also shown is an evolved G type star (star L) believed by both Madore(1975) and Lyngå (1977) to be a cluster member.

The infrared period-luminosity relation and observations of TW Nor (Welch et al. 1984b,1984a) yield a distance modulus of 11.8 ± 0.2. The dashed line in Figure 2 is the zero age main sequence drawn for this assumed distance. Since the infrared Cepheid distance scale appears to be well established, we are then left with two possible conclusions:
1.)Lyngå 6 is at a distance modulus of 11.1 and TW Nor is not a member but rather behind the cluster at a distance modulus of 11.8. or,
2.)The Cepheid distance is the correct distance to the cluster. The stars we have assumed to be main sequence stars are actually slightly evolved B stars and/or the reddening is wrong.

REFERENCES

van den Bergh, S. and Harris, G. 1976, Ap.J., 208, 765
Lynga, G. 1977, Astron. and Astrophys., 54, 311.
Madore, B.F. 1975, Astron. and Astrophys., 38, 471.
Pedreros, M.H. 1984, Ph.D. Thesis, University of Toronto.
Welch,D.L., Wieland,F., McAlary,V.W., McGonnegal,R., Madore,B.F., McLaren,R.A., and Neugebauer,G. 1984a, Ap.J. Supp., 54, 547.
Welch,D.L., McAlary,C.W., Madore,B.F., McLaren,R.A. and Neugebauer,G. 1984b, In preparation.

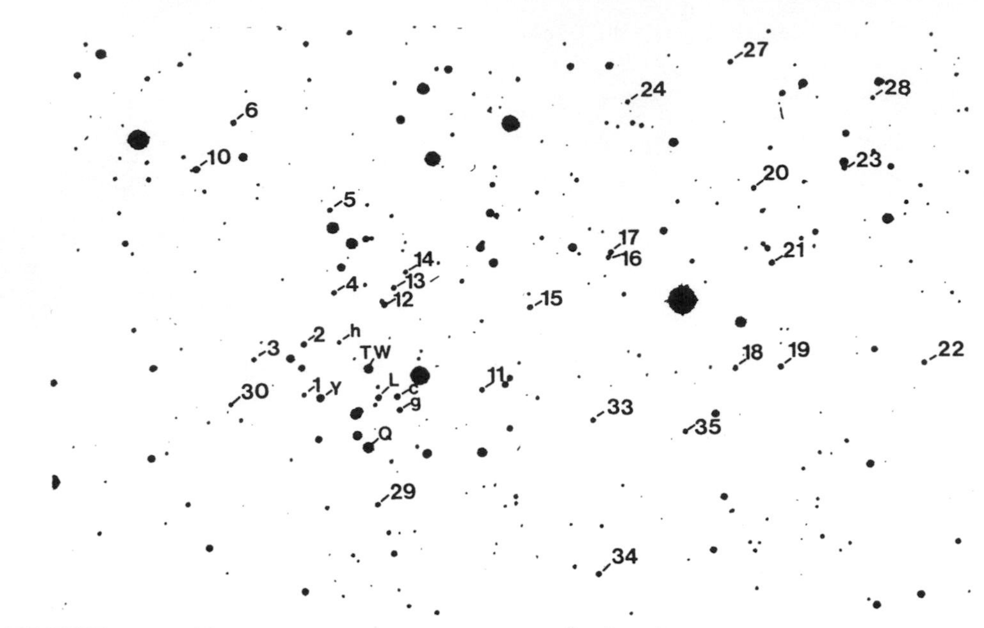

FIGURE 1. The open cluster Lyngå 6. (U photo)

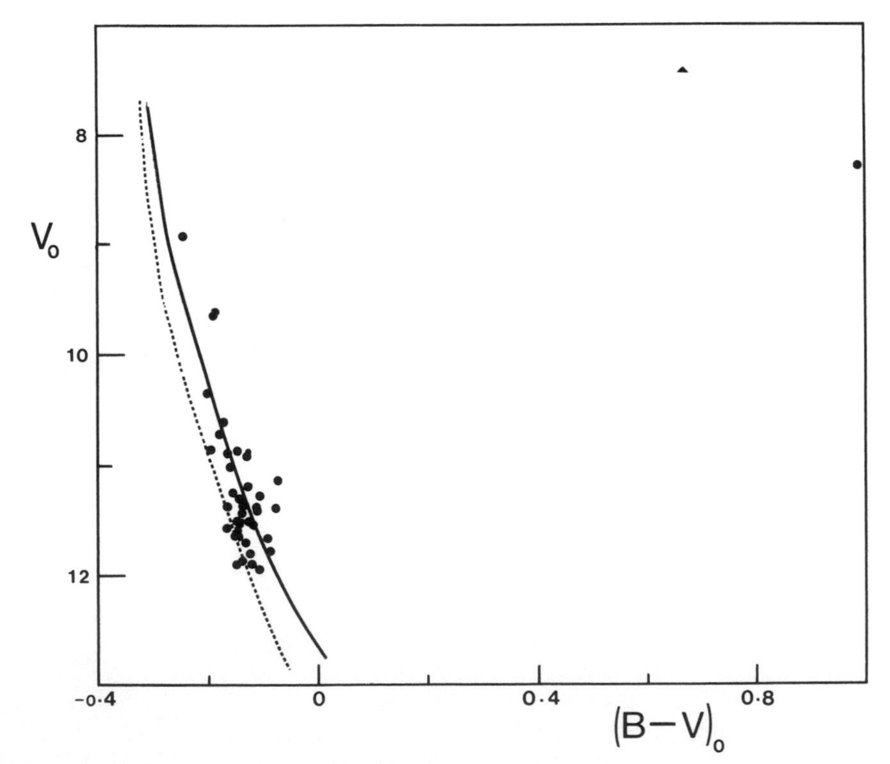

FIGURE 2. The colour-magnitude diagram for Lyngå 6.

SOME REMARKS CONCERNING THE ACCURACY OF PHOTOMETRIC REDDENINGS FOR CLASSICAL CEPHEIDS

D.G. Turner
Department of Astronomy
Saint Mary's University
Halifax, Nova Scotia, B3H 3C3
Canada

Considerable effort has been spent over the years in deriving colour excesses for classical Cepheids using their observed photometric indices. Various broad band, intermediate band, and narrow band systems have been used for this purpose, all with varying degrees of success. Reddenings derived in this manner are extremely useful, and most of the factors which limit their accuracy seem to be reasonably well known. Two others, which often tend to be overlooked but which are addressed here, are the presence of photometric companions to many Cepheids and the tendency to treat interstellar reddening in a prescriptive manner. The point is not necessarily to raise doubts about the validity of published photometric reddenings for classical Cepheids, but rather to examine the possibility that one can derive slightly more reliable colour excesses using the available observational data.

PHOTOMETRIC COMPANIONS

Many Cepheids, probably in excess of 25%, have been found to have unresolved companions in recent years, with blue companions encountered more frequently than red companions. The majority are probably physical, being late B or early A stars formed with the Cepheid, but, being of smaller mass, still residing close to the ZAMS in the H-R diagram. Such companions tend to be faint, and differences of 3^m or 4^m (or more) with respect to the Cepheid primary are not uncommon.

Fig. 1 plots the effect that a blue companion, chosen here to have the colours of an A0 star, has on the mean colours (in this case B-V and R-I) of a typical classical Cepheid with $P \simeq 5^d$. Fig. 1 is intended only for illustrative purposes, but it raises an interesting point, namely that blue companions 4^m or 5^m less luminous than the Cepheid can make the mean colours of the system $0^m.01$ to $0^m.02$ too blue in B-V. A photometric reddening for such a system using UBV colours would tend to be systematically too small by the same amount, and this can be as large as $0^m.05$ for a companion only 3^m fainter.

Similar problems exist for most systems centred on the blue end of the spectrum, although these can be reduced using near infrared photometry. The contamination in R-I, for example, is 3 or 4 times smaller than that in B-V in Fig. 1, although the magnitude range of R-I colours is only half that of B-V colours. In the case of a red companion, however, there may be no advantage gained from using near infrared photometry.

Figure 1.

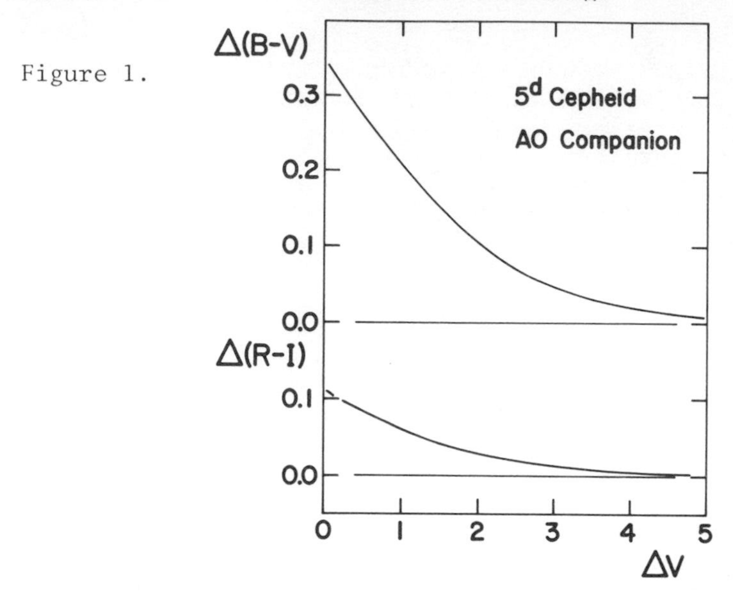

Clearly there is considerable advantage in knowing if a Cepheid has a companion before one derives its photometric reddening. On the other hand, how reliable is a photometric reddening estimate when it is not known if the Cepheid has an unseen companion?

INTERSTELLAR REDDENING

So many theoretical studies have been done of interstellar reddening in the various photometric systems that one tends to overlook the observational results which are available. My own work has involved examining the reddening slope in the UBV system using the best available photometry and MK spectral types published for early-type stars in restricted regions of the galactic plane where the reddening varies by more than a magnitude in E(B-V). Regression fits made to the data for stars in 6 different regions lead to the conclusion that the curvature term in the reddening slope E(U-B)/E(B-V) is constant near +0.02, but the slope itself varies with location in the Galaxy over an extreme range from 0.62 to 0.80. The average slope is close to the canonical value of 0.72, but only in 1 of the 6 regions examined did the calculated slope actually fall within $\pm\sigma$ of this value. The

Table 1. Reddening Parameter Variations

E_{U-B}/E_{B-V}	E_{m_1}/E_{b-y}	E_{c_1}/E_{b-y}	E_{khg_2}/E_{b-y}
0.65	-0.38	0.13	-0.02
0.72	-0.35	0.18	-0.01
0.75	-0.33	0.20	-0.03
0.80	-0.31	0.23	-0.02
Feltz & McNamara	-0.35	0.20	+0.44

interstellar reddening relation appears to vary throughout the Galaxy, and its exact form depends upon the region being examined. Since most photometric analyses of Cepheid reddenings assume that a constant interstellar reddening law is valid, this is bound to introduce some errors into the resulting values of colour excess.

Photometric indices most susceptible to reddening slope variations are those which involve colours in the violet portion of the spectrum, where the reddening law varies most. Table 1 lists for 4 different UBV colour excess ratios the corresponding results calculated for a few specific photometric colour excess ratios, namely those involving the Strömgren m_1 and c_1 indices, and the narrow band khg_2 index defined by McNamara et al. (1970). The Strömgren m_1 and c_1 indices involve violet and ultraviolet colours, respectively, and show distinct variations in reddening slope from the mean values assumed in most studies. The khg_2 index, however, was designed to be almost reddening-independent, and this is confirmed here.

The last line of Table 1 shows the reddening slopes used by Feltz & McNamara (1980) in their examination of photometric reddenings for 41 classical Cepheids using a mixture of Strömgren photometry, Hβ photometry, and khg_2 photometry. The Strömgren m_1 and c_1 indices are somewhat poorly suited for determining Cepheid reddenings, unless one is prepared to treat the reddening law for each Cepheid on an individual basis. Hβ photometry also seems to be ill-suited for this purpose, perhaps because of the magnitude of the scatter in published data. The khg_2 index, however, seems to be ideally suited to the problem, but it is surprising that Feltz & McNamara (1980) found it necessary to use a large reddening correction for what was designed to be a reddening-free index. This is likely to be an error on their part.

Table 2 lists space reddenings recently derived for 5 Cepheids (Turner 1984) using stars in their immediate vicinities which seem to be either physically related to the Cepheids or to lie at roughly the same distance. These reddenings have been combined with the b-y and khg_2 photometry of Feltz & McNamara (1980) for these variables, with the results plotted in Fig. 2. It should be evident from the plotted data that the khg_2 index is an excellent temperature indicator for most Cepheids since it correlates extremely well with unreddened b-y colour. There is a slight difference in this dependence between the declining

Table 2. Cepheid Space Reddenings

Cepheid	P^d	E_{B-V}	Origin of Space Reddening
SZ Tau	3.15	0.30	Nearby Coronal Members of NGC 1647
α UMi	3.97	0.01	Nearby Companion of Similar Distance
BB Sgr	6.64	0.30	Nearby Coronal Members of Cr 394
ζ Gem	10.15	0.04	Nearby Companion of Similar Distance
SV Vul	45.10	0.45	Nearby Members of Vul OB1 Subgroup

Figure 2.

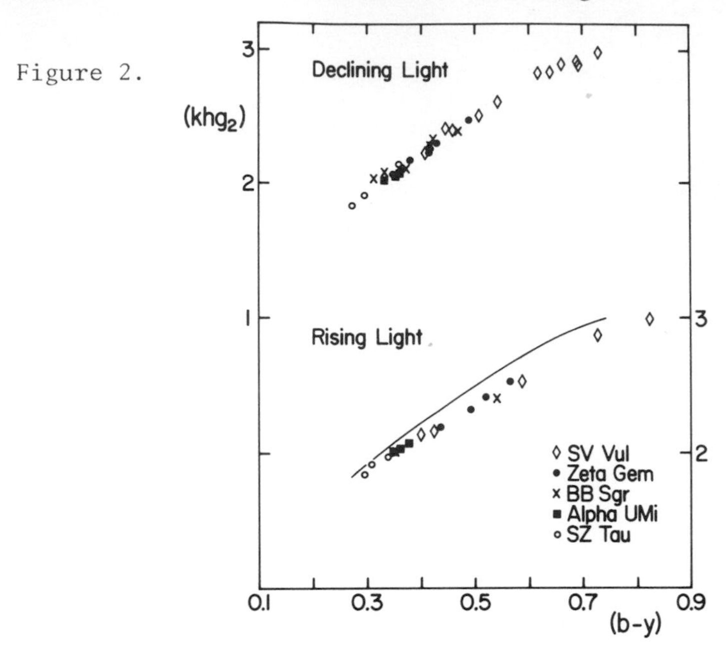

light and rising light portions of the light curves, and blue companions do produce difficulties, as outlined by Feltz & McNamara. However, the khg_2 photometry of Feltz & McNamara (1980) appears to be a better indicator of Cepheid reddening than these authors realize. Regrettably, the reddenings derived by Feltz & McNamara from their data are seriously compromised by their adopted reddening corrections. In particular, their reddening estimates tend to be systematically too large, except perhaps for the least reddened Cepheids. Since the khg_2 photometry of Feltz & McNamara (1980) is itself otherwise ideally suited for the determination of photometric reddenings for Cepheids, one may hope that it will be possible to extend such photometry to additional Cepheids.

References.

Feltz, K.A., Jr. & McNamara, D.H. (1980). Publ. Astron. Soc. Pacific 92, 609.

McNamara, D.H., Helm, T.M., & Wilcken, S.K. (1970). Publ. Astron. Soc. Pacific 82, 293.

Turner, D.G. (1984). J. Roy. Astron. Soc. Can. submitted.

ARE THE CEPHEIDS IN CLUSTER NUCLEI A RARE BREED?

D.G. Turner
Department of Astronomy
Saint Mary's University
Halifax, Nova Scotia, B3H 3C3
Canada

As reviewed by Kholopov (1968), star counts for a variety of open clusters reveal the existence of low density coronal regions surrounding the nuclear concentrations of most star clusters. Such cluster coronae have diameters 2.5 to 5 times larger than the respective nuclear diameters for clusters which are poor to medium-rich in member stars, and have star densities only about 10% those observed in cluster nuclei. Cluster coronae therefore contain roughly 40% to 70% of the stars in an open cluster, and are subsequently a (or, more appropriately, the) major component of most star clusters.

The search for new calibrators for the Cepheid PL relation has yielded many Cepheids which qualify spatially as possible members of cluster coronae. The majority of these can be rejected as bona fide cluster members by such tests as star counts, age differences with respect to the turn-off point ages for the associated clusters, and derived luminosities greatly different from PL relation values when cluster membership is assumed. Specific examples are AB Cam and XZ CMa (cf. Tsarevsky et al. 1966), which fail as possible coronal members of the clusters Tombaugh 5 and Tombaugh 1, respectively, by location outside the cluster corona for AB Cam and by marked discrepancies in age and luminosity for XZ CMa (Turner 1983). Many Cepheids remain, however,

Table 1. Cepheids in Cluster Nuclei

Cepheid	P(days)	Cluster	Class
EV Sct	3.09	NGC 6664	III2m
V1726 Cyg	4.24	Anon	IV2p
CE Cas B	4.48	NGC 7790	II2m
CF Cas	4.88	NGC 7790	II2m
CE Cas A	5.14	NGC 7790	II2m
CV Mon	5.38	Anon	III2p
V367 Sct	6.29/4.38	NGC 6649	II2m
U Sgr	6.74	IC 4725	I2p
DL Cas	8.00	NGC 129	IV2p
S Nor	9.75	NGC 6087	I2p
CPD-53°7400p	11.29	NGC 6067	I2r
S Vul	68.5	Anon	III3m

which do qualify as cluster coronal members from these tests, despite the lack of confirmation which might be possible from radial velocity and proper motion studies. A current listing of these objects and their parent clusters is provided here, along with a comparison of the properties of these clusters with those which contain Cepheids in their nuclear regions.

Table 1 is a summary of Cepheids seen projected against cluster nuclei for which the currently-available membership tests are consistent with cluster membership. The omission of a few notable calibrators such as CS Vel and TW Nor rests upon the results of a number of recent studies which suggest that they may not be cluster members. Table 1 contains 12 Cepheids in 10 different clusters, most of which are either poor or medium-rich in member stars. Trumpler types for these clusters are from the literature or from new classifications derived from photographs of the clusters. Since the clusters are mainly of poor to medium richness, the results given earlier for typical dimensions and star densities lead one to predict from simple geometry that between 21% and 39% (2 to 5) of the Cepheids in Table 1 are coronal members seen in projection against their cluster nuclei. Good candidates are V1726 Cyg and S Vul, which lie near the boundaries of their cluster nuclei, and one or more of the Cepheids seen projected against the nucleus of NGC 7790.

Table 2 is a list of coronal Cepheids for which the presently-available data are consistent with cluster membership, at least with regard to the tests mentioned earlier. The majority of these objects are established as cluster members by photometric studies alone, and a few additional objects such as TV CMa near NGC 2345 may be added later as studies of their possible cluster membership are completed. Two Cepheids, UY Per and RU Sct, are unusual in the sense that they lie roughly midway between two clusters of similar age and distance. Although Table 2 is

Table 2. Cepheids in Cluster Coronae

Cepheid	P(days)	Cluster	Class
SZ Tau	3.15	NGC 1647	II2m
BD Cas	3.65	Czernik 1	IV2m
HD 144972	3.79	NGC 6067	I2r
CG Cas	4.37	Berkeley 58	IV2p
UY Per	5.37 {	King 4	III2p
		Czernik 8	II3p
V Cen	5.49	NGC 5662	II3m
GH Car	5.73	Trumpler 18	III2p
R Cru	5.83	NGC 4349	II2r
BB Sgr	6.64	Collinder 394	IV2m
T Cru	6.73	NGC 4349	II2r
RU Sct	19.70 {	Trumpler 35	II2m
		Dolidze 32	I2p
WZ Sgr	21.85	Anon	IV2p

Figure 1.

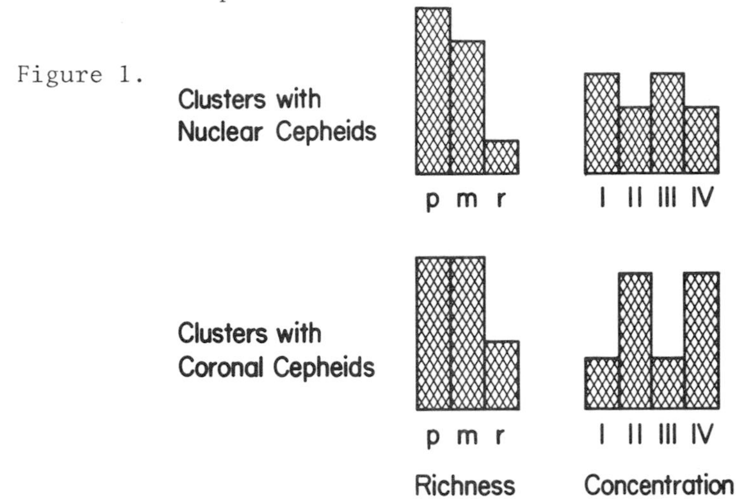

probably incomplete, it already contains as many objects as Table 1. From the arguments presented earlier, it follows that coronal Cepheids must outnumber nuclear Cepheids by a factor of between 1.4 (14/10) and 2.4 (17/7). Coronal Cepheids are therefore probably about twice as abundant as nuclear Cepheids in this region of the Galaxy.

Fig. 1 shows the relative frequencies of the various concentration and richness classes found for clusters in Tables 1 & 2. There is no obvious preference of Cepheids for either very loose, class IV, or very compact, class I, clusters, although they are rare in rich clusters. This can be explained by cluster dynamics, since most star clusters will have undergone a large amount of disintegration through evaporation of member stars by the time they are old enough to produce Cepheids from the post-main-sequence evolution of intermediate mass stars.

It may also be possible to explain with similar arguments the apparent preference of Cepheids for cluster coronae. High density cluster nuclei are ideal sites for the formation, through frequent stellar encounters, of close binary systems in which the components are not sufficiently separated to permit either to evolve to the supergiant dimensions of typical Cepheids. Stone's (1980) H-R diagram for nuclear and coronal members of the cluster NGC 654 exhibits the expected effects of such dynamical evolution, namely larger scatter above the ZAMS for nuclear members than for coronal members. The nucleus of NGC 654 probably contains large numbers of unresolved binary systems with nearly equal components, in contrast to mostly single stars in the corona. If this is also true for the clusters in Tables 1 & 2, then conditions in their nuclei may hinder the natural evolution of intermediate mass stars to the Cepheid state. Lower density cluster coronae may therefore be the best sites to search for new calibrators for the Cepheid PL relation.

References.

Kholopov, P.N. (1968). Astron. Zh. 45, 786.
Stone, R.C. (1980). Publ. Astron. Soc. Pacific 92, 426.
Tsarevsky, G.S., Ureche, V. & Efremov, Y.N. (1966). Astron. Tsirk. 367.
Turner, D.G. (1983). J. Roy. Astron. Soc. Can. 77, 31.

A SEARCH FOR LONG-PERIOD CEPHEIDS IN ASSOCIATIONS

Sidney van den Bergh
Dominion Astrophysical Observatory
Herzberg Institute of Astrophysics
5071 W. Saanich Road, Victoria, B.C. V8X 4M6, Canada

Abstract. Skeleton photoelectric sequences and plates in U, B and V have been obtained for fields surrounding all known Cepheids with P $\gtrsim$ 12 days in the Southern Milky Way. The study of 14 of these fields has been completed and is discussed in this paper.

Cepheid variables remain the most important calibrators of the extra-galactic distance scale. Unfortunately the Cepheids that are presently known in open clusters are intrinsically fainter, and have shorter periods, than do those that can be detected in Galaxies beyond the Local Group. It is therefore important to calibrate long-period Cepheids, similar to those that we observe in distant galaxies, in our own galaxy. At Cerro Tololo I have therefore obtained skeleton UBV sequences around all Population I Cepheids with P $\gtrsim$ 12 days, that are located along the Southern Milky Way. Plates of fields surrounding these same Cepheids have been obtained in ultraviolet, blue and yellow light by Brosterhus and Younger with the 1-m Swope telescope on Las Campanas. Alcaino and Turner have contributed to the measurements (typically 300 - 800 stars in each field) and to the reduction of these data.

To date observations of the following Cepheids have been reduced:

U Car $38^{d}.8$	QY Cen $17^{d}.8$	AA Nor $12^{d}.3$
EZ Vel 34.5	XZ Car 16.6	KK Cen 12.2
VZ Pup 23.2	YZ Car 16.6	VX Cru 12.2
WZ Car 23.0	SV Vel 14.1	UU Mus 11.6
CT Car 18.1	OO Cen 12.9	

Results on WZ Car, YZ Car, KK Cen and OO Cen have been published by van den Bergh, Brosterhus and Alcaino (1982). Those on CT Car, UU Mus, VZ Pup, SV Vel and EZ Vel by van den Bergh, Younger, Brosterhus and Alcaino (1983) and those on U Car, XZ Car, QY Cen, VX Cru and AA Nor by van den Bergh, Younger and Turner (1984). The results obtained to date are summarized below:

U Carinae. This Cepheid is located ~ $1^{\circ}.5$ east of the center of the η Car nebula. The photometry suggests that U Car is somewhat closer to us than the η Car complex.

EZ Velorum. This object is not embedded in a major association that is young enough to contain O-B1 stars. The present observations do not extend faint enough to detect an association containing main-sequence stars with spectral types later than B1.

VZ Puppis. Somewhat surprisingly the low-absorption field surrounding this long-period Cepheid contains no stars with $B-V < 0.0$ or $U-B < -0.50$. The absence of such stars is puzzling because VZ Pup (if it is a Cepheid of Population I) has an age of only $\sim 2 \times 10^7$ years.

WZ Carinae. There is no evidence to suggest that WZ Car is embedded in a major association.

CT Carinae. The UBV colors of stars in this field may be interpreted in two ways. Either CT Car is embedded in a rich association of distant B stars or it is projected on a concentration of foreground F-type stars. Spectroscopy will be required to distinguish between these two alternatives.

QY Centauri. The situation in this field is confused. It is not yet clear if the early-type stars in this area are at the same distance as the Cepheid.

XZ Carinae. This Cepheid lies at a distance comparable to that of the rich Car OB1 complex. Twelve open clusters lie within 1° of this variable. The closest of these is the unstudied cluster Ruprecht 93.

YZ Carinae. The color-color diagram shows no evidence for significant numbers of reddened OB stars in this field.

SV Velorum. Two B-type stars with reddening values similar to the Cepheid are located in this field.

OO Centauri. Three stars in this field have UBV colors indicating that they are OB stars with reddenings similar to that of the Cepheid. A deeper study of this field would be desirable.

AA Normae. This object is probably a Cepheid of Population II.

KK Centauri. No objects with the colors of reddened OB stars are seen in this field.

VX Crucis. This Cepheid is not embedded in an association containing very early-type stars. Deeper photometry (particularly in U) would be required to detect late B stars in this field.

UU Muscae. The color-magnitude and color-color diagrams of the field surrounding this Cepheid show no evidence for a concentration of blue stars that might be associated with UU Mus.

The results that are summarized above are, in some ways, disappointing. Reasons for this are probably that:

1. Most long-period Cepheids are not situated in the cores of rich associations.

2. In some poor associations the surface density of OB stars may be too low for significant numbers of such objects to be situated within 10' of a Cepheid.

3. Associations containing mainly late B-type stars may, in a few cases, have been missed because our photometry did not go deep enough in some reddened fields.

We hope to be able to continue publishing the results of this search for long-period Cepheids in associations at the rate of about half a dozen fields per year.

REFERENCES

van den Bergh, S., Brosterhus, E.B.F. & Alcaino, G. (1982). Ap.J.Suppl., 50, 529.
van den Bergh, S., Younger, P.F., Brosterhus, E.B.F. & Alcaino, G. (1983). Ap.J.Suppl., 53, 765.
van den Bergh, S., Younger, P.F. & Turner, D.G. (1984). In preparation.

LEAVITT VARIABLES: THE BRIGHTEST CEPHEIDS VARIABLES and their IMPLICATIONS for the DISTANCE SCALE

Gerald R. Grieve
Canadian Centre for Remote Sensing, Ottawa, Ontario

Barry F. Madore and Douglas L. Welch
Department of Astronomy, David Dunlap Observatory
University of Toronto, Toronto, Ontario

Abstract. Two low-amplitude variable supergiants in the Large Magellanic Cloud, S65-08 and S65-48 are each found to have periods of approximately 250 days. The optical data suggest that these stars are high-luminosity cepheid variables falling more than one magnitude brighter than any other known Cepheids in the LMC. Confirmation of the cepheid nature of these stars comes from their H-band magnitudes which place them accurately on a simple linear extrapolation of the narrower infrared Period-Luminosity relation. So it appears that the cepheid Period-Luminosity relation extends up to $M_V \sim -8.5$. To honour the astronomer who discovered the first of these highest-liminosity Cepheids, we have sub-classified the variables with log P > 1.8 as being "Leavitt variables". As soon as these long-period variables are discovered in other external galaxies, reliable distances should be possible out to (m-M) $\sim$ 30.

I. Introduction

It is a commonly held belief that Cepheids span a period range extending from 1 to 50 or 60 days (cf. Cox 1980 or Strohmeier 1972) but this is a somewhat constrained view imposed by our galactic perspective. It has in fact been known since the discovery of the PL relation for Cepheids by Leavitt (1907) that the relation extends to periods beyond 200 days. The discovery and calibration of the longest-period (and hence brightest) Cepheids is important since these stars are fundamental to the determination of reliable distances to nearby galaxies.

Comprehensive reviews of the status of the extragalactic sample with the inevitable emphasis on the Magellanic Cloud Cepheids have been published by Madore (1983) and more recently by Feast (1984). It is emphasized by these reviewers and has been noted before (Sandage and Tammann 1968) that in the optical, the longest-period Cepheids (log P > 2.0) are somewhat deviant from a simple linear Period-Luminosity relation as extrapolated from shorter periods. Sandage and Tammann (1968) considered this apparent flattening in the Period-Luminosity relation beyond 100 days to be intrinsic to the calibration. On the other hand, Madore (1982) has argued that based on a reddening-free formulation of the Period-Luminosity relation, the same data indicate no curvature and that excess differential reddening of the four LMC Cepheids might offer a natural explanation. Obviously, small number statistics play an important role in such an evaluation; from all points of view, a larger sample or complimentary observations would be desirable.

II. The New Variables S65-08 and S65-48

Recently, Grieve (1983) has completed a photoelectric program monitoring all of the brightest Magellanic Cloud intermediate-type supergiants using the Las Campanas 61cm reflector of the University of Toronto. The purpose was to discover the frequency and type of light variations to be found in these stars. Low-amplitude variations ($\Delta B < 0.5$ mag) could easily have been missed in the early photographic surveys and extrapolation of the period-amplitude diagram (e.g., Schaltenbrand and Tammann 1970) would suggest that very long-period Cepheids might be expected to have small amplitudes. The recent observations of Eggen (1983) confirm this suspicion.

S65-08 is of spectral type G2 Ia (Ardeberg et al. 1972); and with a radial velocity of +311 km/s (Brunet et al. 1973) it is a confirmed member of the LMC. It is now known that S65-08 is a variable with an amplitude of approximately 0.3 mag at V and 0.1 mag in (B-V). A period gram analysis of the V data yields a best-fitting period of 250 days. A second low-amplitude supergiant variable has been pointed out by Eggen (1983). This star, S65-48, was found to have a V amplitude of 0.3 mag. Grieve (1983) has also observed this star and in combining these data, a period of roughly 250 days has been found. Ardeberg et al. (1972) give the radial velocity of S65-48 as +311 km/sec confirming its membership in the LMC. They also quote a spectral type of F8 Ia while Morgan and Keenan (1973) consider it to define the super-supergiant type F8 0.

Fig. 1. The <V> Period-Luminosity relation for 33 LMC Cepheids for which both V and H observations are available.

Fig. 2. The <H>Period-Luminosity relation for 33 LMC Cepheids. including the two new Leavitt variables at 250 days.

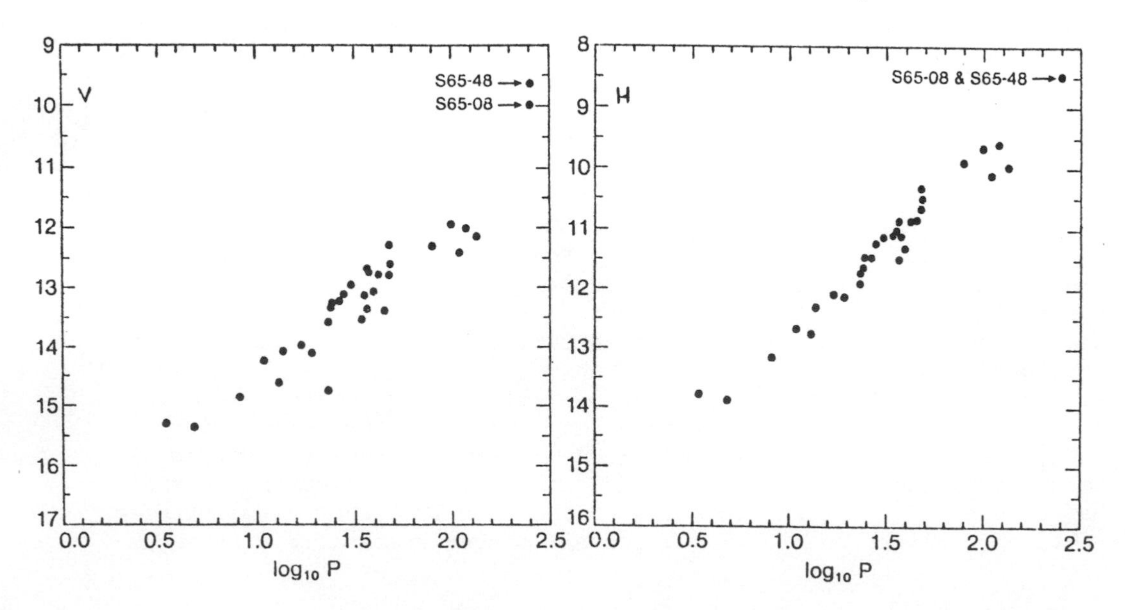

In Figure 1, we present the V Period-Luminosity relation for the photoelectrically observed LMC Cepheids as compiled by Madore (1984). Neither individual nor statistical corrections for foreground or external reddening have been applied to the data. Also plotted are the intensity-averaged <V> magnitudes of S65-08 and S65-48. As can be seen, the V Period-Luminosity relation does in fact continue linearly from the shorter periods to the region unambiguously defined by these two new Cepheids. The curvature in the Period-Luminosity relation suggested by the previous data sample is not confirmed for substantially longer periods.

As a further powerful check in establishing that the intrinsic Period-Luminosity relation extends in a linear fashion out to 250 days, we obtained JHK observations for S65-08 and S65-48 using the Dupont 2.5m at Las Campanas, Chile in January, 1983. Figure 2 shows the position of these stars in the infrared H-band relation. The data defining this relation for the previously catalogued Cepheids are intensity-mean averages obtained by the techniques outlined in Welch et al. (1984) using the data in McGonegal et al. (1982) and additional data to be published. As can be seen the new Cepheids fall precisely where they would be expected to be based on a simple linear extrapolation of the H Period-Luminosity relation.

III. Implications and Discussion

Sandage (1984) has suggested that the flattening in his calibration of the apparent <B> Period-Luminosity relation can be put to good practical use, in the sense that only approximate periods are needed at the long-period end in order to obtain reasonable distances. Inspection of his Figure 1 or equivalently Figure 5 in Sandage and Tammann (1968) reveals that in practice this apparent flattening could be used to give distance moduli good to $\pm$0.2 mag (rms) for periods in the interval 60 to 160 days (i.e., log P = 2.0 $\pm$ 0.2).

However, using a strictly linear Period-Luminosity relation has a two-fold advantage. Firstly, the scatter in the infrared Period-Luminosity relation is so small that periods, in fact, need not be determined with great accuracy anywhere along the relation. For instance, for a Cepheid whose true period is 100 days, any determination between 80 to 125 days will also suffice to give and H-band distance modulus good to $\pm$0.2 mag (rms). Secondly, and perhaps more importantly, a linear Period-Luminosity relation greatly extends the potential of Cepheids as distance indicators.

When Henrietta Leavitt (1907) plotted the periods and magnitudes of the variable stars that she had been studying in the Magellanic Clouds "... she realized that her discovery could be used as an indicator of intrinsic brightness, but was prevented from pursuing the subject any further ..." (Berendzen, Hart and Seeley 1976). Not only did she discover the Period-Luminosity relation but she discovered the longest-period members of that class -- none of which were known locally. Others soon identified Leavitt's Harvard variables with galactic Cepheids and so the relation that she found does not bear her name. Now as a tribute to the discoverer of the objects that have and will continue to play a key role in the extragalactic distance scale, we plan to refer to the sub-class of these variables beyond log P $\sim$ 1.8 as Leavitt variables. With these objects in the calibration, galaxies at least an extra magnitude further in distance modulus can now be calibrated by traditional methods.

References

Ardeberg, A., Brune , J.P., Maurice, E., and Prevot, L. 1972, Astr . Ap. Suppl., 6, 249.
Berendzen, R., Hart, R., and Seeley, D. 1976, "Man Discovers the Galaxies", (New York, Science History Publications).
Bruent, J.-P., Prevot, L., Maurice, E., and Muratorio, G. 1973, Astr. Ap. Suppl., 9, 447.
Cox, J.P. 1980, "Theory of Stellar Pulsation", (Princeton, Princeton University Press).
Eggen, O.J. 1983, A. J., 88, 1458.
Feast, M.W. 1984, in I.A.U. Symposium #108, Structure and Evolution of the Magellanic Clouds, ed. S. van den Bergh and K. de Boer, (Dordrecht, Reidel), p 157.
Grieve, G. 1983, Unpublished Doctoral Thesis, University of Toronto.
Leavitt, H. 1907, Harvard Ann., 60, 87.
Lindblad, B. 1922, Ap. J., 55, 85.
Madore, B.F. 1982, Ap. J., 253, 575.
Madore, B.F. 1983, Highlights in Astronomy, Vol 6, ed. R.M. West, Dordrecht, Reidel), p 217.
Madore, B.F. 1984, in I.A.U. Colloquium No. 82., Cepheid Variables: Theory and Obwervations, ed B.F. Madore, (Cambridge Univ. Press).
McGonegal, R., McLaren, R.A., McAlary, C.W., Madore, B.F. 1983, Ap. J. (Letters), 257, L33.
Morgan, W.W. and Keenan, P.C. 1973, Ann. Rev. Astr. Ap., 11, 29.
Sandage, A. 1984, (preprint).
Sandage, A. and Tammann, G.A. 1968, Ap. J., 151, 531.
Schaltenbrand, R. and Tammann, G.A. 1970, Astr. Ap., 7, 289.
Strohmeier, W. 1972, Variable Stars, (New York, Pergamon Press)
Welch, D.L., Wieland, F., McAlary, C.W., McGonegal, R., Madore, B.F., McLaren, R.A., Neugebauer, G. 1984, Ap. J. Suppl., 54, 547.

THE INFRARED DISTANCE SCALE: THE GALAXY AND THE MAGELLANIC CLOUDS

D.L. Welch
Dept. of Astronomy, University of Toronto

C.W. McAlary
Steward Observatory, University of Arizona

R.A. McLaren and B.F. Madore
Dept. of Astronomy, University of Toronto

Abstract. The advantages of using near-infrared photometry of Cepheids to determine distances to nearby galaxies are now well known. In this paper we summarize the current state of the infrared period-luminosity (P-L) relations for the Galaxy, the LMC, and the SMC and present composite P-L relations derived from all available photometry. We give distance moduli for the LMC and SMC based on these data and briefly report on the status of other work in progress.

Introduction

Magellanic Cloud Cepheids have always played a central role in our understanding of Cepheids and the determination of cepheid P-L relations. It is particularly important to examine the behaviour of these P-L relations in the infrared, as the Cepheid-based Local Group distance scale is best determined at these wavelengths.

We present P-L relations determined from intensity-mean JHK magnitudes of 40 LMC and 28 SMC Cepheids. At present we have multiple observations for three-quarters of the LMC Cepheids and one-third of the SMC Cepheids. Variables which are now known to be optical doubles, overtone pulsators, and those lacking a published ephemeris, were excluded from the sample. The individual observations will be published at a later date.

Reduction

To transform random-phase observations to intensity-mean average magnitudes, we adopt the procedure described in Welch et al. (1984). The recent ephemerides of van Genderen (1983a) were used where possible. Predicted lightcurves for each Cepheid were examined visually for goodness-of-fit. Significant phase adjustments were necessary for 50% of the LMC sample and 10% of the SMC sample. The vast majority of adjustments resulted in mean magnitudes differing by less than 0.03 mag from those derived from uncorrected curves. In 10% of the combined sample, an amplitude adjustment was also necessary. Particularly noteworthy is HV 2883 in the LMC, whose infrared amplitude is twice that predicted from the scaling relations of Welch et al. (1984).

Period- Luminosity Relations

The observational P-L relations derived from our data are as follows:

LMC	SMC
$J = 16.14 - 2.96 \log P$	$J = 16.70 - 3.09 \log P$
$\pm 0.04 \quad \pm 0.03$	$\pm 0.07 \quad \pm 0.04$
$H = 15.87 - 3.04 \log P$	$H = 16.45 - 3.19 \log P$
$\pm 0.04 \quad \pm 0.03$	$\pm 0.07 \quad \pm 0.05$
$K = 15.80 - 3.06 \log P$	$K = 16.50 - 3.28 \log P$
$\pm 0.04 \quad \pm 0.03$	$\pm 0.10 \quad \pm 0.06$

The largest source of error in using these same data to derive LMC and SMC distance moduli is the uncertainty in the galactic P-L zero-points. Unfortunately, the final distance moduli are sensitive to the choice of galactic calibrators and Hyades modulus. We justify our choice of calibrators in a paper on the improved calibration of the galactic near-infrared P-L relations (in preparation). In this work we use the calibrators EV Sct, CF Cas, CV Mon, V Cen, V367 Sct, U Sgr, DL Cas, S Nor, and RS Pup.

Adopting the galactic calibration data of Caldwell (1983), the distance of RS Pup from Havlen (1972), and the ratios of total-to-selective absorption from McGonegal et al. (1983), the apparent distance moduli to the LMC and SMC can be determined using the method described in McAlary et al. (1983). Briefly, the apparent moduli for both the LMC and SMC are solved for by assuming a common slope for the P-L relations in all three galaxies. We present the results of these calculations in Table 1. The composite H P-L relation is displayed in Figure 1. We have assigned an uncertainty of 0.2 mag to all galactic cepheid distance moduli.

Table 1

Composite P-L Relations and Apparent Distance Moduli for the LMC and SMC

Filter	Zero-point	Slope	(m-M) mag LMC	(m-M) mag SMC
J	-2.41 ±0.07	-3.00 ±0.02	18.60 ±0.07	18.97 ±0.07
H	-2.56 ±0.07	-3.08 ±0.02	18.49 ±0.07	18.86 ±0.07
K	-2.62 ±0.07	-3.10 ±0.03	18.48 ±0.07	18.84 ±0.07

Fig. 1 The composite H P-L relation for the galactic calibrators, the LMC, and the SMC. Symbols are as follows: diamonds - galactic calibrators, squares - LMC, circles - SMC.

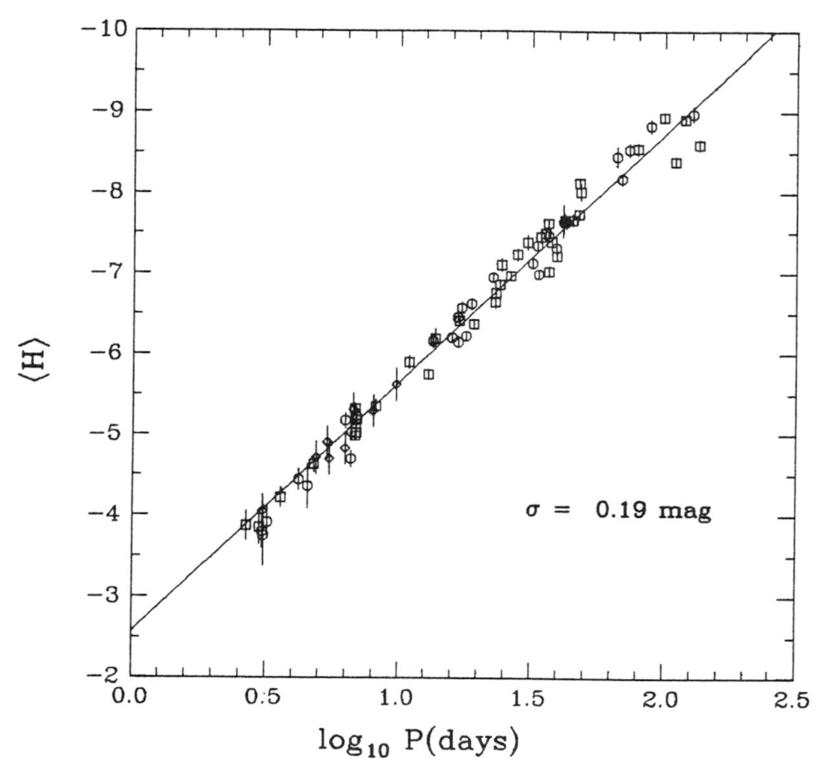

We adopt the average reddenings found by van Genderen (1983b): E(B-V)= 0.14 mag for the LMC and 0.11 mag for the SMC. The J, H, and K absorptions for the LMC are then 0.12, 0.07, and 0.04 mag, respectively, and for the SMC, 0.09, 0.06, and 0.03 mag, respectively. We obtain a true distance modulus of 18.45 for the LMC and 18.83 mag for the SMC.

Discussion

Welch and Madore (1984) and Feast (1984) have both presented H-band P-L relations. Feast's report of the unpublished work of Laney and Stobie cites a common slope of -3.34 for the LMC and SMC P-L relations. The SAAO work encompasses a smaller number of stars and a shorter baseline in log P, but uses more complete lightcurves. The slope difference may be accounted for by the exclusion of the 100-day Cepheids from the SAAO work. Welch and Madore (1984) reported observational H-band P-L relations for the LMC and SMC which differ only marginally from those reported here (as a result of more data, phasing and amplitude corrections, and a slightly different sample).

The standard deviation of one point about the observational and composite P-L relations is only ±0.2 mag. It is clear from Figure 1 that this width is very nearly constant for the entire range of observed periods.

Conclusions

The composite P-L relations reported above clearly illustrate the numerous advantages of infrared photometry for distance scale work. There is a great deal of information yet to be gleaned from these data.

Important applications of these new P-L relations are:

1) The recalibration of the Local Group distance scale.
2) The possible depth and structure of the SMC.
3) Constraints on the slope of the period-radius relation resulting from the slope of the infrared P-L relations.
4) The determination of infrared period-color relations, using more complete photometry.

Work on these problems is now in progress.

This work was supported by the Natural Sciences and Engineering Research Council of Canada and the National Science Foundation.

Caldwell, J.A.R. 1983, Observatory, 103, 244.
Feast, M.W. 1984, IAU Symp. 108, Structure and Evolution of the Magellanic Clouds, ed. S. van den Bergh and K.S. de Boer, (Dordrecht: D. Reidel), p. 157.
van Genderen, A.M. 1983a, Astr. Ap. Suppl., 52, 423.
van Genderen, A.M. 1983b, Astr. Ap., 119, 192.
Havlen, R.J. 1972, Astr. Ap., 16, 252.
McAlary, C.W., Madore, B.F., McGonegal, R., McLaren, R.A., and Welch, D.L. 1983, Ap. J., 273, 539.
McGonegal, R., McAlary, C.W., McLaren, R.A., and Madore, B.F. 1983, Ap. J., 269, 641.
Welch, D.L., and Madore, B.F. 1984, IAU Symp. 108, Structure and Evolution of the Magellanic Clouds, ed. S. van den Bergh and K.S. de Boer, (Dordrecht: D. Reidel), p. 221.
Welch, D.L., Wieland, F., McAlary, C.W., McGonegal, R., Madore, B.F., McLaren, R.A., and Neugebauer, G. 1984, Ap. J. Suppl., 54, 547.

NEW DISTANCES TO THE MAGELLANIC CLOUDS

N. Visvanathan
Mount Stromlo and Siding Spring Observatories, Australia.

Abstract. The mean phase magnitudes at the IV waveband (1.05 micron) of thirteen Cepheids in the SMC and the LMC and nine Cepheids in groups and clusters in the Galaxy, are used in conjunction with periods, to construct P-L(IV) relations in these galaxies. The slopes and the dispersions of the relations are nearly the same. We derive a distance modulus of 19.11±0.07 for the SMC and 18.82±0.07 for the LMC.

The classical Cepheid has the unique property that the absolute luminosity is related to its period. This property has been used to determine distances to external galaxies in the Local Group. In this paper, we derive distances to the LMC and the SMC using IV mag (1.05 micron luminosity) corrected to the mean phase of the Cepheid, with the aid of V mag. The extinction in the IV waveband is four times lower than that in the B waveband. Further, the spread in the P-L(IV) relation due to the width of the instability strip in the H-R diagram as well as the metallicity variations from galaxy to galaxy, is smaller by a factor of ~2, compared to that in the P-L relation in optical wavelengths. Hence we believe that the P-L(IV) relation will lead to better distances to external galaxies compared to those derived using optical luminosities.

The absolute calibration of the P-L(IV) relation has been achieved through the observations of Cepheids in galactic clusters whose distances are known. On total, nine Cepheids in the Galaxy, thirteen Cepheids in the SMC and fourteen Cepheids in the LMC have been observed at the one metre telescope and the Anglo-Australian Telescope at the Siding Spring Observatory. A Varian photomultiplier tube with an InGaAsP cathode which is sensitive to IV(1.05 micron) and V wavebands has been used. For each Cepheid, both, IV and V magnitudes have been measured. The V mag has been used to identify the phase of the Cepheid from the published V light curve. Assuming the ratio of the amplitudes of Cepheids from V to IV, as ~3, the observed IV mag has been corrected to the mean phase.

The mean phase mag <IV>, of galactic Cepheids have been corrected further for the extinction and their absolute magnitudes in IV and $M_{<IV>_0}$ have been derived using the distance moduli given by Fernie and McGonegal (1983). A value of 0.02 mag and 0.04 mag for the reddening in the direction of the SMC and the LMC has been used. The mean phase and reddening corrected absolute mag $M_{<IV>_o}$ for the galactic Cepheids and mag $m_{<IV>_o}$ for the LMC and the SMC are combined with their periods to form a P-L(IV) relation for the Cepheids in the Galaxy, the LMC and the SMC. The least-squares regressions of these relations are given below:

Galaxy

$$M_{<IV>_o} = -2.21 - 3.136 \log P; \quad \sigma = 0.15 \text{ mag}$$

LMC

$$m_{<IV>_o} = 16.56 - 3.089 \log P; \quad \sigma = 0.14 \text{ mag}$$

SMC

$$m_{<IV>_o} = 16.97 - 3.195 \log P; \quad \sigma = 0.18 \text{ mag}$$

It can be seen that the slope and the dispersion in the P-L(IV) relation is nearly the same in the SMC, the LMC and the Galaxy. Hence we take the mean of the three slopes to represent the best slope of the P-L(IV) relation (-3.14). If we fit this slope to Cepheids in the LMC and the SMC, we derive a distance modulus of 18.82 for the LMC and 19.11 for the SMC. These distances are based on a distance modulus of 3.29 for the Hyades.

Our distance modulus for the LMC (18.82) is in agreement with that derived using 1.6 micron luminosity (18.71; McAlary et al. 1983) and with the extinction and abundance corrected distance modulus computed by Martin et al. (1979; 18.95). As regards the SMC, our distance modulus (19.11) is again in agreement with that corrected for extinction and metallicity difference between our Galaxy and the SMC by Gascoigne (1974) and strengthens the fact that the SMC has a lower metal content than our Galaxy. However, our moduli are higher by ~0.4 mag than those derived by de Vaucouleurs (1978). A part of the difference ~0.2 mag, is due to the high reddening correction adopted by de Vaucouleurs.

References:

de Vaucouleurs, G. (1978). The extragalactic distance scale II. Distances of the nearest galaxies from primary indicators, Ap. J. 223, pp. 730-739.

Fernie, J.D. & McGonegal, R. (1983). Cepheids in open clusters and associations, Ap.J., 275, pp. 732-736.

Gascoigne, S.C.B. (1974). Metal abundance and the luminosities of Cepheids, Mon. Not. R. astr. Soc., 166, Short comm. pp. 25-27.

McAlary, C.W., Madore, B.F., McGonegal, R. McLaren, R.A., & Welch, D.L. (1983). The distance to NGC 6822 from infrared photometry of Cepheids, Ap.J. 273, pp. 539-542.

The Distance to M33 Based on BVRI CCD Observations of Cepheids

Wendy L. Freedman
Department of Astronomy, University of Toronto
Toronto, Canada

Introduction

As part of a long-term program to study the distance and stellar content of M33, CCD frames of a number of fields centered on known Cepheids in this galaxy (Hubble 1926, Sandage and Carlson 1983) have been obtained, in collaboration with Madore, McAlary, and Davis. The observations were made at the prime focus of the Kitt Peak 4m telescope, and reduced using the Kitt Peak RICHFLD profile fitting programs. These data are still preliminary, but quite sufficient to illustrate the power in the application of our method.

Figure 1. Differential (M33-LMC) P-L Relations

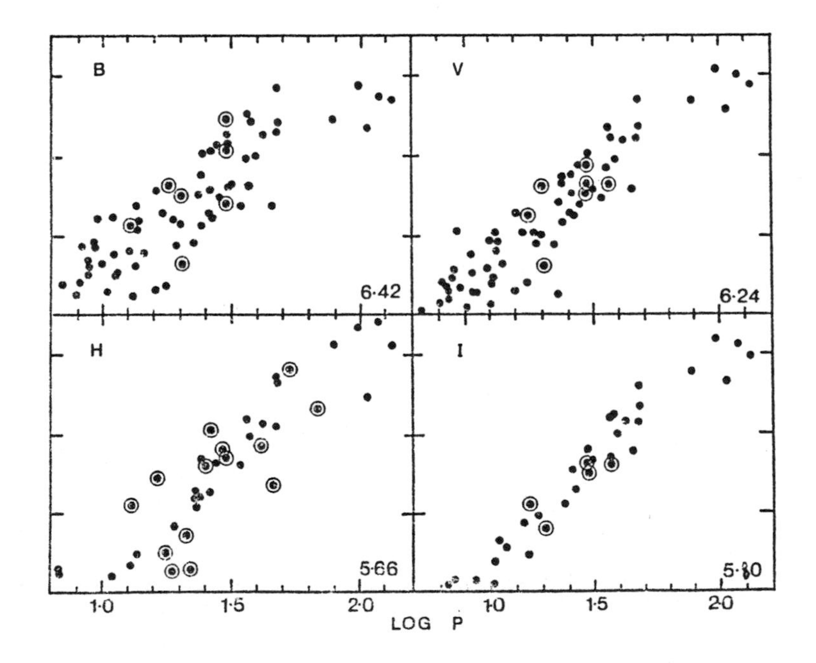

The Application of I-band Photometry of Cepheids to the Determination of Distances

McGonegal et al. (1982) have illustrated the advantages of near-infrared photometry applied to the cepheid Period-Luminosity (PL) relation. The effects of reddening and metallicity are reduced at longer wavelengths, and the amplitude of the Cepheids themselves decreases toward the red, so that the scatter in the PL relation is significantly reduced compared to the optical, even for random phase observations. However, the main disadvantage of the method is the following: aperture photometry, which is presently the only readily available technique for near-infrared observations, is limited by crowding in the main beam and by background contamination in the comparison beams.

Panoramic detectors such as CCD's, however, offer the advantage of seeing-limited photometry while retaining a linear response. With these detectors the stellar magnitude, locally corrected for background light can be well-determined using profile-fitting techniques. Working with stellar profiles, rather than fixed apertures, has a distinct advantage for the more distant systems. The following study was undertaken in order to combine the advantages offered by both techniques: the CCD observations were extended as far to the red as possible (0.9μm). The R and I bands are relatively insensitive to the radiation redistribution effects due to possible metallicity variations from galaxy to galaxy (see McGonegal et al. 1982, their Figure 1). Furthermore, the effects of extinction are more than a factor of 2 lower at I with respect to B.

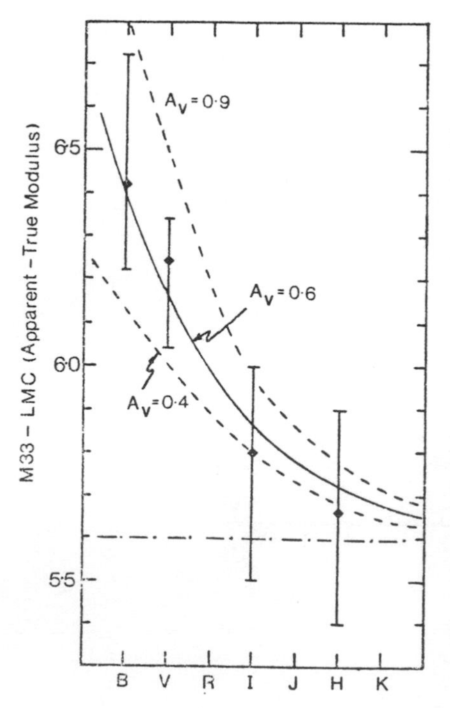

Figure 2. Differential moduli versus wavelength.

Distance to M33

The Cepheids in Field No. 1 in M33 have periods ranging from 18 to 42 days. The solid dots in Figure 1 show PL relations in BVI obtained from photoelectric photometry of Cepheids in the LMC (Martin, Warren and Feast 1979; Grieve 1983). H-band observations from the study by McGonegal et al. (1982) are also plotted (lower left panel). Differential reddening has not been corrected for in any of the plots. Superimposed on the LMC mean-light relations are the CCD data for the M33 Cepheids (circled dots), each shifted in distance modulus by an amount which gives the minimum scatter between the LMC and M33 P-L relations for BV and I. Also plotted for comparison, are the H-Band observations of Madore et al. (1984) for M33.

The scatter in the M33 H data is appreciably larger than the scatter in the LMC H-band data. This is likely the result of contamination problems. At the increased distance of M33 relative to the LMC, crowding in the main beam and comparison fields is more severe. The scatter of the random-phase I observations on the other hand is much lower than at H indicating that the CCD method can likely be applied with success to distance determinations further than the infrared aperture photometry technique. A single CCD observation, which can be obtained at R or I, is therefore a highly competitive alternative to classical B photometry, which requires many nights of observing to follow the variation.

The shifts in magnitude which result in the minimum scatter between the LMC and M33 data at each of the various wavelength bands are: 6.42 at B, 6.24 at V, 5.80 at I and 5.66 at H. These differences in apparent moduli as a function of wavelength are plotted in Figure 2. The shape of this relation is the same as that expected for the extinction as a function of wavelength. If we assume that the true distance modulus is being approached as the wavelength increases, and choose a true differential distance modulus of 5.6, then the extinction at V and A_V = 0.6. Computing the extinction at the various other wavelengths using the recent calibration of Leitherer and Wolf (1984), the solid curve drawn in Figure 2 is obtained. Lower and upper limit extinctions which bracket the data are shown as dashed lines at values of A_V = 0.4, and 0.9 mag, respectively.

An extinction of E(B-V)=0.08 has been applied to the LMC data, and a distance modulus of 18.5 adopted (McGonegal et al. 1982). This results in a new preliminary distance modulus of 24.1 for M33, 1.2 mag closer than the recently determined apparent blue modulus of 25.3 preferred by Sandage (1983). Part of this difference may be attributed to the inclusion of E(B-V)=0.2 mag of internal absorption in M33 itself, and part to the higher LMC distance modulus adopted by Sandage.

References

Grieve, G. 1983, (Unpublished Ph.D. thesis: University of Toronto)
Hubble, E. 1926, Ap. J., 63, 236.
Leitherer, C. and Wolf, B. 1984, Astr. and Ap., 132, 151.
McGonegal, R. McLaren, R. A., McAlary, C.W. and Madore, B.F. 1982, Ap. J. (Letters), 257, L33.
Madore, B.F., McAlary, C.W., McLaren, R.A., Welch, W.L. and Neugebauer, G. 1984,(in preparation)
Madore, B.F., Freedman, W.L., Grieve, G., McAlary, C., and Davis, L. 1985, (in preparation).
Martin, W.L., Warren, P.R. and Feast, M.W., 1979, M.N.R.A.S., 188, 139.
Sandage, A.R. 1983, A.J., 88, 1108.
Sandage, A.R. and Carlson, G. 1983, Ap. J. (Letters), 267, L25.

THE DISTANCES TO NEARBY GALAXIES FROM NEAR-INFRARED PHOTOMETRY OF CEPHEIDS

Christopher W. McAlary
Steward Observatory, University of Arizona

Douglas L. Welch
Deparment of Astronomy, University of Toronto

Abstract. Near-infrared photometry of Cepheid variables in the Local Group galaxies, NGC 6822, IC 1613, and M33, and in the M81 Group galaxy, NGC 2403 has been used to determine new, independent distances to these objects, which are almost unaffected by dust extinction and by differences in metallicities among the galaxies.

Introduction

The advantages inherent in using the near-infrared Period-Luminosity (P-L) relation have been discussed in some detail by McGonegal *et al*. (1982) and in various papers at this meeting. For the purpose of using the technique for more distant galaxies, it should be stressed that it is the smaller variation of surface brightness with temperature that allows near-infrared observations of Cepheids in external galaxies to be practical. The intrinsic width of the *H* band (1.65 um) P-L relation for random-phase observations has been shown to be approximately the same as that for fully phase-averaged *V* (0.55 um) measurements. Thus, single-phase near-infrared observations of Cepheids in external galaxies can lead to apparent distance moduli which have about the same intrinsic uncertainty as the best available in the optical; however, the large telescope time required for such measurements is an order of magnitude less, and it is primarily this consideration that makes the near-infrared technique so attractive. In addition, the smaller extinction in the near-infrared means that the uncertainty in the true distance modulus to a galaxy is likely to be smaller than in the optical.

Over the past three years, observations of Cepheids in six nearby galaxies have been carried out. Welch *et al*. (1984) have already discussed the continuing program to monitor the Cepheids in the Magellanic Clouds. This paper will discuss four more distant objects observed.

Observations

More than 90% of the near-infrared observations of extragalactic Cepheids outside the Magellanic Clouds have been obtained with the Multiple Mirror Telescope (MMT), operated by the Smithsonian Institution and the University of Arizona. The telescope has a collecting area equivalent to a 4.5m conventional telescope, and the specialized near-infrared photometer built for the MMT is the most sensitive such system in the world. As an example of the sensitivity of the photometer, a signal-to-noise ratio of one can be obtained in one hour of integration for an object of J = 22.3. The mount for the MMT is altitude-azimuth, and chopping is performed in elevation. This provides a singular advantage when observing Cepheids in the crowded fields of external galaxies. Multiple observations of a given Cepheid, when obtained several hours apart, will be made at significantly different position angles, and the effects of field crowding can be minimized.

The four galaxies which have been observed are the Local Group members NGC 6822, IC 1613, and M33. In addition, several Cepheids in NGC 2403, which is probably a member of the M81 group, were also measured. Further descriptions of the observations can be found in McAlary *et al.* (1983), McAlary, Madore, and Davis (1984), McAlary and Madore (1984), and Madore *et al.* (1984). For the Local Group members, the Cepheids are all brighter than 19th magnitude in the near-infrared and were therefore observed in the H band so as to minimize reddening and intrinsic strip width. The extreme faintness of the Cepheids in NGC 2403 necessitated observations in the J band, where the MMT photometer is about 0.7 mag more sensitive. The observations are shown in Fig 1, where the solid line indicates the composite P-L relation determined by Welch *et al.* (1984).

Fig. 1 Near-infrared P-L relations for a) NGC 6822, b) IC 1613, c) M33, and d) NGC 2403. Photometric uncertainties for the Local Group galaxies are about twice the size of the symbols. The solid line is the composite P-L relation determined from Cepheids in the Magellanic Clouds and galactic clusters.

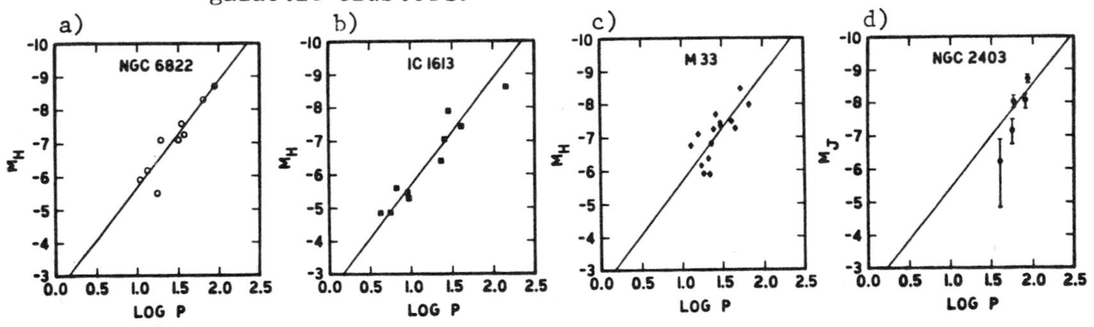

The distances to the four galaxies were determined by incorporating the observations into a multiple linear regression fit which also included the galactic cluster Cepheids (Caldwell 1983) and those in the Magellanic clouds. The distances are given in Table 1. These compare favorably with most other determinations with the exception of NGC 6822, where the extinction appears to have been underestimated in most optical studies, and for NGC 2403, where our distance is even larger than that found by Tammann and Sandage (1968). Recently, however, Sandage (1984)

has estimated the distance modulus to be 27.9, in good agreement with our result. Our result for M33 also disagrees with that of Sandage (1983). At least part of this must be related to his low value for the foreground and internal extinction toward the Cepheids in the galaxy, but there may also be residual zero-point errors in his re-analysis of the photographic photometry of Hubble (1926).

The Red Supergiant Calibration

Beyond the realm of the Cepheids, distance moduli are not well determined. It is now generally conceded that the diameters of HII regions are not reliable indicators, and Sandage (1984a) has shown that the brightness of the blue supergiants within a galaxy is a strong function of the galaxy's intrinsic luminosity. In a series of papers on the supergiants of nearby galaxies Humphreys (1984, and references therein) has asserted that the absolute visual magnitude of the brightest red supergiants is a constant for all galaxies, with $M_V=-8.0$. Since we now have a consistent set of distances for 6 galaxies with detailed searches for supergiants, this hypothesis can be properly tested. Fig. 2 shows how the *V* and *K* magnitudes of the 3 brightest supergiants, corrected for extinction, relate to the absolute *B* magnitude of the galaxy, corrected for foreground extinction and inclination effects. It is immediately apparent that, while the mean *V* magnitude is approximately -8.0, there is definately a trend toward brighter supergiant magnitudes with higher luminosity galaxies. This effect is even more pronounced in the *K* band.

This should not be viewed as discouraging for distance work. The slope of the relation is very shallow, and far removed from that of a distance degenerate case. When a least-squares solution is run on the data, it gives hope that the dispersion of the fit may not exceed 0.15 mag. Since the red supergiants can be detected out to the distance modulus of the Virgo cluster, such an uncertainty would be a great improvement over present techniques.

Table 1 Distance Moduli of Nearby Galaxies

Galaxy	No. Cepheids	$(m-M)_H$	E(B-V)	$(m-M)_o$
LMC	39	18.54 ± 0.06	0.08 ± 0.04	18.50 ± 0.07
SMC	25	18.91 ± 0.06	0.08 ± 0.04	18.89 ± 0.07
NGC 6822	9	23.49 ± 0.13	0.36 ± 0.06	23.30 ± 0.13
IC 1613	10	24.10 ± 0.14	0.03 ± 0.03	24.08 ± 0.14
M 33	15	24.28 ± 0.14	0.2: ± 0.1	24.17 ± 0.15
NGC 2403	5	28.14 ± 0.21[1]	0.06 ± 0.03	28.09 ± 0.21

[1] $(m-M)_J$

Fig. 2 Absolute magnitude of the three brightest red supergiants in a galaxy in the V and K bands as a function of the absolute B magnitude of the galaxy. The open circle is represents the local galactic neighborhood. The two points shown for NGC 2403 (connected by a line) indicate the range in absolute luminosity possible if the candidates proposed by Sandage (1984) are shown to be supergiants.

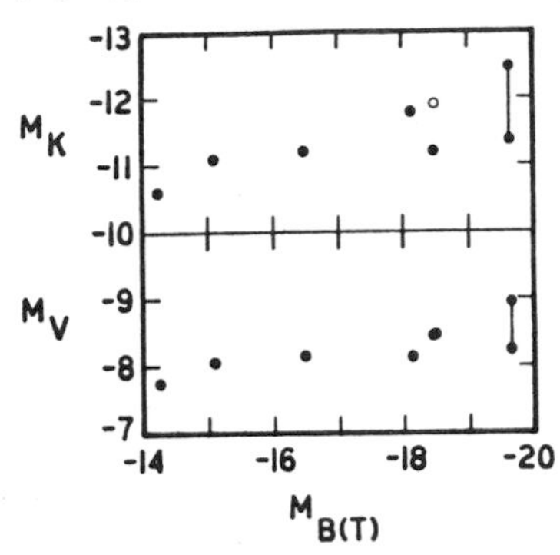

Conclusions

The advent of modern infrared techniques has allowed a significant improvement in our estimates of the distances to nearby galaxies. There are only two more objects, M31 and NGC 300, which can be observed in this manner, and observations of both are planned. The major problem with photometry of Cepheids in external galaxies is field crowding, and this accounts for more than one-half of the uncertainty in distance modulus. In the next year the infrared Astronomy group at Steward Observatory hopes to have in operation a near-infrared CCD. If the Cepheids can be detected with this device, distance moduli for galaxies out to 4 Mpc will be obtainable with 0.1 mag accuracy.

This work was supported by the National Science Foundation and the Natural Sciences and Engineering Research Council of Canada.

Caldwell, J.A.R. 1983, The Observatory, 103, 244.
Hubble, E. 1926, Ap.J., 63, 236.
Humphreys, R.M. 1984, A.J., in press.
Madore, B.F., McAlary, C.W., McLaren, R.A., Welch, D.L., and Neugebauer, G. 1984, in preparation.
McAlary, C.W., and Madore, B.F. 1984, Ap.J., in press.
McAlary, C.W., Madore, B.F., and Davis, L.E. 1984, Ap.J., 276, 487.
McAlary, C.W., Madore, B.F, McGonegal, R., McLaren, R.A., and Welch, D.L. 1983, Ap.J., 273, 539.
McGonegal, R., McLaren, R.A., McAlary, C.W., and Madore, B.F. 1982, Ap.J. (Letters), 257, L33.
Sandage, A. 1984, preprint.
Tammann, G.A., and Sandage, A. 1968, Ap.J., 151, 825.
Welch, D.L., McAlary, C.W., Madore, B.F., and McLaren, R.A. 1984, IAU Colloquium 82.

POPULATION II CEPHEIDS

H.C. Harris
McMaster University, Hamilton, Ontario, Canada

Abstract. A review of Cepheids in globular clusters finds 46 stars that are highly probable members of 21 clusters, all of intermediate to low metallicity with blue horizontal branches. In order to separate Type II Cepheids in the field from classical Cepheids, an analysis is made of the distribution of all Cepheids: 144 are sufficiently far from the galactic plane to be considered Type II. Their properties are compared with the cluster Cepheids. Some recent studies are reviewed that are improving our understanding of low-mass Cepheids.

INTRODUCTION

The distinction between young, massive, classical Cepheids and old, low-mass, Population II Cepheids has been recognized for more than three decades. Cepheids in globular clusters have been carefully studied and compared with field Cepheids near the sun. A simple (and in some respects traditional) view of the Cepheids in our Galaxy leaves them divided cleanly into the two types with no overlap. This view is based on the idea that stars in our Galaxy with intermediate masses (1 to 2 $M_{\odot}$) are sufficiently metal-rich to avoid making loops through the instability strip and becoming Cepheids in appreciable numbers during their helium burning evolution. It is supported by the absence of Cepheids in metal-rich globular clusters.

As in most research, however, further study reveals complications. The old, low-mass, field Cepheids include some metal-poor analogs to those in globular clusters, but many come from a metal-rich, old-disk population, for which the term "Type II" seems more appropriate than "Population II". These old-disk stars and the Anomalous Cepheids in dwarf spheroidal galaxies challenge our simple picture of two types of Cepheids. Type II Cepheids are often separated into long-period W Vir stars and short-period BL Her stars (with the division at a period of about 10 days). This separation is useful because the evolutionary state of the two groups is different, and these terms will be used occasionally throughout this paper. Various aspects of these stars have been reviewed previously (Joy 1949; Arp 1955; Payne-Gaposchkin 1956; Plaut 1965; Kukarkin 1975; Wallerstein & Cox 1984). This paper will discuss the present state of both the cluster and field Type II Cepheids, as well as current research that is shedding new light on this late stage of evolution for old stars.

CEPHEIDS IN GLOBULAR CLUSTERS

The list of known Cepheids in globular clusters continues to grow slowly. The presently known members are listed in Table 1, including 23 BL Her stars and 23 W Vir and/or RV Tauri stars. This list is based on the data from Sawyer Hogg (1973), updated by numerous more recent papers. References can be found in Clement _et al_. (1984a) and Harris _et al_. (1983). The magnitudes marked as approximate are based on B or pg magnitudes and colours estimated from the P-C relation, so they cannot be considered reliable. In addition, the following stars are possible cluster Cepheids, but will require further study: NGC 3201-V65 (probable eclipsing field star), NGC 4833-V9 (probable SR), NGC 6093-V2 (uncertain type), NGC 6293-V2 and NGC 6522-V8 (probable field Cepheids), NGC 6626-V21 (possible field Cepheid), NGC 6626-V22 (possible field RR Lyrae, but uncertain period), and NGC 7492-V4 (probable red variable). However, the following stars are probably _not_ cluster Cepheids: NGC 362-V8 and V10 (Cepheids in the SMC), NGC 5024-V24 and NGC 6626-V9 (RR Lyrae stars).

Included in Table 1 are the 6 known cluster stars that show the RV Tauri characteristic of alternating deep and shallow minima. These are

Table 1. Cepheids in Globular Clusters

CLUSTER	STAR	PERIOD	<V>	<B>-<V>
NGC 2419	V18	1.58	18.8	0.35
NGC 5139	V1*	29.34	10.85	0.68
(w Cen)	V29	14.73	11.82	0.95
	V43	1.16	13.36	0.46
	V48	4.47	12.69	0.63
	V60	1.35	13.45	0.39
	V61	2.23	13.40	0.63
	V92	1.35	13.93	0.53
NGC 5272 (M3)	V154	15.29	12.32	0.51
NGC 5466	V19	0.82	14.75	0.20
NGC 5904	V42*	25.74	11.3	0.52
(M5)	V84*	26.42	11.5	0.50
NGC 6093 (M80)	V1	16.30	13.42	0.76
NGC 6205	V1	1.45	14.0	0.26
(M13)	V2	5.11	13.1	0.43
	V6	2.11	14.1	0.42
NGC 6218 (M12)	V1	15.51	12.0:	...
NGC 6229	V8	14.85	15.65:	...
NGC 6254	V2	18.73	11.76	0.74
(M10)	V3	7.91	12.69	0.68
NGC 6273	V1	16.92	13.0:	...
(M19)	V2	14.14	13.3:	...
	V3	16.5	12.9:	...
	V4	2.43	14.1:	...
NGC 6284	V1	4.48	15.2:	...
	V4	2.82	15.4:	...
NGC 6333 (M9)	V_	short	...	...
NGC 6402	V1	18.73	14.06	1.22
(M14)	V2	2.79	15.64	0.75
	V7	13.60	14.80	1.22
	V17	12.08	14.81	1.14
	V76	1.89	15.84	0.76
NGC 6626	V4	13.46	13.5:	...
(M28)	V17*	46.0	12.5:	...
NGC 6656 (M22)	V11	1.69	12.6	0.8
NGC 6715 (M54)	V1	1.35	17.0:	...
NGC 6752	V1	1.38	12.97	0.40
NGC 6779	V1	1.51	15.3:	...
(M56)	V6*	45.01	13.24	0.83
NGC 7078	V1	1.44	14.89	0.34
(M15)	V72	1.14	15.08	0.47
	V86	17.11	13.7:	...
NGC 7089	V1	15.58	13.46	0.53
(M2)	V5	17.61	13.34	0.52
	V6	19.30	13.18	0.54
	V11*	33.6	12.11	0.56

*Shows RV Tauri characteristics.

included for two reasons. The identification of a star as an RV Tauri star rather than a Cepheid can be difficult, because detecting the alternating minima sometimes requires more complete and accurate observations than are available. In fact V1 in NGC 5139 and V42 in NGC 5904 have been called both types. Second, there is probably no qualitative difference between the evolutionary state of RV Tauri stars in clusters and other long-period Cepheids (although the situation for field RV Tauri stars is more complicated). Both types have pulsation driven by the same mechanism (the opacity of the hydrogen and helium ionization zones), and some models of long-period Cepheids display RV Tauri behavior (Bridger 1984). Both types lie together in the instability strip, in contrast to yellow semiregulars (SRd stars), which have lower luminosities, and red semiregulars, irregulars, and long-period variables, all of which are cooler (Rosino 1978, Wehlau & Sawyer Hogg 1977, Lloyd Evans 1977). The periods listed in Table 1 are the times between successive minima.

Rosino (1978), Lloyd Evans (1983a), and others have pointed out the usefulness of infrared colours and magnitudes for studying cluster variables. However, the UBV system remains the only system for which data are available for many of these stars. The P-L and P-C relations

Figure 1. The period-luminosity relation (a) and the period-colour relation (b) for the cluster Cepheids from Table 1 with measured V magnitudes and B-V colours. The line in the P-L plot is taken from Harris (1981). The line in the P-C plot is a least-squares fit to the data:
$(B-V)_o = 0.275 + 0.206 \log P$.

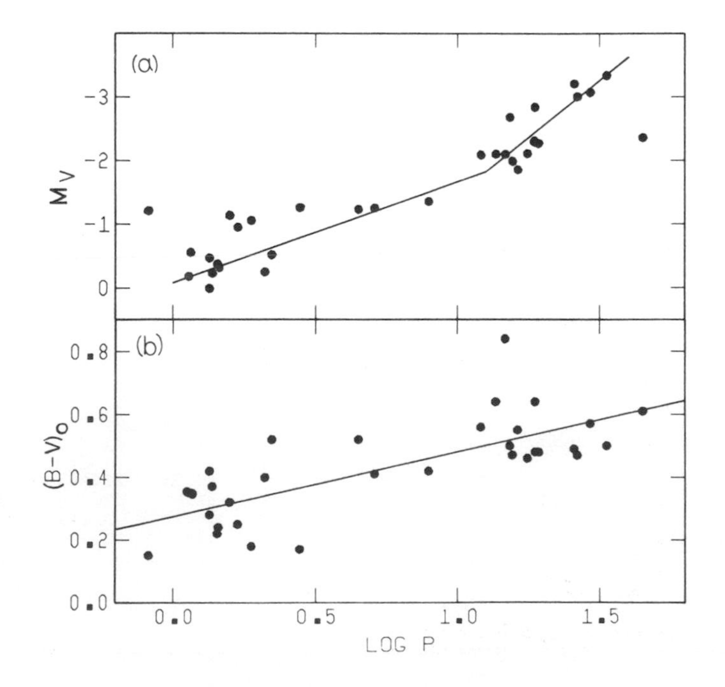

are shown in Figure 1, using distances and reddenings taken from Harris and Racine (1979). The mean deviations from the lines are 0.21 in V and 0.08 in B-V, and they probably include a significant contribution from observational errors for these faint, often crowded, variable stars. These deviations are not very large, to my eye (compared to similar plots for classical Cepheids by Feast at this conference, for example), and show that these relations can be used to determine distances and reddenings to globular cluster Cepheids without introducing excessive errors. The P-C relation in B-V for Type II Cepheids in the field has a larger scatter than appears in Figure 1 (Demers & Harris 1974), related to the larger range of abundances among the field stars. However, data presently available in V-K, R-I, and T_1-T_2 indicate that the use of red or infrared colours reduces the scatter. Further work is needed on field stars to determine whether the instability strip for Type II Cepheids is much wider in temperature than it is for classical Cepheids.

The distribution of periods is shown in Figure 2(a). The bimodal distribution noted by Kraft (1972) is obvious, with the gap around P = 8 to 10 days separating BL Her stars from W Vir stars. However, Clement *et al*. (1984b) suggest that instead the gap at P = 4 days should be used

Figure 2. (a) The distribution of periods for the cluster Cepheids from Table 1. Stars with RV Tauri characteristics are shown shaded. (b) The distribution for field Type II Cepheids. The cross-hatched region shows stars within a cylinder of radius 5 kpc centred on the sun.

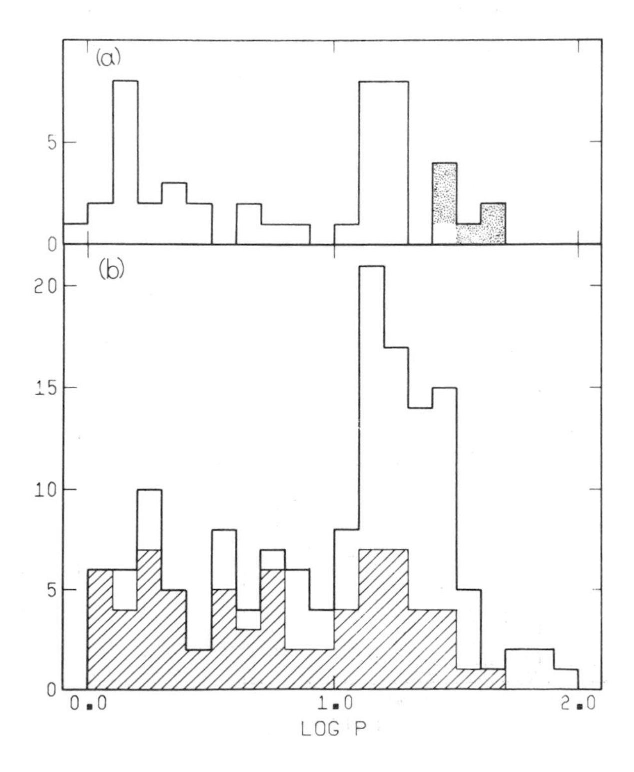

as the upper limit defining the BL Her class. Similar period distributions are found for subsamples of metal-poor or intermediate metallicity clusters.

Cepheids appear only in clusters with blue horizontal branches, a fact demonstrated by Wallerstein (1970) and supported with today's more complete data. More generally, UV-bright stars are most frequent in clusters with blue horizontal branches (Harris _et al_. 1983), and Cepheids are simply the UV-bright stars that happen to be passing through the instability strip. BL Her stars, in particular, may be most frequent in clusters such as M15 that have an extended blue tail on the horizontal branch. The metal abundance [A/H] of clusters containing Cepheids is generally low. The abundance distribution is shown in Figure 3(a), where the abundances are taken to be those of the parent clusters from Pilachowski (1984). (Use of a different source such as Zinn makes little difference for these clusters.) The spread of abundances in ω Cen makes the values for its Cepheids more uncertain than for others.

The evolution of globular cluster stars into Cepheids is generally understood (Schwarzschild & Harm 1970; Strom _et al_. 1970; Newell 1973; Zinn 1974; Gingold 1976). Stars from the blue end of the horizontal branch evolve through the instability strip with M_V = 0 to −1 and P = 1 to 5 days. Stars evolving up the asymptotic giant branch make loops into the instability strip (with M_V = −2 to −3 and P = 15 to 30 days)

Figure 3. The histogram of metal abundances for (a) cluster and (b) field Type II Cepheids. RV Tauri stars are shaded and BL Her stars are cross-hatched.

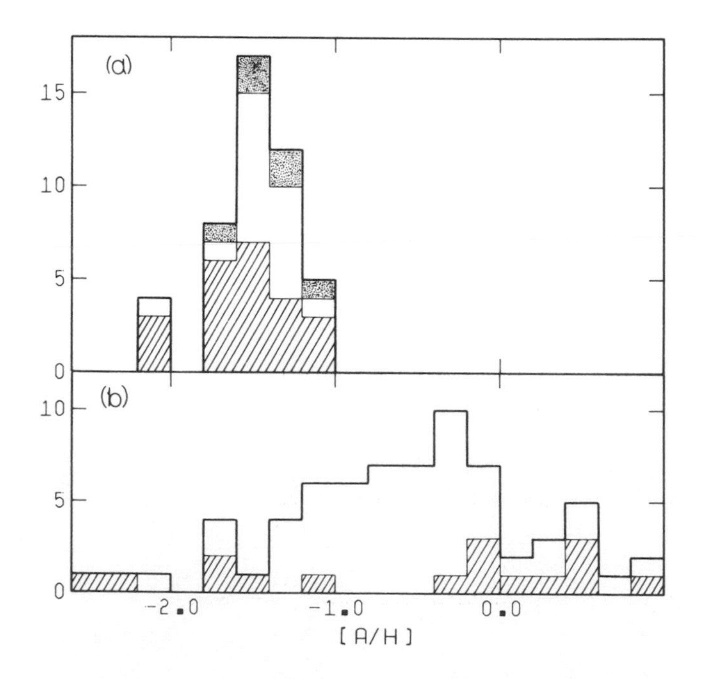

due to helium shell flashes and the final transition toward becoming a white dwarf. The bimodal distribution of periods seen in Figure 2 is predicted, but the distribution of periods, particularly for long periods, differs from that predicted (Gingold 1976). One factor that might help to explain the discrepancy is mass loss by AGB stars, limiting the luminosity attained by W Vir stars.

FIELD CEPHEIDS

Separation of Classical and Type II Cepheids

Classifying individual Cepheids that are not members of clusters as Type I or Type II is, at best, difficult. The problem arises because Type II Cepheids not only have smaller masses than classical Cepheids, but also smaller radii, so their surface gravities at a given period are almost identical; they likewise have similar temperatures at a given period, so most spectroscopic indicators cannot distinguish the two types. Some Type II Cepheids are metal-poor and have weak-lined spectra, but metallicity is a characteristic we want to measure, so it should not be used to define our sample. Classification methods must use a criterion sensitive to mass or radius. Several have been devised, based on different observational characteristics: Type II Cepheids sometimes exhibit open loops in color-color plots (Mianes 1963; Kwee 1967; Nikolov & Kunchev 1969) or color-magnitude plots (Warren & Harvey 1976); they often are bluer at maximum light (Walraven _et al._ 1958); they are more prone to variations in period or phase jitter (Hoffleit 1976; Szabados 1983); for periods of 13 to 28 days, they almost always show strong hydrogen emission lines during rising light (Joy 1949; Wallerstein 1958; Harris & Wallerstein 1984); they often have distinctive bumps in their light curves, and so show a Hertzsprung progression different from classical Cepheids (Petit 1960; Plaut 1965; Stobie 1973; Warren & Harvey 1976; Szabados 1977). Unfortunately, a comparison of results from these methods shows considerable disagreement for various reasons. For example, the presence of a hot main-sequence companion can cause a classical Cepheid to execute loops, mimicking a Type II Cepheid (Madore 1977). If good photometric data is available, the light curve shape is perhaps the most useful criterion for periods of 1 to 3 days and 12 to 20 days, and the agreement of observed and model light curves (Carson & Stothers 1982; Bridger 1984) adds confidence in this method of classification. However, for other periods for metal-rich Cepheids, there is at present no reliable way of separating the two types. When more accurate velocity curves and temperature indicators are available for many Cepheids, direct determinations of their radii through the Baade-Wesselink method will allow better identification of the Type II's.

A different approach to separating Type II Cepheids relies on their larger distances from the Galactic plane. Fernie (1968) found that classical Cepheids are distributed exponentially from the galactic plane with a scale height of only 70 pc. Hence any Cepheids farther than about 500 pc from the plane are very likely Type II. (The distances can first be estimated assuming the P-L relation for classical Cepheids. Then for the more distant stars, now assumed to be Type II, the distances are recalculated with the fainter Type II P-L relation, giving distances from the plane larger than about 200 pc.) This method was used by Smith et al. (1978) to identify BL Her stars. However, it may not always give a correct result. Runaway OB stars defy the exponential distribution, so some runaway Cepheids might be expected. Note, however, that runaway OB stars are rare and OB stars outnumber Cepheids in the Galaxy by almost two orders of magnitude, so runaway Cepheids should be very rare. Might the classical Cepheid disk be thicker toward the galactic center? Probably not, because the young disk of HI and molecular clouds is not thicker there, and because scattering out of the disk (perhaps by molecular clouds) is probably not effective during the short (typically 10^8 years) Cepheid lifetimes. At very large distances away from the galactic center, this approach should take into account the increasing thickness and warp of the galactic disk, but in practice these effects are usually negligible.

I have carried out a new analysis of the distribution of Cepheids using this approach. Working from a tape of the General Catalogue of Variable Stars, edited to include the Third Supplement (1976), I find 771 Cepheids, including 5 Cepheids with periods near one day classified as RR Lyraes. Removing Magellanic Cepheids, some non-Cepheids, and some without determined periods leaves 708 stars. Reddenings and distances are calculated from the P-C and P-L relations of Fernie & McGonegal (1983) for the 450 stars with measured B-V colours. (B-V is used rather than some other colour only because it is available for most stars.) Fernie (1968) had only 212 stars with measured colours, and the improved reliability of these reddenings constitutes the major improvement over Fernie's study. Reddenings are taken from Fernie & Hube (1968) for 80 more Cepheids, estimated from nearby field stars. Finally, reddenings are taken from Burstein & Heiles (1982) for 35 stars with $|b| > 10°$ from observed HI column densities. Therefore 553 Cepheids are included in the analysis.

The distribution of $|z|$ distances in this sample is peaked close to the Galactic plane. (Half are within 150 pc of the plane.) Before using the data to find the distribution of classical Cepheids, we must limit the sample to a region near the sun where the incompleteness caused by dust absorption is not too great. Using a cylinder of radius 2 kpc centred on the sun, we obtain a density in the central plane N_o = 69 kpc^{-3}, and a scale height h = 71 pc. These data are shown in Figure 4. Similar results are obtained for a cylinder of radius 3 kpc (N_o = 57 kpc^{-3}, h = 69 pc). These are least-squares fits to the stars with $|z| <$ 350 pc. Relatively few Type II Cepheids will be contaminating this restricted sample. The scale height found here for classical Cepheids agrees very well with the value (70 $\pm$ 10 pc) found by Fernie.

A total of 144 Cepheids are calculated to lie beyond $|z|$ = 600 pc, and these are all presumably Type II Cepheids. (In the GCVS, 14 are classified as Cδ, 81 as CW or CW?, and 49 as Cep or Cep?.) Their reddenings and distances have been recomputed using the P-C and P-L relations in Figure 1. After restricting the sample to a cylinder of radius 5 kpc centred on the sun, the distribution of $|z|$ distances is shown in Figure 5. The first bin is very incomplete, of course, because stars with small $|z|$ have gone into the classical Cepheid group. The distribution is probably not exponential (probably due to the mixture of old-disk and halo stars), but the number of stars is too small to draw a firm conclusion. The line drawn by eye in Figure 5 has $N_o = 1.5$ kpc^{-3}, h = 500 pc. This estimate of N_o should be considered a lower limit because of incompleteness for these faint stars; comparison with counts in smaller cylinders suggests that $N_o = 2.2$ kpc^{-3} is more realistic.

These distributions in $|z|$ for classical and Type II Cepheids suggest several conclusions: Classical Cepheids outnumber Type II Cepheids in the Galactic plane near the sun by about 30 to 1. Type II Cepheids dominate at $|z| > 300$ pc. When projected onto the Galactic plane, the densities in the solar neighborhood of classical and Type II Cepheids are 9.8 kpc^{-3} and 2.2 kpc^{-3}, respectively. We do not know if these relative numbers are also valid near the Galactic centre and toward the anticentre. The above estimate for N_o for Type II Cepheids derived statistically is consistent with the numbers of individual stars identified in the solar neighborhood (Harris 1981). The relative numbers of Type II Cepheids and RR Lyraes appear to differ in clusters, in the halo field, and in the solar neighborhood (Harris & Wallerstein 1984), probably due to real population differences.

Figure 4. The distribution in 50 pc bins of Cepheids within a cylinder of radius 2 kpc centred on the sun.

Figure 5. The distribution in 200 pc bins of Type II Cepheids within a cylinder of radius 5 kpc.

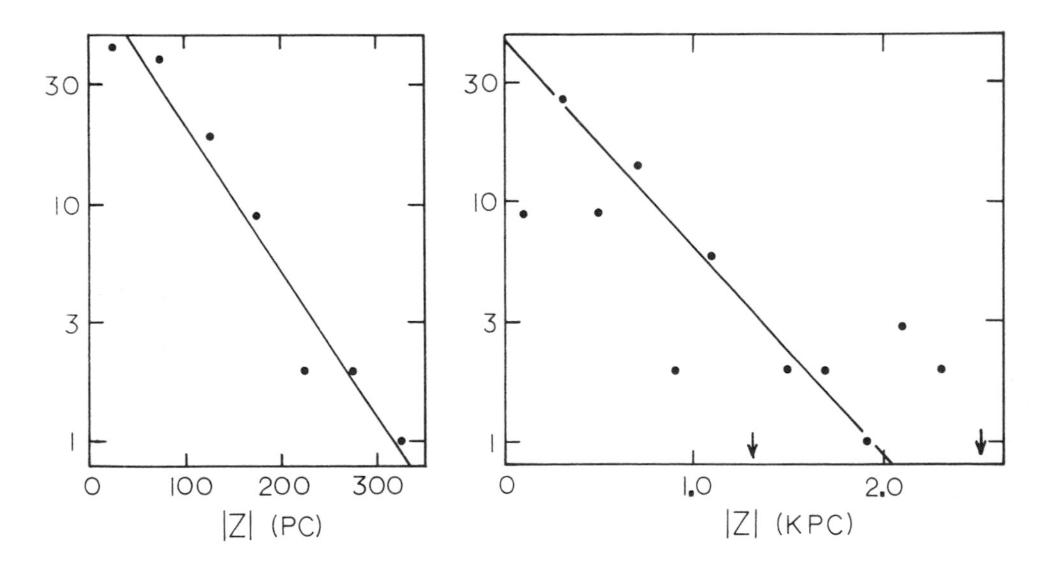

Properties of Field Type II Cepheids

The period distribution for the 144 field Type II Cepheids is shown in Figure 2(b). To minimize biases caused by selection effects, we should restrict the sample to a limited volume. The cross-hatched histogram in Figure 2(b) shows the bimodal distribution seen in cluster Cepheids in Figure 2(a), but the short-period peak is not as prominent in the field star sample as in clusters. This may be due to a selection effect (because short-period stars are fainter and discovered less completely), but I suspect that it shows a real difference. Probably the field population does not have a horizontal branch as extremely blue as the mean for clusters, and so contributes relatively fewer BL Her stars. Also the metal-rich field Cepheids (not present in clusters) may tend to fill in the 6- to 12-day range of periods. An earlier plot by Kraft (1972) included stars already classified as Type II, probably biasing his sample toward those periods where the light curve is a good discriminator of Cepheid type. Kraft's result that long-period Cepheids dominate in the nuclear bulge is supported here: over 80% of stars in this sample of Type II Cepheids that lie within 15° of the Galactic centre have periods longer than 10 days. Some of this excess must be due to greater completeness for brighter stars, but some of the excess is probably real.

Until recently, abundances of only five Type II Cepheids have been measured using high-resolution spectroscopy (Rodgers & Bell 1963; Barker *et al.* 1971; Anderson & Kraft 1971; Caldwell & Butler 1978; Cottrell 1979; Wallerstein *et al.* 1979), too few to draw conclusions about their origins. Recently, however, the abundances of 70 field Cepheids have been measured by Harris (1981) and Harris & Wallerstein (1984) using the Washington system, a broad-band system efficient for a survey project of this type. Similar studies are now in progress by McNamara, Kwee and Diethelm, and others using other observing systems (and these narrow-band or spectroscopic systems offer potential advantages over a broad-band system), but it is too early to tell if the Washington-system results will be confirmed. The abundance histogram for all observed field stars is shown in Figure 3(b). It differs markedly from the histogram for cluster Cepheids in Figure 3(a), and it demonstrates that the solar abundance of BL Her is not unusual.

The radial velocities of field Type II Cepheids have been analyzed by Harris & Wallerstein (1984) improving earlier studies based on small samples (Woolley 1966; Plaut 1965). Their kinematic properties correlate with their abundances in the sense that metal poor stars have a low velocity of rotation around the Galaxy and a high velocity dispersion, while metal-rich stars have mostly normal, circular motions. The kinematics of the metal-rich Type II Cepheids are consistant with their origin in the old-disk, confirming this result already suggested from their spatial distribution and their abundances.

Radial velocity curves have been used to determine radii for seven field Type II Cepheids using the Baade-Wesselink method: BL Her (Abt & Hardie 1960), κ Pav (Rodgers & Bell 1963), W Vir and TW Cap (Bohm Vitense 1974), XX Vir (Wallerstein & Brugel 1979), AU Peg (Harris *et al.* 1984), and SW Tau (Burki 1984). The distances and luminosities derived for these stars are consistent with the P-L relation for cluster Cepheids in Figure 1. However, these results are not very precise for several stars. Better quality velocity curves should be obtained for more stars, particularly the metal-rich Type II Cepheids, to check their luminosities.

FURTHER INVESTIGATIONS

Cepheids should change their pulsation periods by about 40% as a result of changing their radii as they evolve through the instability strip. The changes for Type II Cepheids are predicted by evolutionary models to occur in about 10^3 to 10^5 years, depending on the model. These changes are sufficiently rapid to be detectable with presently available observations covering about 80 years, allowing a check on the applicability and correctness of the models. Unfortunately, W Vir stars have quite unstable periods, often exhibiting apparently random changes larger than are seen in classical Cepheids (Coutts 1973; Szabados 1983). The cause of these period fluctuations is not known, but the fluctuations are large enough that so far they appear to have masked the systematic changes that might be present due to evolution.

BL Her stars show less extreme random period changes, however, and the studies by Wehlau & Bohlender (1982) and Wehlau & Sawyer Hogg (1984) have detected changes that are probably caused by evolution. Among 13 Cepheids with periods from 1 to 5 days in six globular clusters, ten had increasing periods and three had constant periods, but none showed decreasing periods, suggesting that most of these stars are evolving through the instability strip toward cooler temperatures. The rates of period change observed were from 0.4 to 18 days/10^6 years. Changes of this size are predicted by the evolutionary tracks of Gingold (1976) and Sweigart and Gross (quoted by Wehlau & Bohlender), supporting the proposal that the changes are indeed caused by evolution. If so, the six stars with fast changes (of the order of 10 days/10^6 years) are probably in their first crossing of the instability strip, while the seven stars with slow changes (less than 1 day/10^6 years) are probably in their third crossing. By combining observed period changes with evolutionary models, we can map the part of the horizontal branch that pumps the BL Her-star domain.

Pulsation models for BL Her stars constructed for a variety of masses, periods, and luminosities have successfully reproduced many of the features observed in the light and velocity curves (Carson & Stothers 1982; Hodson *et al.* 1982; Cox & Kidman 1984). For example, the bump around minimum light that distinguishes 1.2- to 2.4-day BL Her stars from classical Cepheids is seen in the model light curves. The most important result of the models is that the masses of all the BL Her stars may be close to 0.6 $M_\odot$. There is some evidence that metal-rich BL Her stars are less luminous than metal-poor ones (by perhaps 0.3 magnitudes), but have nearly the same mass. Considerable structure appears in the light curves of some stars when accurate observations are obtained with good phase coverage, so accurate light curves for a large sample of stars combined with the models is likely to provide further constraints on M, L, Y and Z. The same remarks apply to W Vir stars for which recent models also indicate masses close to 0.6 $M_{\odot}$ (Bridger 1983, 1984).

The study of Type II Cepheids has revealed some with unusual properties. For example, CC Lyr and ST Pup may be unusually hot for their periods, while AU Peg is unusually cool. Several are carbon stars (Lloyd Evans 1983b) and show erratic behavior, the most notable being RU Cam. Detailed analyses of temperature and abundances have been carried out for only a few stars and are needed for more. Three Type II Cepheids, AU Peg, TX Del, and IX Cas, all metal-rich Cepheids with periods in the 6-12 day gap, are now known to have close binary companions that have likely influenced their evolution by mass transfer (Harris *et al*. 1984; Harris, in preparation). This may provide an explanation of how some of these stars have evolved into the instability strip, since evolutionary tracks do not predict such Cepheids. However, other similar Cepheids have not shown, with presently available data, velocity variations indicating a binary companion, so it is not yet clear whether some other explanation is needed for their existence.

One explanation for the presence of metal-rich old-disk stars in the instability strip is large mass loss during hydrogen-shell burning evolution. This produces unusually blue stars on the horizontal branch, and might give rise to both BL Her and W Vir stars. Large mass loss can also explain the presence of metal-rich RR Lyrae stars in the solar neighborhood (Taam *et al*. 1976). However, the cause of the mass loss in old-disk field stars and its absence in metal-rich globular cluster stars remains to be explained.

Outside our Galaxy, the Magellanic Clouds are the only systems with much data on Type II Cepheids available. Twenty are known in the LMC: HV2351, 5598, 5690, 13063, 13064, 13065 (Hodge & Wright 1969; Wright & Hodge 1971; Connolly 1975; Butler 1978), and fourteen other stars (Payne-Gaposchkin 1971). Four are known in the SMC: HV206, 1828, and 12901 (Payne-Gaposchkin & Gaposchkin 1966) and probably Star 48 (Wesselink & Shuttleworth 1965). The magnitudes to which the variable star searches at Harvard are considered to be complete suggest that these lists are largely complete outside of crowded regions for Type II Cepheids with $P > 15$ days, but that virtually all of the BL Her stars in the Magellanic Clouds remain undiscovered. Only a few small areas have been searched for RR Lyrae stars at fainter magnitudes. In some cases variables have been found which might include Type II Cepheids, but they have not been investigated further. The stars with $P = 1$ to 3 days found in the SMC by Wesselink and Shuttleworth (1965) probably represent the short-period, low-mass end of the sequence of classical Cepheids. Cepheids with $M \sim 2\ M_{\odot}$ and $P \sim 1$ day are predicted to occur in large numbers in a metal-poor population (Becker *et al*. 1977). Stars with $P <$ 1 day but brighter than the RR Lyraes in the SMC (Wesselink & Shuttleworth 1965; Graham 1975; Butler *et al*. 1982) and the LMC (Connolly 1984) may include classical Cepheids, Anomalous Cepheids, normal Type II Cepheids, and foreground RR Lyraes. The status of most of these short-period variables is not yet clear.

Anomalous Cepheids in the dwarf spheroidal galaxies (anomalously bright BL Her stars) have received a great deal of attention during the last decade (Wallerstein & Cox 1984, and references therein). The only Anomalous Cepheid known in our Galaxy, V19 in NGC 5466 (Zinn & Dahn 1976), stands out in Figure 1, but we should remember that others may exist among field Type II Cepheids and RR Lyraes. Anomalous Cepheids are more massive than either the majority of the stars in the dwarf spheriodals or the normal BL Her stars found in Galactic globular clusters. Possible explanations for their presence are that they are younger than the RR Lyraes in these galaxies, that they have gained mass by exchange from a binary companion, or that they result from a coalescence of two stars in a binary system. The apparent lack of normal Type II Cepheids in these galaxies probably can be understood as a small numbers effect: most of these galaxies have red horizontal branches and so should produce almost no Type II Cepheids, while Sculptor and Ursa Minor (with blue horizontal branches) have not yet had a sufficient number of variables investigated. Many variables remain to be investigated in several dwarf spheroidals (van Agt 1973), and Demers is studying Fornax variables. I hope that all types of variables will be pursued to allow comparison of their relative numbers and properties.

I wish to thank George Wallerstein for developing my interest in Type II Cepheids and for numerous conversations on their various aspects. Much of this research has been supported by the Natural Sciences and Engineering Research Council of Canada at the Dominion Astrophysical Observatory and at McMaster through a grant to William Harris, to whom I am most grateful.

REFERENCES

Abt, H.A. & Hardie, R.H. (1960). Astrophys. J. 131, 155.
Anderson, K.S. & Kraft, R.P. (1971). Astrophys. J. 167, 119.
Arp, H. (1955). Astron. J. 60, 1.
Barker, T., Baumgart, L.D., Butler, D., Cudworth, K.M., Kemper, E., Kraft, R.P., Lorre, J., Rao, N.K., Reagan, G.H. & Soderblom, D.R. (1971). Astrophys. J. 165, 67.
Becker, S.A., Iben, I. & Tuggle, R.S. (1977). Astrophys. J. 218, 633.
Bohm-Vitense, E. (1974). Astrophys. J. 188, 571.
Bridger, A. (1983). Ph.D. thesis, University of St. Andrews.
Bridger, A. (1984). This conference.
Burki, G. (1984). This conference.
Burstein, D. & Heiles, C. (1982). Astron. J. 87, 1165.
Butler, C.J. (1978). Astron. Astrophys. Suppl. 32, 83.
Butler, D., Demarque, P. & Smith, H.A. (1982). Astrophys. J. 257, 592.
Caldwell, C.N. & Butler, D. (1978). Astron. J. 83, 1190.
Carson, R. & Stothers, R. (1982). Astrophys. J. 259, 740.
Clement, C., Sawyer Hogg, H. & Lake, K. (1984b). This conference. p.260.
Clement, C., Sawyer Hogg, H. & Wells, T. (1984a). This conference. p.264.

Connolly, L.P. (1975). Ph.D. thesis, University of Arizona.
Connolly, L.P. (1984). Preprint.
Cottrell, P.L. (1979). Mon. Not. R. Astron. Soc. 189, 13.
Coutts, C.M. (1973). In Variable Stars in Globular Clusters and in Related Systems, ed. J.D. Fernie, p. 145. Reidel: Dordrecht.
Cox, A.N. & Kidman, R.B. (1984). This conference. p.250.
Demers, S. & Harris, W.E. (1974). Astron. J. 79, 627.
Fernie, J.D. (1968). Astron. J. 73, 995.
Fernie, J.D. & Hube, J.O. (1968). Astron. J. 73, 492.
Fernie, J.D. & McGonegal, R. (1983). Astrophys. J. 275, 732.
Gingold, R.A. (1976). Astrophys. J. 204, 116.
Graham, J.A. (1975). Publ. Astron. Soc. Pac. 87, 641.
Harris, H.C. (1981). Astron. J. 86, 719.
Harris, H.C., Nemec, J.M. & Hesser, J.E. (1983). Publ. Astron. Soc. Pac. 95, 256.
Harris, H.C., Olszewski, E.W. & Wallerstein, G. (1984). Astron J. 89, 119.
Harris, H.C. & Wallerstein, G. (1984). Astron J. 89, 379.
Harris, W.E. & Racine, R. (1979). Ann. Rev. Astron. Astrophys. 17, 241.
Hodge, P.W. & Wright, F.W. (1969). Astrophys. J. Suppl. 17, 467.
Hodson, S.W., Cox, A.N. & King, D.S. (1982). Astrophys. J. 253, 260.
Hoffleit, D. (1976). Inf. Bull. Var. Stars No. 1131.
Joy, A.H. (1949). Astrophys. J. 110, 105.
Kraft, R.P. (1972). In The Evolution of Pop II Stars, ed. A.G.D. Philip, p. 69. Dudley Obs.: Albany.
Kukarkin, B.V. (1975). In Variable Stars and Stellar Evolution, ed. V.E. Sherwood & L. Plaut, p. 511. Reidel: Dordrecht.
Kwee, K.K. (1967). Bull. Astron. Inst. Neth. 19, 260.
Lloyd Evans, T. (1977). Mon. Not. R. Astron. Soc. 178, 353.
Lloyd Evans, T. (1983a). Mon. Not. R. Astron. Soc. 204, 945.
Lloyd Evans, T. (1983b). Observatory 103, 276.
Madore, B.F. (1977). Mon. Not. R. Astron. Soc. 178, 505.
Mianes, P.M. (1963). Ann. d'Astrophys. 26, 1.
Newell, E.B. (1973). Astrophys. J. Suppl. 26, 37.
Nikolov, N. & Kunchev, P. (1969). Astrophys. Sp. Sci. 3, 46.
Payne-Gaposchkin, C. (1956). Vistas Astron. 2, 1142.
Payne-Gaposchkin, C. (1971). Smith. Cont. Ap. 13, 1.
Payne-Gaposchkin, C. & Gaposchkin, S. (1966). Smith. Cont. Ap. 9, 1.
Petit, M. (1960). Ann. d'Astrophys. 23, 681.
Pilachowski, C.A. (1984). Astrophys. J. 281, in press.
Plaut, L. (1965). In Galactic Structure, ed. A. Blaauw & M. Schmidt, p. 267. Univ.Chicago: Chicago.
Rodgers, A.W. & Bell, R.A. (1963). Mon. Not. R. Astron. Soc. 125, 487.
Rosino, L. (1978). Vistas Astron. 22, 39.
Sawyer Hogg, H. (1973). Publ. David Dunlap Obs. 3, No. 6.
Schwarzschild, M. & Harm, R. (1970). Astrophys. J. 160, 341.
Smith, H.A., Jacques, J., Lugger, P.M., Deming, D. & Butler, D. (1978). Publ. Astron. Soc. Pac. 90, 422.
Stobie, R.S. (1973). Observatory 93, 111.
Strom, S.E., Strom, K.M., Rood, R.T. & Iben, I. (1970). Astr. Ap. 8, 243.
Szabados, L. (1977). Mitt. Stern. Ungar. Akad. Wissen. No. 70.

Szabados, L. (1983). Astrophys. Sp. Sci. 96, 185.
Taam, R.E., Kraft, R.P. & Suntzeff, N. (1976). Astrophys. J. 207, 201.
van Agt, S. (1973). In Variable Stars in Globular Clusters and in Related Systems, ed. J.D. Fernie, p. 35. Reidel: Dordrecht.
Wallerstein, G. (1958). Astrophys. J. 127, 583.
Wallerstein, G. (1970). Astrophys. J. Lett. 160, L345.
Wallerstein, G., Brown, J.A. & Bates, B.A. (1979). Publ. Astron. Soc. Pac. 91, 47.
Wallerstein, G. & Brugel, E.W. (1979). Astron. J. 84, 1840.
Wallerstein, G. & Cox, A.N. (1984). Publ. Astron. Soc. Pac. in press.
Walraven, Th., Muller, A.B. & Oosterhoff, P.Th. (1958). Bull. Astron. Inst. Neth. 14, 81.
Warren, P.R. & Harvey, G.M. (1976). Mon. Not. R. Astron. Soc. 175, 129.
Wehlau, A. & Bohlender, D. (1982). Astron. J. 87, 780.
Wehlau, A. & Sawyer Hogg, H. (1977). Astron. J. 82, 137.
Wehlau, A. & Sawyer Hogg, H. (1984). This conference.
Wesselink, A.J. & Shuttleworth, M. (1965). Mon. Not. R. Astron. Soc. 130, 443.
Woolley, R. (1966). Observatory 86, 76.
Wright, F.W. & Hodge, P.W. (1971). Astron. J. 76, 1003.
Zinn, R. (1974). Astrophys. J. 193, 593.
Zinn, R.J. & Dahn, C.C. (1976). Astron. J. 81, 527.

THEORETICAL MODELS OF W VIRGINIS VARIABLES

Alan Bridger
Royal Greenwich Observatory, Herstmonceux Castle, Hailsham, East Sussex, BN27 1RP, England.

INTRODUCTION

W Virginis variables are the population II counterparts of the classical cepheids, although they do not show quite the same trends as are seen in the latter. Theoretical studies of the population II cepheids have not been very extensive until recent studies of the shorter period variables (BL Herculis variables, with periods between 1 and 10 days). The variables with periods above 10 days (up to about 50 days) have only been studied by a few authors, modelling the prototype star (W Vir) (.g. Christy 1966; Davis 1974). Although these models qualitatively reproduced the observations they were not very successful, and were based on a stellar mass (0.88 $M_{\odot}$) that now seems likely to be too high.

This paper summarises the results of a series of hydrodynamic (non-linear) models constructed in an attempt to reproduce the observed characteristics of these stars. The full results and analysis are given elsewhere (Bridger 1984). Conventional non-linear methods are used for the modelling, including attempts to improve on previously used outer dynamic boundary conditions. Convection is ignored. A realistic equation of state is used, with opacities calculated by Carson (1976) for a composition of Y = 0.25, Z = 0.005.

OBSERVATIONAL DATA

Observations of W Virginis variables are unfortunately rather scarce, This is especially the case for velocity observations. Very few observed stars have well defined velocity curves, only W Virginis itself being well observed. The major thing that the velocity observations do tell us is that the stars with periods above 13 days have highly asymmetric velocity curves frequently accompanied by hydrogen emission lines at rise to maximum light. This is probably caused by strong outward moving shock waves.

Kwee (1967) gives some good light curves of some field W Vir stars, and splits the curves into two classes which he calls "crested" (C-type) and "flat-topped" (F-type). The stars with periods less than 13 days usually fit neither class, having rather featureless light curves (X-type).

A study has been made of the best available observations, producing new estimates of luminosity ($\log(L/L_\odot)$) and effective temperature ($\log T_e$) for the stars. The study shows that the C- and F-type variables (hereinafter the variables are typed according to their light curve) occupy different positions in the HR diagram. Following Kwee two almost parallel Period-Radius (P-R) relations can be derived for the two classes. Figure 1 is an HR diagram showing the positions of the variables, observed blue and red edges (Demers & Harris 1974), a theoretical blue edge based on the Carson opacties (Worrell, private communication) and a red edge estimated from static models.

This division may turn out to be due to different evolutionary tracks. The mechanism for feeding stars into the population II instability strip at this point seems to be "blue-looping" from the Asymptotic Giant Branch (AGB) caused by thermal instabilities in the helium burning shell (e.g. Mengel 1973). Making simple assumptions (that metallicity varies little or that it has little effect on the evolution), then it may be that a slightly higher mass star moving up the AGB executes this blue-loop at a later stage, causing the star to have a higher luminosity when passing through the instability strip, producing a C-type variable. Thus we see that the C-type variables may be slightly more massive than the F-type. Assuming the two classes follow parallel P-R relations then the mass ratio may be estimated, giving $M_C \sim 1.3 M_F$. This fits nicely into the range of possible masses given by Bohm-Vitense et al (1974) of $0.5 < M/M_\odot < 0.75$.

SURVEY RESULTS

The survey consisted of a series of 25 models constructed to cover the Instability Strip, aiming at periods in the range 10 - 20 days. A mass of 0.6 $M_\odot$ was used, with a few additional models constructed with masses of 0.5 $M_\odot$ and 0.8 $M_\odot$. The survey models were then looked at in the same way as the observational data.

Figure 1. HR diagram

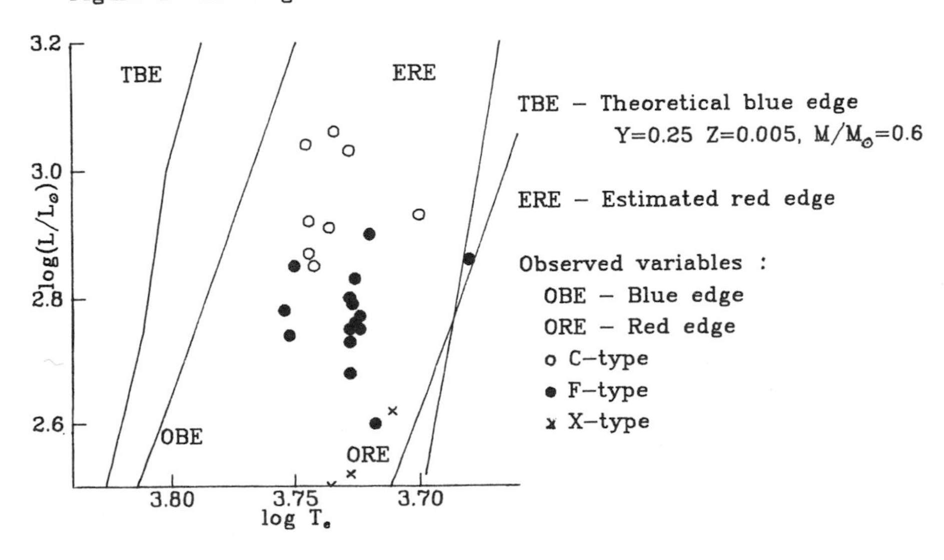

The modelled light curves also show the split into C- and F-type curves, at the approximate expected stellar parameters, with the derived P-R relation agreeing well with the observed relations, to within estimated observational errors. The general agreement with the observed curves is good, although it did prove more difficult to reproduce the F-type curves (figure 2). Using a mass of 0.5 $M_\odot$ seemed to make little difference to the general pulsational characteristics (except, of course, to the period), but using 0.8 $M_\odot$ produced very different curves. It seemed likely that that the observed light curves could be reproduced using this mass and reasonable values of the stellar parameters. Thus 0.8 $M_\odot$ seems to be too high to be the mass of most of these variables. This is quite reasonable as it is at the upper limit of the possible evolutionary masses in any case. The amplitudes of the velocity curves are quite reasonable, and the curves for many models show the high asymmetry and strong shocks that presumably accompany the hydrogen emssion lines seen in some stars.

A few models showed very slight "RV Tauri behaviour" in their light and/or velocity curves. This is the alternation of larger and smaller amplitudes, and possibly also longer and shorter periods, causing a doubling of the repetition period. In particular one model showed this behaviour to a large extent (figure 3), and also was capable of switching to a more normal variation and back again.

Further models were calculated using Los Alamos opacities. Two of these used the same composition as the survey, and the same parameters as two of the survey models. The light curves computed were not as close to those observed as the survey models were, being significantly different in shape and the position of the secondary bump. A model using Y = 0.299 (the King Ia mix, Cox & Tabor 1976) showed a light curve

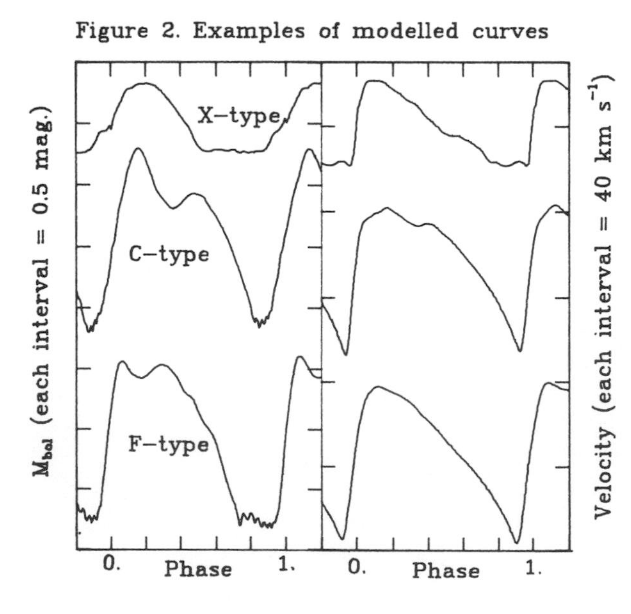

Figure 2. Examples of modelled curves

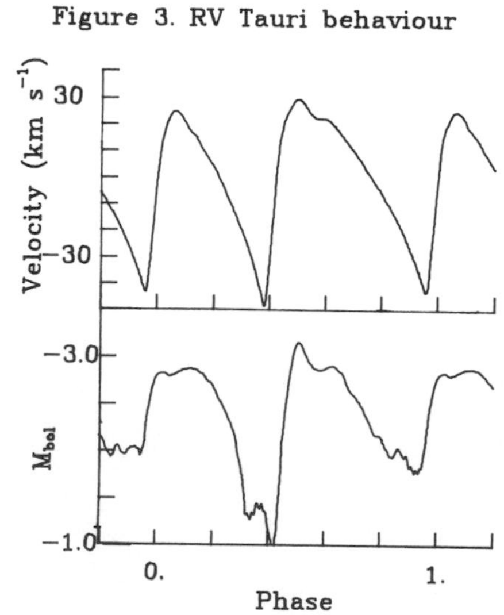

Figure 3. RV Tauri behaviour

completely unlike those observed, suggesting that 30% Helium is too high for these stars.

Finally, three of the survey models proved to be very good models of individual stars, even without attempting to tune the parameters to improve the fit. One of these reproduces the light curve and general properties of the variable CS Cas very well. Table 1 shows the parameters of model and star and figure 4 compares their light curves.

CONCLUSION

This study shows that the W Virginis variables can be modelled using suitable input physics and stellar parameters derived from evolution or observation. Indeed the modelling can be very good. However it also shows that convection is probably very important for the cooler stars.

REFERENCES

Bohm-Vitense, E., Szkody, P., Wallerstein, G., and Iben, I., (1974). Astrophys. J., 194, 25.
Bridger, A. (1984). Ph. D. Thesis, University of St. Andrews.
Carson, T.R. (1976). Ann. Rev. Astron. & Astrophys., 14, 95.
Christy, R.F., (1966). Astrophys. J., 145, 337.
Cox, A.N., & Tabor, J.E., (1976). Astrophys. J. Suppl., 31, 271.
Davis, C.G., (1974). Astrophys. J., 187, 175.
Demers, S. & Harris, W.E., (1974). Astron. J., 79, 627.
Kwee, K.K., (1967). Bull. Astron. Netherlands, 19, 260.
Mengel, J.G., (1973). In Variable Stars in Globular Clusters and related systems, ed. J.D. Fernie. IAU Coll. no. 21. Reidel.

Table 1. Comparison of CS Cas and model star.

ϕ_B Phase of secondary bump
$\phi_B-\phi_M$ Phase of bump from maximum

Parameter	Model	CS Cas
$\log(L/L_\odot)$	2.9	2.91
$\log T_e$	3.75	3.74
Period (days)	14.6	14.7
Amplitude (mag.)	1.8	1.44
Asymmetry in light curve	3.2	2.0
ϕ_B	0.40	0.48
$\phi_B-\phi_M$	0.27	0.25

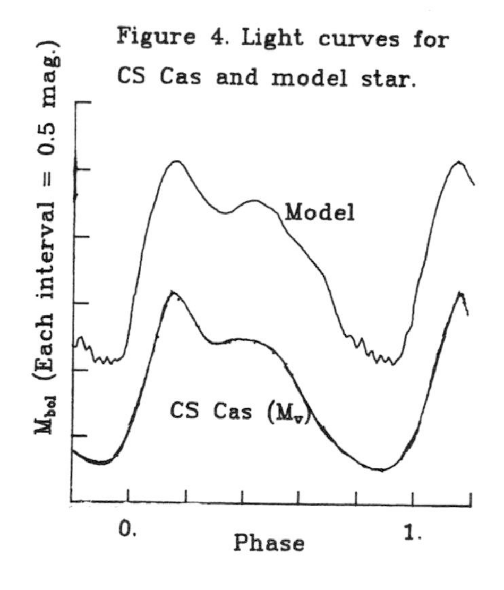

Figure 4. Light curves for CS Cas and model star.

A NEW POSSIBLE RESONANCE FOR POPULATION II CEPHEIDS

A. N. COX
Theoretical Division, Los Alamos National Laboratory,
University of California

R. B. KIDMAN
Theoretical Division, Los Alamos National Laboratory,
University of California

Light and velocity curves of some radial mode variable stars seem to indicate a resonance where the second overtone has a period exactly half that of the fundamental mode. The two classes of stars that show this resonance by bumps in their light curves are the classical Cepheids and the population II BL Her variables. We here propose that there is another resonance for the population II W Vir variables where the ratio of the first overtone to the fundamental periods is 0.5.

Observations by Kwee (1967) show that these population II Cepheids, have frequently a crested or a flattop shape of their light curve. Some light curves seem to be featureless, but that may be only the result of too few observations resulting in too few phase points.

Crested light curves such as for W Vir show a bump or standstill on descending light rather like the classical Cepheids with periods between 5 and 10 days. The flattop type of light curves sometimes have bumps on rising light as for Cepheids with periods between 10 and 20 days. The problem with these variable stars is that they do not have a correlation of these light curve shapes with period as long-ago demonstrated by Hertzsprung (1926) for the classical Cepheids. The two types of light curves both exist all the way from about 10 to 20 day periods.

BL Her variables that show bumps in their light and velocity curves have some correlation of the light curve shape with period, but it is not a perfect one. King, Cox and Hodson (1981) have shown that on the Hertzsprung-Russell diagram, lines of constant period (slightly steeper than lines of constant radius) have smaller slopes than the slightly curved lines of constant period ratio P_2/P_0. This has also been shown again by Hodson, Cox and King (1982) using the Carson opacities. In this latter case, the period ratios are a bit smaller, but that is a direct result of the larger Carson opacities that deconcentrate the outer envelope structure in a way that P_0 is increased more than P_2. As is well known, this smaller period ratio for the BL Her variables does not affect the bump mass determination in any significant way, but the smaller period ratio using the Carson opacities for the classical Cepheids can compensate for otherwise smaller bump masses (Vemury and Stothers, 1978).

The easy visibility of the Hertzsprung sequence for the classical Cepheids is because the lines in the Hertzsprung-Russell diagram of constant period and those of constant period ratio P_2/P_0 are parallel, resulting in an almost perfect correlation between P_0 and P_2/P_0. Both red and blue Cepheids with a given P_0 show a bump at the same phase.

The difference in slope between the lines of constant period and constant period ratio P_2/P_0 for the BL Her variables means that at a given period, a range of period ratios is possible. Then the phase of the bumps is not fixed, but it is confined to a small range. This explains why BL Her itself has a bump on falling light with a period ratio inferred to be 0.52 or 0.53, whereas XX Vir at the same 1.3 day period has no bump, implying a period ratio of greater than 0.53 which can occur if XX Vir is bluer, fainter, or more massive than BL Her.

Simon and Schmidt (1976) showed that this correlation of bump phase with period can be seen in the results of the nonlinear hydrodynamic integrations that have been done by Stobie (1969) and many others. A unique relation was found that made it possible to predict the bump phase as seen in the theoretical light and velocity curves by knowing the period ratio P_2/P_0, if this ratio was between 0.46 and 0.53. From the recent work of Buchler (1984), we see that the bump is actually the result of a resonance between these two modes.

Standard models with masses and luminosities that are consistent with evolution theory give linear theory period ratios too large for a given period. For example, for a 7 $M_\odot$ Cepheid and the evolution theory luminosity of just over 3700 solar luminosities, on the blue side of the instability strip this star would have a period of 6 days while on the red side of the strip its period would be 12 days. In these cases the predicted period ratio would range from about 0.57 to 0.53, but the appearance of observed light curve bumps indicate period ratios ranging from about 0.52 to 0.49. The original way to reconcile this discrepancy was to lower the mass to about 4 $M_\odot$'s at this luminosity. Then the period ratios from linear theory and the bumps seen in nonlinear hydrodynamic integration light and velocity curves fit the data. Actually this has recently been considered more of a period ratio anomaly than a mass anomaly, and ways to deconcentrate the Cepheid envelope and reduce the period ratio have been sought.

Of the three currently viable ways of deconcentrating the outer 0.01 to 0.1 per cent of the stellar mass, perhaps only one is now possible. The large magnetic fields postulated by Stothers (1979) can strongly affect the stellar structure, but their existence to give a pressure as large as the gas pressure has not been (and maybe cannot be) measured. If the required fields are generated by a convective dynamo, they are many powers of ten too large. A primordial field may be strong enough, but then all the Cepheids in this mass range of 5 to 9 $M_\odot$ need to have had a rather uniform early formation history.

The excellent suggestion by Simon (1982) that the envelope can be deconcentrated with a higher metal opacity must be rejected because such an effect does not seem possible from the viewpoint of atomic physics and the compositions that much exist in Cepheid envelopes. Actually some of this very effect has been seen earlier because the higher Carson opacities at temperature above 5×10^5 K did slightly deconcentrate the stellar models, especially for the population I case where there is a significant CNONe abundance. The discovery that the Carson opacities are inaccurate in this high temperature region by Carson, Huebner, Magee, and Merts (1984) has removed the earlier mass anomaly improvements, and the parallel result by Magee, Merts, and Huebner (1984) that no other opacity source is available, seem to eliminate the enhanced opacity explanation for the period ratio anomaly.

We therefore want to reemphasize that an enhanced helium abundance in the Cepheid envelope can deconcentrate the density structure and reconcile both the bump and beat Cepheid mass anomalies. Cox, Michaud, and Hodson (1978) have shown that a Cepheid wind, if it is anything like the solar wind, can blow away more hydrogen than helium, leaving a very thin homogeneous convective zone that is enhanced to a mass fraction as large as 0.65 or even possibly 0.75.

This helium enhancement takes time. A 7 $M_\odot$ star evolving on its blue loops from the red giant region with a very deep convective envelope has only 1-10 million years before it becomes a second crossing Cepheid. At 8 $M_\odot$ this time is reduced so much that very little wind enhancement is possible. Therefore the very blue 8 $M_\odot$ Cepheid with a period of 10 or more days should not be deconcentrated, and it should not show any period ratio anomaly for periods up to 20 days as it evolves back to the red edge of the instability strip.

Recent observations and interpretations by Davis, Moffett, and Barnes (1981) show that X Cyg at 16.4 days shows little or no period ratio (bump phase) anomaly. Thus enhanced helium is not needed or predicted for the masses above about 8 $M_\odot$. Unlike the tangled magnetic field or the increased opacity proposals, which do not have a reversion to standard models at greater than 8 $M_\odot$ or a period of about 15 days, the enhanced helium model slowly merges with non-enhanced models at this mass and period. Bumps on rising light are seen for periods as long as 26 days for the Cepheid X Pup as observed by Pel (1976).

Returning to the population II Cepheids, we see in Figure 1 that lines of constant P_1/P_0 near 0.5 are in the very region where the crested and flattop type light curves occur. In this figure we show only globular cluster variables where their luminosities are rather well determined. The symbols b and B stand for, respectively, BL Her variables in ω Cen and in all the other globular clusters. The f and F symbols refer to again, respectively, the ω Cen and other globular cluster variables with flattop light curves. The c and C symbols refer to the crested light curves and the r and R symbols refer to two RV Tau variables.

Constant period lines for the BL Her variables have been computed for a mass of 0.65 $M_\odot$ with the King Ia table with a Z composition of 0.001. The Stellingwerf (1975ab) fit for the opacity and equation of state has been used for the 0.60 $M_\odot$ W Vir lines at higher luminosities. The location of the blue edge and the curvature of the constant period lines may be changed when the actual tabular data are used. It has been verified that both the fit and the tablular material properties give constant P_1/P_0 lines that go right up the instability strip at least approximately as drawn. Thus the flattop curves, with often bumps on rising light and maybe a period ratio of less than 0.5 should be always redder than the crested light curve cases as observed.

Figure 1. On the Hertzsprung-Russell diagram lines of constant period and lines of constant period ratio P_1/P_0 are drawn for 0.65 $M_\odot$ for the BL Her luminosities and for 0.60 $M_\odot$ for the W Vir variables. Individual stars in globular clusters and their periods are also plotted.

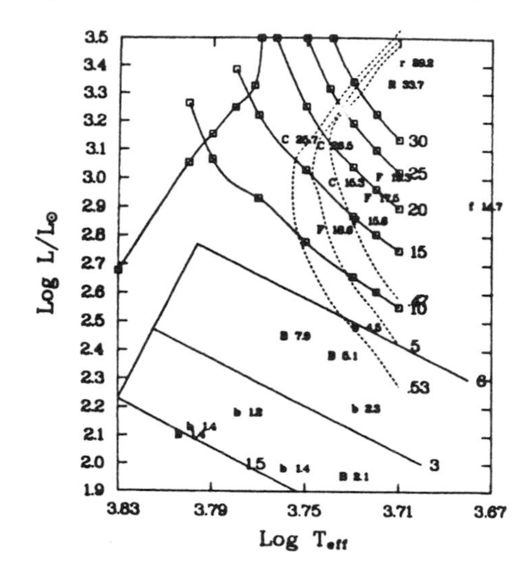

REFERENCES

Buchler, J.R. (1984). Ap. J., in press.
Carson, T.R., Huebner, W.F., Magee, N.H., & Merts, A.L. (1984). Ap. J., (Aug 15).
Cox, A.N., Michaud, G., & Hodson, S.W. (1978). Ap. J., 222, 621.
Davis, C.G., Moffett, T.J., & Barnes, T.G. (1981). Ap. J., 246, 914.
Hertzsprung, E. (1926). Bull. Astr. Int. Netherlands, 3, 115.
Hodson, S.W., Cox, A.N., & King, D.S. (1982). Ap. J., 253, 260.
King, D.S., Cox, A.N., & Hodson, S.W. (1981). Ap. J., 244, 242.
Kwee, K.K. (1967). Bull. Astr. Int. Netherlands, 19, 260.
Magee, N.H., Merts, A.L. & Huebner, W.F. (1984). Ap. J., (Aug 15).
Pel, J.W. (1976). Astr. Ap. Suppl., 24, 413.
Simon, N.R. & Schmidt, E.G. (1976). Ap. J., 205, 162.
Simon, N.R. (1982). Ap. J., Lett 260, L87.
Stellingwerf, R.F. (1975a). Ap. J., 195, 441.
Stellingwerf, R.F. (1975b). Ap. J., 199, 705.
Stobie, R.S. (1969). M.N.R.A.S., 144, 485.
Stothers, R. (1979). Ap. J., 234, 257.
Vemury, S.K. & Stothers, R. (1978). Ap. J., 225, 939.

HYDRODYNAMIC MODELS OF POPULATION II CEPHEIDS

Yu.A.Fadeyev, A.B.Fokin
Astronomical Council of the Academy of Sciences of the USSR
48 Pjatnitskaya St. Moscow 10917, USSR

This is a short report on hydrodynamic modelling for BL Her and W Vir pulsating variables. We present only a qualitative summary of the results obtained, whereas detailed paper will appear soon.

We studied radial self-exciting stellar oscillations with convection ignored. The Massevitch I mixture (X=0.7, Z=0.004) was assumed for calculating the opacity coefficient. All models have a mass $M=0.6M_\odot$, but their absolute magnitudes are in the range from -0.5 to -3 mag. The study covers the period range from 1.3 to 20 days. The models under investigation revealed the following properties:

Increase of the luminosity is accompanied with increasing nonadiabaticity of stellar pulsation. This leads to both increase of the growth rate while pulsations are exciting and increase of the oscillation amplitude of the limit cycle. The relative oscillatory moment of inertia of the radiative damping region was found to decrease with increasing luminosity. As a result, in the models with $L \gtrsim 800L_\odot$ amplitude growth ceases due to strong dissipation of the mechanical energy by shocks in the stellar atmosphere. The relative amplitude of oscillations in photospheric layers of such stars is $\Delta R/R \gtrsim 0.5$.

Within the period range from 1.3 to 3 days the theoretical light curves have a bump the phase of which depends on the pulsation period. Detailed analysis of the non-linear models confirmed a dualistic pulse-reasonance approach proposed by Whitney (1983) to interpretation of the Hertzsprung progression. The main conclusions are as follows: at maximum compression two pulses are generated at the antinode of the second overtone. One of the pulses moves inward and exactly after a half of the period reflects off the stellar core. The pulse arrives at the antinode exactly one period after its generation. The other pulse propagating outwards arrives at the photosphere δt after its generation. It is the dependence of the time delay δt on the pulsation period that is responsible for the existence of the Hertzsprung progression. The reflection coefficient of the photosphere for the outward pulse does not exceed 0.1, so that the pulse goes on to propagate through the stellar atmosphere as a shock. The pulse reflected off the photosphere is very weak but arrives at the antinode together with that reflected off the stellar core. The time delays estimated for the pulses obey to pulse-reasonance condition (Aikawa & Whitney 1983). Thus, the mechanism of propagating pulses and that of modal resonance are complementary.

For long-period models we tried to find a connection between W Vir and RV Tau pulsating variables. Estimates of adiabatic periods showed that the resonance condition $P_1/P_0=1/2$ begins to be fulfilled when the period of the fundamental mode $P_0=10$ days. Non-linear analysis of the model with the pulsation period 10 days revealed alternation of deep and shallow minima on the radial temporal dependences of the layers lying near the antinode of the first overtone. However, the amplitude of first overtone at the photosphere is not perceptible and the photospheric layers do not show the alternation. The depth of the shallow minimum increases with increasing pulsation period and the model having P=20 days reveals slight alternation on the temporal dependence of the photospheric radius. Thus, our non-linear calculations confirm the assumption proposed by Takeuti & Petersen (1983) about the resonance between the fundamental mode and first overtone in RV Tau pulsating variables.

The Population II Cepheids are known to have a degenerate carbon-oxygen core surrounded by a double shell source. Fadeyev (1982) proposed the period-luminosity relation for stars with degenerate cores and showed that this relation predicts the observer period increase for FG Sge. Expressing the pulsation constant Q as a function of the ratio M/R we used this dependence to determine the luminosities of the Population II Cepheids having known effective temperatures. We found that within errors of observations the theoretical period-luminosity relation is in good agreement with the observable one.

References

Aikawa, T. & Whitney, C.A. (1983). Stellar acoustics II. Pulse resonance in giant star models (Preprint).

Fadeyev, Yu.A. (1982). Models of pulsating low-massive yellow supergiants. Astrophys. Space Sci., 86, No.3,143-155.

Takeuti, M. & Petersen, J.O. (1983). The reasonance hypothesis applied to RV Tauri stars. Astron. Astrophys., 117, No.2, 352-356.

Whitney, C.A. (1983). Stellar acoustic. I. Astrophys. J., 274, No.2, 830-839.

SOME MASSES FOR POPULATION I AND II CEPHEIDS

R. B. Kidman
Theoretical Division, Los Alamos National Laboratory,
University of California

A. N. Cox
Theoretical Division, Los Alamos National Laboratory,
University of California

The masses of Cepheids can be obtained in several ways. If a Cepheid luminosity is known from membership in a galactic cluster, the mass-luminosity relation obtained from stellar evolution theory gives its mass. This evolution mass depends slightly on the composition, that is, the mass fraction of helium, Y, and on the the mass fraction of all the heavier elements, Z, but as we shall see later, the composition dependence is small.

A mass of a Cepheid that is based entirely on pulsation theory is called the pulsation mass. Here the needed observations are: the luminosity, the color (giving a surface effective temperature by use of a conversion formula), and the easily observed pulsation period. In this report the observed period is assumed to always be the fundamental mode for the classical Cepheids. The mass, M, the pulsation constant, Q, and the radius, R, are unknowns that are solved for using the three equations

$$L = 4\pi R^2 \sigma T_e^4 \quad , \quad Q = P(M/R^3)^{1/2} \quad , \text{ and } Q = Q(M,R,L,T_e) \quad .$$

These equations are: the definition of the effective temperature, the period-mean density relation, and an expression fitting the pulsation theory values for the pulsation constant Q obtained from a large number of models covering a large range of stellar parameters.

A third mass can be derived based on both evolution and pulsation theories. The required observations are only the well determined period and a surface effective temperature. The above three equations, plus the evolution mass-luminosity relation, are used to simultaneously solve for the unknowns M, R, L, and Q. Due to the strong influence of the evolution theory mass-luminosity relation for the theoretical mass, this mass agrees to within a few % with the evolution mass for almost all Cepheids. The major value of the theoretical mass is its availability when a luminosity has not been observed, and one cannot get an evolution mass.

These three masses have been calculated for 29 Cepheids recently listed by Fernie and McGonegal (1983). The luminosities of these Cepheids have been uniformly set by assuming a distance modulus of 3.29 for the Hyades cluster. Table 1 gives our evolution, theoretical, and pulsation masses. Fernie and McGonegal have suggested that CS Vel and V810 Cen are not in the cluster as supposed, and therefore their luminosities are incorrectly stated. Due to the large discrepancy between the evolution and pulsation masses for V Cen, GY Sge, and S Vul, we have rejected them also. We find only a small difference between the masses for TW Nor, and therefore we do not reject this Cepheid even though there is a cluster membership question about it. The ratios of these masses relative to the theoretical masses average to the values at the bottom of the evolution and pulsation mass columns. The usual scatter in the pulsation masses is present, but on the average, the masses seem in accord with evolution theory.

Table 2 gives similar results with only the distance scale to the Hyades changed to 3.45, a value recently suggested by Vandenberg and Bridges (1984). With the increased luminosities, evolution masses increase a bit, while the luminosity-independent theoretical masses remain unchanged. However, pulsation masses are greatly increased so that the average is about 1.22 above the theoretical masses. For both these tables we have converted the dereddened (B-V) colors to effective temperatures by use of the Kraft (1961) formula. As discussed by Cox (1979) and many others such as Pel (1978) these temperatures may need to be cooled by 0.01 to 0.03 in log T_e. That would greatly alleviate the mass discrepançy. It appears that a distance scale which has the Hydade distance modulus at 3.45 is too large, but part of the change from the currently used 3.29 may be acceptable from the viewpoint of Cepheid masses.

Other recent data on Cepheid luminosities have been prepared for publication by Schmidt (1984). Table 3 shows at the top section the evolution, theoretical, and pulsation masses for 7 Cepheids using his luminosities and dereddened colors. The pulsation masses are so low that for conventional compositions, stars would not evolve into the instability strip. The possible cooling of the temperature scale is investigated in the second section. Even though we have reduced the log

Table 1

Fernie and McGonegal Data
$(m-M)_{Hyades}=3.29$

Cepheid	P(d)	T_e(K)	$L/L_\odot$	M_{ev}	M_{th}	M_Q
SU CAS	1.95	6288	791	4.6	4.6	4.8
EV SCT	3.09	6212	1176	5.1	5.4	4.1
CE CAS B	4.48	6088	2447	6.3	6.2	6.6
CF CAS	4.88	5848	2278	6.1	6.0	6.5
CE CAS A	5.14	5895	2673	6.4	6.3	7.1
UY PER	5.37	5848	2568	6.3	6.3	6.5
CV MON	5.38	6113	2676	6.4	6.7	5.4
V CEN	5.50	6138	2061	6.0	6.8	3.7
VY PER	5.53	5895	3690	7.0	6.4	9.8
CS VEL	5.90	5991	828	4.7	6.8	1.3
V367 SCT	6.30	6138	3484	6.9	7.2	5.9
U SGR	6.75	5801	3566	6.9	6.8	7.3
DL CAS	8.00	5824	3723	7.0	7.4	5.8
S NOR	9.75	5685	4233	7.3	7.7	5.8
TW NOR	10.79	5594	4444	7.4	7.9	5.8
VX PER	10.89	5731	5012	7.6	8.2	5.8
SZ CAS	13.61	5872	8400	8.7	9.2	7.2
VY CAR	18.92	5309	11389	9.5	9.1	10.9
RU SCT	19.68	5685	15087	10.2	10.2	10.4
RZ VEL	20.42	5549	14532	10.1	10.0	10.6
WZ SGR	21.83	5266	8000	8.6	9.6	6.0
SW VEL	23.44	5549	14532	10.1	10.6	8.7
T MON	27.04	5161	17763	10.7	10.1	13.3
KQ SCO	28.71	5058	19950	11.1	10.1	15.8
RS PUP	41.40	5224	28161	12.1	12.1	12.1
SV VUL	44.98	5099	33526	12.7	12.1	15.3
GY SGE	51.05	4898	41139	13.5	12.0	20.7
S VUL	67.61	4819	71024	15.6	13.1	32.4
V810 CEN	130.32	6088	154420	19.3	23.1	10.4
				.993		.999
				±.052		±.247

T_e by only 0.01, it is apparent that more than three times this amount is needed to reconcile Schmidt's low pulsation masses. With the use of the Becker, Iben and Tuggle (1977) fits for the second crossing luminosities as a function of composition and mass, we give the lower two sections for, respectively, higher Y and lower Z. These extreme composition changes for population I Cepheids do not lower the evolution masses enough to match the Schmidt pulsation masses. We feel that his luminosities are very much too low.

Anomalous Cepheids have periods like RR Lyrae variables, but they are typically 5 times more luminous. Zinn and King (1982) have proposed that variable V19 in NGC 5466 at 0.82 day is pulsating in the first radial overtone mode, but even with that mode, its mass is about 1.4 $M_\odot$. This is surprising for a population II star because it is believed that stars

Table 2 Fernie and McGonegal Data
$(m-M)_{Hyades}=3.45$

Cepheid	P(d)	T_e(K)	$L/L_\odot$	M_{ev}	M_{th}	M_Q
SU CAS	1.95	6288	917	4.8	4.6	5.9
EV SCT	3.09	6212	1362	5.3	5.4	4.9
CE CAS B	4.48	6088	2836	6.5	6.2	8.1
CF CAS	4.88	5848	2640	6.4	6.0	7.9
CE CAS A	5.14	5895	3097	6.7	6.3	8.7
UY PER	5.37	5848	2976	6.6	6.3	8.0
CV MON	5.38	6113	3100	6.7	6.7	6.6
V CEN	5.50	6138	2389	6.2	6.8	4.4
VY PER	5.53	5895	4276	7.3	6.4	12.2
CS VEL	5.90	5991	959	4.8	6.8	1.5
V367 SCT	6.30	6138	4038	7.2	7.2	7.1
U SGR	6.75	5801	4132	7.2	6.8	9.0
DL CAS	8.00	5824	4314	7.3	7.4	7.0
S NOR	9.75	5685	4905	7.6	7.7	7.0
TW NOR	10.79	5594	5149	7.7	7.9	6.9
VX PER	10.89	5731	5808	7.9	8.2	7.0
SZ CAS	13.61	5872	9734	9.1	9.2	8.7
VY CAR	18.92	5309	13198	9.9	9.1	13.3
RU SCT	19.68	5685	17482	10.7	10.2	12.7
RZ VEL	20.42	5549	16839	10.6	10.0	12.9
WZ SGR	21.83	5266	9270	9.0	9.6	7.2
SW VEL	23.44	5549	16839	10.6	10.6	10.5
T MON	27.04	5161	20583	11.2	10.1	16.3
KQ SCO	28.71	5058	23118	11.5	10.1	19.5
RS PUP	41.40	5224	32632	12.6	12.1	14.7
SV VUL	44.98	5099	38849	13.3	12.1	18.6
GY SGE	51.05	4898	47671	14.0	12.0	25.8
S VUL	67.61	4819	82302	16.2	13.1	42.7
V810 CEN	130.32	6088	178939	20.1	23.1	12.3
				1.036		1.218
				±.054		±.312

Table 3

Schmidt Photometry and Kraft Temperatures

Cepheid	P(d)	T_e(K)	$L/L_\odot$	M_{ev}	M_{th}	M_Q
		Y=0.28 Z=0.02				
EV SCT	3.09	6364	1176	5.1	5.6	3.6
CF CAS	4.88	5685	997	4.9	5.8	2.6
CV MON	5.38	5824	1352	5.3	6.3	2.9
U SGR	6.75	5943	2682	6.4	7.1	4.5
DL CAS	8.00	5848	3233	6.7	7.4	4.8
S NOR	9.75	5572	2786	6.5	7.5	3.8
TW NOR	10.79	5395	1568	5.5	7.5	2.0
		T_e cooler by 0.01 in $\log T_e$ Y=0.28 Z=0.02				
EV SCT	3.09	6219	1176	5.1	5.4	4.0
CF CAS	4.88	5556	997	4.9	5.6	2.9
CV MON	5.38	5692	1352	5.3	6.1	3.3
U SGR	6.75	5808	2682	6.4	6.8	5.0
DL CAS	8.00	5715	3233	6.7	7.2	5.4
S NOR	9.75	5445	2786	6.5	7.3	4.3
TW NOR	10.79	5272	1568	5.5	7.2	2.2
		Y=0.35 Z=0.02				
EV SCT	3.09	6364	1176	4.3	4.6	3.6
CF CAS	4.88	5685	997	4.1	4.7	2.6
CV MON	5.38	5824	1352	4.5	5.1	2.9
U SGR	6.75	5943	2682	5.5	5.9	4.5
DL CAS	8.00	5848	3233	5.8	6.2	4.8
S NOR	9.75	5572	2786	5.6	6.3	3.8
TW NOR	10.79	5395	1568	4.7	6.2	2.0
		Y=0.28 Z=0.01				
EV SCT	3.09	6364	1176	4.3	4.5	3.6
CF CAS	4.88	5685	997	4.1	4.7	2.6
CV MON	5.38	5824	1352	4.5	5.1	2.9
U SGR	6.75	5943	2682	5.5	5.8	4.5
DL CAS	8.00	5848	3233	5.8	6.1	4.8
S NOR	9.75	5572	2786	5.5	6.2	3.8
TW NOR	10.79	5395	1568	4.7	6.2	2.0

more massive than about 0.8 $M_{\odot}$ should have long ago died. We here confirm the proposal that this star is the result of a coalescence of two stars in a binary system that had an initial separation smaller than the red giant radius of the more massive star.

Figure 1 gives the work done in each of the 195 zones over each pulsation cycle to cause pulsations in each of the three lowest radial modes. The mass used for this nonadiabatic pulsation analysis is 1.4 $M_{\odot}$, and the luminosity is the observed 257 suns. The typical population II composition used is the King Ia table with Z=0.001. Here at 7500K, hotter than NGC 5466 V19, all modes are stable because the radiative damping of the interior (lower zone number) layers is greater than the hydrogen and helium driving in the surface layers.

Figure 2 gives the same plot for the effective temperature of 6500K. In this case it is apparent that the first and second overtone modes are driven more than they are damped, and linear theory predicts growing pulsations. Actually the first overtone is driven the strongest, and at the observed effective temperature of 7000K, probably only the first overtone is unstable to pulsations.

Fig. 1. The work per pulsation cycle to drive pulsations in a NGC 5466 V19 model at 7500K is plotted for each of the 195 zones.

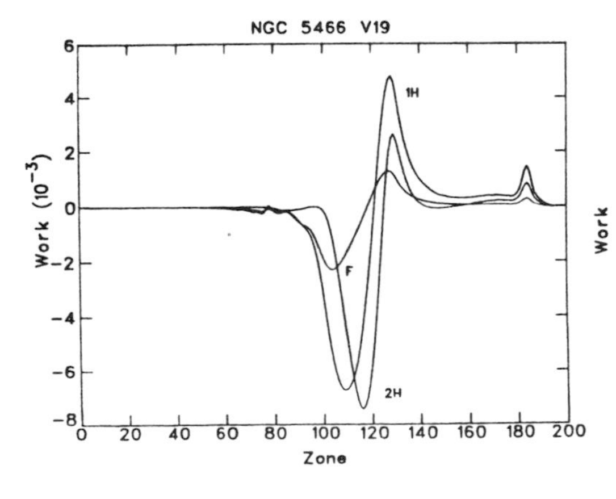

Fig. 2. The work per pulsation cycle to drive pulsations in a NGC 5466 V19 model at 6500K is plotted for each of the 195 zones.

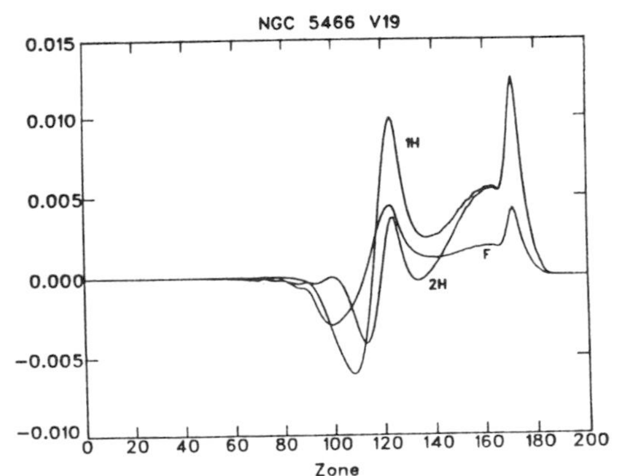

REFERENCES

Becker, S.A., Iben, I., and Tuggle, R.S. (1977). Ap. J. 218, 633.
Cox, A.N. (1979). Ap. J. 229, 212.
Fernie, J.D. & McGonegal, R. (1983). Ap. J. 275, 732.
Kraft, R.P. (1961). Ap. J. 134, 616.
Pel, J.W. (1978). Astr. Ap. Suppl. 31, 489.
Schmidt, E.G. (1984). Ap. J., in press.
Vandenberg, D.A. & Bridges, T.J. (1984). Ap. J. 278, 679.
Zinn, R. & King, C.R. (1982). Ap. J. 262, 700.

THE TWO POPII CEPHEIDS IN THE GLOBULAR CLUSTER MESSIER 10

C.Clement, H.Sawyer Hogg
David Dunlap Observatory, Univ. of Toronto, Toronto M5S 1A7

K.Lake
Dept. of Physics, Queen's University at Kingston,Ont.K7L 3N6

The globular cluster Messier 10 has three known variables. The first two of these were discovered by one of us (Sawyer 1933) and the third by Arp (1955). Two of the variables, V2 (P=18.7226) and V3 (P=7.831), are population II cepheids while V1 appears to be an irregular variable. Another star which lies in the Schwarzschild gap on the horizontal branch is a suspected variable (Voroshilov 1971).

In this investigation, we examine the variations in the periods of the two cepheids over the interval 1912 to 1983 (for V2) and 1931 to 1983 (for V3). The study is based on photographs obtained with seven different telescopes - the Mt.Wilson 100-inch and 60-inch (1912 to 1919), the Dominion Astrophysical Observatory 72-inch, the David Dunlap 74-inch and 19-inch, the 16-inch at the University of Toronto downtown campus and the University of Toronto 24-inch at the Las Campanas Observatory of the Carnegie Institution of Washington. Some of our magnitudes have already been published (Sawyer 1938) and the remaining ones will be submitted to the Astronomical Journal for publication. We have also included material published by Arp (1955,1957) in our study.

Our phase shift diagrams are shown on the next page. To derive these, we have plotted light curves at various epochs for each star and then measured the shift relative to Arp's 1952 observations (Arp 1955). From the diagrams, we can see that the periods for both stars have fluctuated during the observed interval so that we can not determine a value for β, the rate of period change.

V2, with P=18.7 days, is a 'loop' cepheid, i.e. one which enters the instability strip when thermal instabilities in the helium burning shell of an asymptotic red giant branch star cause it to loop to the left in the HR diagram (Schwarzschild and Harm 1970). The period for V2 has been constant since 1945, but before that, it varied. Because of the gap in observations between 1919 and 1931, we can not determine the nature of this variation.

The period of V3, 7.8 days, is very unusual for a population II cepheid in a globular cluster. This makes it difficult to decide whether it is a 'loop' cepheid or a 'suprahorizontal branch' cepheid, i.e. one which passes through the instability strip on its way to the asymptotic giant branch after the exhaustion of helium in its core (Strom et al. 1970). However, the random character of its period fluctuations resembles the variation observed in the 'loop' cepheid V84 in

Messier 5 (Clement and Sawyer Hogg 1977). Furthermore, in Norris and Zinn's (1975) table of luminosities for population II cepheids in globular clusters, there is a gap between 2.16 and 2.43 solar luminosities and V3 belongs to the brighter group. We therefore conclude that V3 is a 'loop' cepheid.

Fig.1 Phase-shift diagrams for the two cepheids in M10:O-C (in fractions of a period) vs. time (in years)

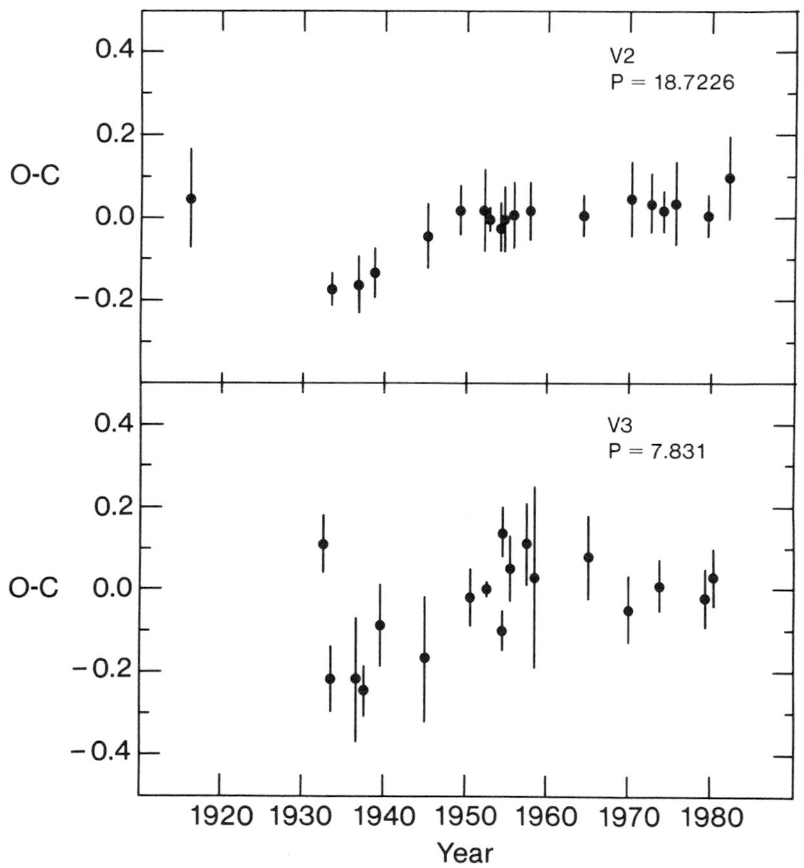

References

Arp, H.C. (1955). Astron.J.60, 1-17.

Arp, H.C. (1957). Astron.J.62, 129-36.

Clement,C. and Sawyer Hogg,H. (1977). J.Roy.Astron.Soc.Can.71, 281-97.

Norris,J. and Zinn,R. (1975). Astrophys.J.202, 335-45.

Sawyer,H.B. (1933). Publ.Amer.Astron.Soc.7, 185-6.

Sawyer,H.B. (1938). David Dunlap Publ.1, 59-68.

Schwarzschild,M. and Harm,R. (1970). Astrophys.J.160, 341.

Strom,S.E.,Strom,K.M.,Rood,R.T.,and Iben,I.Jr. (1970). Astron. and Astrophys.8, 243-50.

Voroshilov,Y.V. (1971). Astron.Circ.623, 7-8.

EIGHT POPII CEPHEIDS RECENTLY IDENTIFIED IN GLOBULAR CLUSTERS

C. Clement, H. Sawyer Hogg and T. Wells
David Dunlap Observatory, Univ. of Toronto, Toronto M5S 1A7

The University of Toronto 24-inch telescope at the Las Campanas Observatory of the Carnegie Institution of Washington has been used for the study of variables in four southern globular clusters: NGC 6273=M19, NGC 6284, NGC 6293 and NGC 6333=M9.

The first three of these clusters lie about 2° from one another in the sky at southern declinations ranging from about 24.5° to 26.5°. They were investigated by one of us (Sawyer 1943) using photographs obtained at the Steward Observatory in Arizona in 1939. A number of variables were discovered, but even with additional David Dunlap plates, it was not possible to determine any periods because of the large southern declinations. At Las Campanas (latitude 29°S), they pass close to the zenith and therefore the periods are more readily determined. From our Las Campanas data, we have found that there are four Cepheids in NGC 6273, two in NGC 6284 and one in NGC 6293.

NGC 6333 with its declination of about 18°S can be studied more easily from the northern hemisphere than the others. Accordingly, one of us (Sawyer 1951) studied the variables in this cluster using plates obtained at the Steward Observatory in 1939 and at the David Dunlap Observatory during the period 1940 to 1949. Eleven periods were determined and all were thought to be of the RR Lyrae type. However, new data from Las Campanas have shown that one of the stars, V12 is a population II cepheid. At northern observatories, this cluster can be observed over such a limited range in hour angle that a spurious period was obtained for V12. However, at Las Campanas, it can be observed for as much as nine hours in one night. Our new period for this star, 1.34 days is related to the old by the relation: $1/P(new) = 1/P(old) - 1$.

Elements for the eight cepheids are presented in the table on the following page. The values of E(B-V) for the clusters were taken from Zinn (1980). The other information is taken from the cited references where light curves and epochs of maximum light may also be found. We hesitate to quote absolute magnitudes for any of these stars because for some, the mean B magnitudes are uncertain due to the contamination of light by nearby stars and for all of them, the distances are uncertain. Of the four clusters, NGC 6273 is the only one with a published colour-magnitude diagram (Harris et al 1976). The diagram has a lot of scatter which the authors attribute partly to contamination from field stars, but mainly to differential reddening. The cluster has the largest ellipticity of any in the galaxy and they suggest that this is caused by a thin absorbing lane on its east side. It is therefore difficult to estimate the distance modulus of NGC 6273 to an accuracy of better than 0.3 magnitudes. In the other three clusters, Racine (Diamond 1976) has made a few photoelectric observations on the B,V system, but in each case, there are only one or two standards in the appropriate magnitude range.

Elements for the Eight Variables

Cluster	Var	P(days)	logP	<B>	E(B-V)	Reference
NGC 6273	1	16.92	1.23	13.73:	0.36	Clement & Sawyer-Hogg (1978)
	2	14.139	1.15	14.17		
	3	16.5	1.22	13.61:		
	4	2.4326	0.39	14.76:		
NGC 6284	1	4.48121	0.65	15.88	0.29	Clement et al(1980)
	4	2.81873	0.45	16.04		
NGC 6293	2	1.1575	0.06	16.91:	0.37	Clement et al(1982)
NGC 6333	12	1.340204	0.13	16.62	0.34	Clement & Ip(1984)

Cluster Membership

The cepheids in NGC 6273 and NGC 6284 have B magnitudes appropriate for cluster membership, when compared with the RR Lyrae variables in these clusters. However, both V2 in NGC 6293 and V12 in NGC 6333 have B magnitudes comparable to those of the RR Lyrae variables, and normally, we would expect them to be about a magnitude brighter if they were cluster members. Nevertheless, we believe that V12 is a member of NGC 6333 because it lies at the edge of a region of considerable obscuration southwest of the cluster. This obscuration is very marked on plates E and O1160 of the Palomar Sky Survey and so it is not surprising that V12 is faint. The case for V2 in NGC 6293 is more complicated. In our earlier paper (Clement et al. 1982), we assumed that it was not a cluster member, but close examination of plates E and O192 of the Palomar Sky Survey indicates that there may be a small region of obscuration to the northwest of NGC 6293 where V2 lies.

Since all of these newly identified cepheids are in and around highly reddened clusters, it would be useful to study them in the infrared and Welch (1984) is planning to do this.

References

Clement,C.M. and Ip,P. (1984). To be submitted to the Astron.J.
Clement,C.M.,Panchhi,P.S. and Wells,T.R. (1982). Astron.J.87,1491-6.
Clement,C. and Sawyer Hogg,H. (1978). Astron.J.83,167-71.
Clement,C.M.,Sawyer Hogg,H. and Wells,T.R. (1980). Astron.J.85,1604-11.
Diamond,G.E. (1976). M.Sc. thesis, Univ. of Toronto.
Harris,W.E.,Racine,R. and de Roux,J. (1976). Astrophys.J.Suppl.31,13-31.
Sawyer,H.B. (1943). David Dunlap Publ. 1,285-93.
Sawyer,H.B. (1951). David Dunlap Publ. 1,511-19.
Welch,D.L. (1984). Private communication.
Zinn,R. (1980). Astrophys.J.Suppl.42,19-40.

TWO MID-GIANT BRANCH VARIABLE STARS IN M15

Chu, Y-H
Purple Mountain Observatory, Nanking, China

C.M. Clement, H. Sawyer Hogg and T.R. Wells
David Dunlap Observatory, Univ. of Toronto, Toronto M5S 1A7

Abstract. High precision photographic photometry indicates that two stars lying on the giant branch in the C-M diagram of M15 are small amplitude ($\sim$0.2 mag) variables. The two stars are Kustner 64 and 152. This investigation is based on plates taken with three telescopes: the Dominion Astrophysical Observatory 1.8-metre reflector, the David Dunlap 1.9-metre reflector and the Yunnan 1-metre reflector in China. The existing data is not sufficient for period determination.

Introduction

Recently, there has been considerable interest in the possibility of the existence of small amplitude variation in globular cluster stars lying outside the instability strip in the colour-magnitude diagram. Before this conference, a number of these variables were suspected in M15 (Chu 1975 & 1977, Mosley & White 1975, Kraft 1981 and Sandage et al 1981). A systematic study of the globular clusters M3 and M13 was conducted by White (1981) who made photoelectric observations of asymptotic giant branch (AGB) stars brighter than V=14 during three observing seasons. He found that some, but not all, of these stars varied in their apparent light and determined periods ranging from 59 to 225 days for six of the variables in M3. In another investigation, Russeva et al. (1982) found a period of 21 days for an M13 star which lies between the instability strip and the AGB. As a result of these findings, we decided that it would be appropriate to make a systematic study of AGB stars in the globular cluster M15.

Investigation

Our study is based on 19 plates taken with the 1.8-metre reflector of the Dominion Astrophysical Observatory during the years 1932 and 1934, 25 plates taken with the David Dunlap 1.9-metre reflector in 1969 and 1971 and 22 plates taken with the Yunnan 1-metre reflector in 1979, 1981 and 1982. The original purpose was to investigate all the AGB stars with B magnitudes in the range 15.5 to 17.0. Accordingly, from the PDS photometry of Buonanno et al.(1983), we selected about 25 giant stars with B magnitudes in the appropriate range to use as comparison stars. The plates were all measured on a Cuffey iris astrophotometer and the data from the different telescopes were reduced separately using a procedure similar to that of Chu et al.(1982). We did not make any background measurements. Instead we closed the iris diaphragm to the minimum size possible around the stars to avoid background contamination.

We computed the mean B and its σ for all the stars on each set of plates and found that the σ values for the stars with Kustner (1921) numbers 64 and 152 ($\sigma \cong 0.05$) were greater than for other stars with similar magnitude, colour and position in the cluster (for which $\sigma \simeq 0.03$). A C-M diagram is shown in Figure 1 and our newly suspected variables are indicated as open triangles. For K64, Sandage (1970) gives photographic V=15.27, B-V=0.83, while Buonanno et al (1983) give V=15.22, B-V=0.87 from PDS photometry based on Sandage's photoelectric standards. For K152, Buonanno et al give V=15.24, B-V=0.81, similar to values determined in earlier work on the m_{pg} system. It is therefore clear that neither of these stars lies in the instability strip of the C-M diagram. It is interesting to note, however, that both stars lie near gaps discovered by Sandage et al.(1968) in the giant branch. Cudworth (1976) has found that the membership probability is 99% for both stars. At this stage, it is difficult to determine periods or even the pattern of variation. However, for K64, the David Dunlap observations indicate rapid variation on two consecutive nights in 1969.

Figure 1. Colour-magnitude diagram for M15 plotted from the data of Buonanno et al (1983). The plus signs represent the RR Lyrae variables (Sandage et al 1981). Our new variables K64 and 152 are indicated by the open triangles. Other suspected variables are indicated: open circle (K1082, Chu 1977), the crosses (Sandage et al 1981) and the open square (Fillipenko's variable, Kraft 1981).

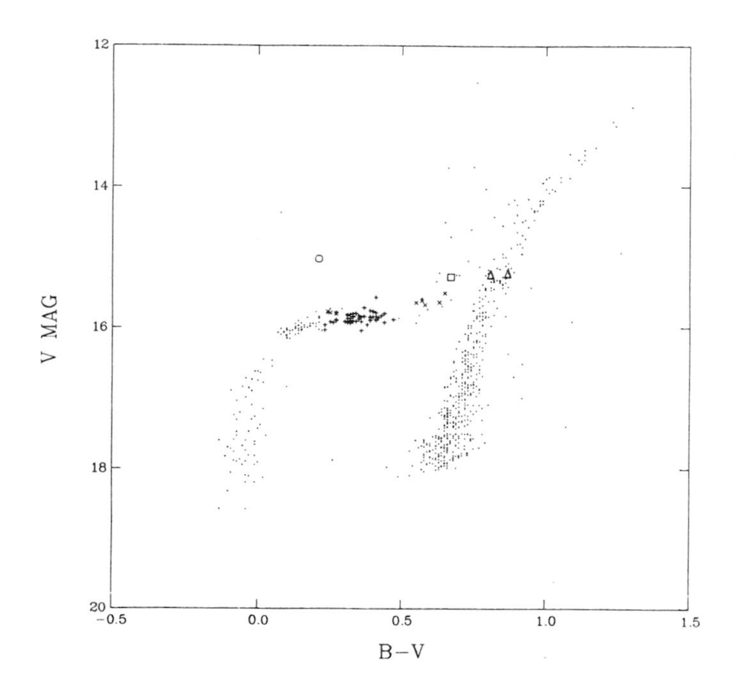

Figure 2, is a print of a plate taken with the Yunnan 1-metre reflector. Our two new suspected variables K64 and K152 are labelled, along with K202 a star we consider to be non-variable. Also indicated is K1082 a a star previously suspected to be a variable by Chu (1977). K1082 has subsequently been studied by Smith et al.(1979) and by Liller & Schommer (1980) who agree that it probably varies. From Figure 2, it can be readily seen that both K64 and K152 lie in uncrowded areas and so we can have confidence in the iris photometry.

Figure 2. A 25 minute exposure of M15 taken with the Yunnan 1-metre reflector on September 19, 1979. Our new variables K64 and 152 are labelled, along with K202, a star we consider to be non-variable (σ_B=0.02). Also indicated is K1082, a variable previously discussed by Chu (1977).

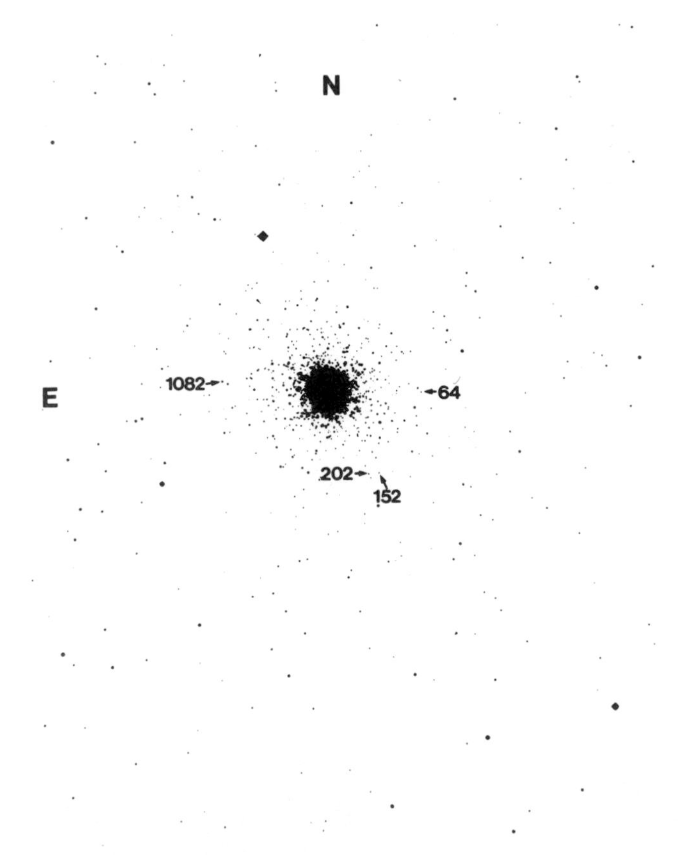

Acknowledgements

This work has been supported by a grant through the Natural Sciences and Engineering Research Council of Canada. We would also like to thank Wm. G. Weller for modifying the iris photometer for high precision work. One of us (Y-H.Chu) is extremely grateful to Dr.J.D. Fernie for offering him a Visiting Professorship for six months at the University of Toronto.

References

Buonanno, R., Buscema, G., Corsi, C.E., Iannicola, G. & Fusi Pecci, F. (1983) Astron. Astrophys. Suppl. 51, 83-92.

Chu, Y-H. (1975) Purple Mountain Observatory Preprint.

Chu, Y-H. (1977) Chinese Astron. 1, 302-8.

Chu, Y-H., Shen, C-j., Di, X-h., Bao, M-x., Deng, L-w., & Wu, X-f. (1982) Chinese Astron. Astrophys. 6, 204-8.

Cudworth, K. (1976) Astron. J. 81, 519-26.

Kraft, R.P. (1981) In Astrophysical Parameters for Globular Clusters, ed. A. G. D. Philip, Comment on p. 165.

Kustner, F. (1921). Bonn Veroff. No. 15.

Liller, M. H. & Schommer, R. A. (1980). Inf. Bull. Var. Stars. No. 1752.

Mosley, D. R. & White, R.E. (1975). Bull Amer. Astron. Soc. 7, 535.

Russeva, T., Iliev, L. & Russev, R. (1982) Inf. Bull Var. Stars No. 2223.

Sandage, A. (1970). Astrophys. J. 162, 841-70.

Sandage, A., Katem, B. & Kristian, J. (1968). Astrophys. J. 153, L129-33.

Sandage, A., Katem, B. & Sandage, M. (1981) Astrophys. J. Suppl. 46, 41-74.

Smith, H. A., Millis, A.J. & Liller, M.H. (1979) PASP 91, 671-3.

White, R. E. (1981) In Astrophysical Parameter for Globular Clusters, ed. A. G. D. Philip, pp. 161-165.

CURRENT PROBLEMS IN HORIZONTAL-BRANCH THEORY: SOME IMPLICATIONS FOR RR LYRAE VARIABLES.

Pierre Demarque
Yale University Observatory, New Haven, CT 06511 USA

I. INTRODUCTION

There are a number of good reasons for wanting to improve our understanding of the structure and evolutionary status of horizontal-branch (HB) stars. To mention a few:

1) The HB morphology of globular clusters can be used as a tool for studying the evolution of the galactic halo (Searle and Zinn 1978) and the time scale for halo collapse (Demarque 1980; Zinn 1980).

2) HB stars offer an opportunity to evaluate the helium abundance of globular clusters. This can be done either by using the R-method (Cole et al. 1983) or the width of the RR Lyrae instability strip [Deupree 1977; but see also the paper by Stellingwerf (1984) in these preceedings which casts doubt on the validity of previous calculations of the dependence of the blue edge on helium content].

3) HB stars may play an important role in understanding the integrated light of old stellar systems. In particular, blue HB stars seem to make a significant contribution to the ultraviolet light of elliptical galaxies. (Ciardullo and Demarque 1978; Gunn, Tinsley and Stryker 1981). At the same time, HB stars are believed to be the direct progenitors of asymptotic giant branch stars which have proved to be powerful tracers of stellar populations in nearby external systems (Blanco et al. 1980)

4) The RR Lyrae variables are part of the HB population. One would wish to relate the observable properties of RR Lyrae variables to their chemical compositions and ages so as to use them: a) as distance indicators for galactic globular clusters (Sandage 1982a,b) and in the Magellanic Clouds; b) as tracers of stellar populations. This has been attempted for the galactic halo and disk by Sandage (1982a,b) and for the Magellanic Clouds by Butler et al. (1982).

The current theory of the HB [see e.g. the grid of HB evolutionary tracks by Sweigart and Gross 1976 (SG)], explains many features of HB morphology. It can, for example, be used to model the stellar distribution of the HB of most observed globular clusters and to describe the effects of age and chemical composition on HB morphology (Demarque 1980).

However, in the light of the new observational and theoretical information of the last few years a reevaluation of some aspects of the standard picture of stellar evolution in globular clusters is

needed. The aim of this paper is:
1) to draw attention those features of HB theory which will require revision. Some of these changes are suggested by more refined observations; others by advances in stellar structure theory (Section II);
2) to discuss briefly the implications of recent observations of clusters main sequences in the Magellanic Clouds which have a direct bearing on the problem of the cosmological distance scale and should serve as a check of RR Lyrae absolute magnitudes and the zero-point of the Cepheid period-luminosity relation (Fernie and McGonegal 1983) (Section III).

II. DEFICIENCIES IN THE STANDARD THEORY

A growing number of questions are being asked about the HB which cannot be answered by the standard theory. In some cases, there are apparent inconsistencies; in other cases, the theory is incomplete. The list of these problems includes:
1) the anticorrelation of Y and Z among globular cluster variables of different chemical compositions discussed by Sandage (1982a,b), which is nearly certainly spurious.
2) our inability to understand the evolutionary status of the bluest HB stars, which are not found in the most metal-poor globular clusters, but rather in systems of intermediate metallicity (Sweigart et al. 1974; Caloi et al. 1984).
3) the problem of the evolutionary status of metal-rich RR Lyrae variables (Taam et al. 1976) which is still very uncertain, and the related more general question of the expected range in ages of RR Lyrae stars in different metallicities.
4) the origin of bimodal stellar distributions on the HB's of some globular clusters and the relation that these bimodal distributions have to similar bimodal distributions in the chemical composition observed on the giant branch of these clusters (Harris 1974; Freeman and Norris 1981). It has been suggested that internal stellar rotation plays an important role in this problem. This suggestion gains additional support from recent observations of surface rotation among blue HB stars (Peterson 1983).

The solution to problem 1) may be found in improved interior opacities (Renzini 1983). Another possibility is mixing of heavy elements produced at a particularly violent core helium flash (Deupree and Cole 1983). Still another is the possibility of a range in core masses (possibly due to internal rotation) among HB stars of the same composition.

Problems 2) and 3) seem primarily due to our inability to predict mass loss rates and their dependence, if any, on metallicity for late-type stars. Recent work by Dupree et al. (1984) suggests that previously derived mass loss rates for metal-poor giants were overestimates. This is one of several hints that one may have to have recourse to a mass ejection mechanism effective in the subgiant region to explain the low masses of HB stars compared to their main sequence progenitors

(Dearborn et al. 1976; Corbally 1983; King et al. 1984).

On the theoretical side, arguments have been presented which cast doubt on the treatment of semi-convection first introduced by Robertson and Faulkner (1972) and used in the SG models (Arimoto 1980). At the same time, the rapid advances in numerical fluid dynamics have made it possible to reconsider the development of the helium core flash using a 2D and 3D description of convection (Deupree and Cole 1983; Deupree 1984). Although still subject to considerable uncertainties in their detailed predictions, the hydrodynamic core flash calculations demonstrate the need for a revision of the structure of ZAHB models. Some of their implications on HB Lifetimes and trach morphology have been discussed by Demarque (1981) and Cole and Demarque (1984).

III. STELLAR EVOLUTION, H_o AND THE DISTANCE TO THE MAGELLANIC CLOUDS.

Finally, I wish to discuss briefly recent observations of star clusters in the Small and Large Magellanic Clouds which, when interpreted with theoretical isochrones, lead to an apparent inconsistency with similar results from our own Galaxy in estimating the cosmic distance scale and the corresponding value of H_o. The current controversy between proponents of the "short" and "long" distance scales of the Universe is well known (Hodge 1981). It is also well known that a fit of galactic globular cluster c-m diagrams to theoretical isochrones yield ages which agree with the "long" estimate of the distance scale and are inconsistent with the "short" distance scale (or $H_o \cong 100$ km/sec Mpc) (Janes and Demarque 1983; VandenBerg 1983).

On the other hand, the recent c-m diagrams of the intermediate age clusters Kron 3 (Rich et al. 1984) and Lindsay 113 (Mould et al. 1984), both in the SMC, which include a sufficient portion of the main-sequence to achieve a good fit to the Yale isochrones (Ciardullo and Demarque 1979), are compatible with a distance modulus of 18.8 (i.e. the "short" distance scale). Similar work on two LMC clusters (NGC2162 and NGC2190) by Schommer et al. (1984), yields $(m-M_o)=18.2 \pm 0.2$ for the LMC distance modulus, also in agreement with the "short" distance scale using both the Yale isochrones and the work of VandenBerg and Bridges (1984).

We are thus left with the paradoxical situation that stellar models, based on the same theoretical assumptions, when applied to globular star clusters in the galactic halo on the one hand, and to old disk clusters in the Magellanic Clouds on the other hand, yield apparently inconsistent results for the age of the Universe, i.e. a high nuclear age, and a low expansion age.

REFERENCES

Arimoto, N. 1980 Sci. Rep. Tohoku Univ., Ser. 1, 62, 134.
Blanco, V.M., McCarthy, M.F. and Blanco, B.M. 1980, Astrophys. J., 242, 938.
Butler, D., Demarque, P. and Smith, H.A. 1982 Astrophys. J. 257, 592.
Caloi, V., Castellani, V., Danziger, J., Gilmozzi, R., Cannon, R.D., Hill, P.W. and Boksenberg, A. 1984 ESO Sci. Preprint No. 318.
Ciardullo, R.B. and Demarque, P. 1978, IAU Symp. No. 88, ed. A.G.D. Philip and D.S. Hayes, p. 345.
Ciardullo, R.B. and Demarque, P. 1979, Dudley Obs. Rep. No. 14, p. 317.
Cole, P.W. and Demarque, P. 1984 Astrophys. J. to be published.
Cole, P.W., Demarque, P. and Green, E.M. 1983, ESO Workshop on Primordial Helium, ed. P.A. Shaver, D. Kunth and K. Kjär, p. 235.
Corbally, C. J. 1983 Ph. D. Thesis, University of Toronto.
Dearborn, D.S.P., Eggleton, P.P. and Schramm, D.N. 1976 Astrophys. J. 203, 455.
Demarque, P. 1980, IAU Coll. No. 85, ed. J.E. Hesser, p. 281.
Demarque, P. 1981, IAU Coll. No. 68, ed. A.G.D. Philip and D.S. Hayes, p. 301.
Deupree, R.G. 1977 Astrophys. J. 214, 502.
Deupree, R.G. 1984 Astrophys. J. in press.
Deupree, R.G. and Cole, P.W. 1983 Astrophys. J. 269, 676.
Dupree, A.K., Hartmann, L. and Avrett, E.H. 1984 Center for Astrophysics Preprint No. 1984.
Fernie, J.D. and McGonegal, R. 1983 Astrophys. J. 275, 732.
Freeman, K.C. and Norris, J. 1981 Ann. Rev. Astr. Ap. 19, 319.
Gunn, J.E., Tinsley, B.M. and Stryker, L. 1981 Ap.J. 249, 48.
Harris, W.E. 1974 Astrophys. J. Letters, 192, L161.
Janes, K.A. and Demarque, P. 1983 Astrophys. J. 264, 206.
King, C.R., Da Costa, G.S. and Demarque, P. 1984 Astrophys. J. to be published.
Mould, J.R., Da Costa, G.S. and Crawford, M.D. 1984 Astrophys. J. in press.
Peterson, R.C. 1983, Astrophys. J. 275, 737.
Renzini, A. 1983, Mem. Soc. Astr. Italiana, 54, 335.
Rich, R.M., Mould, J.R. and Da Costa, G.S. 1984, IAU Symp. No. 108, ed. S. van den Bergh and K.S. de Boer, p. 45.
Robertson, J.W. and Faulkner, D.J. 1972 Astrophys. J. 171, 309.
Sandage, A. 1982a, Astrophys. J. 252, 553.
Sandage, A. 1982b, Astrophys. J. 252, 574.
Schommer, R.A., Olszewski, E.W. and Aaronson, M. 1984 preprint.
Searle, L. and Zinn, R. 1978 Astrophys. J. 225, 357.
Stellingwerf, R.F. 1984 These Proceedings.
Sweigart, A.V. and Gross, P.C. 1976 Astrophys. Suppl., 32, 367.
Sweigart, A.V., Mengel, J.G. and Demarque, P. 1974 Astr. Ap. 30, 13.
Taam, R.E., Kraft, R.P. and Suntzeff, N. 1976 Astrophys. J., 207, 201.
VandenBerg, D.A. 1983 Astrophys. J. Suppl. 51, 29.
VandenBerg, D.A. and Bridges, T.R. 1984 Astrophys. J. 278, 679.
Zinn, R. 1980 Astrophys. J. Suppl. 42, 19.

FOURIER DECOMPOSITION PARAMETERS FOR THE HALO RR LYRAE VARIABLES U AND V CAELI

L. Hansen and J.O. Petersen
Copenhagen University Observatory, Øster Voldgade 3
DK-1350 Copenhagen K, Denmark

Abstract. UBVRI light curves are obtained for the two halo RR Lyrae variables U Caeli with period 0.420 days (73 observations) and V Caeli with period 0.571 days (42 observations). It is shown that their light curve characteristics are very similar to those of field RR Lyrae stars.

Fourier decompositions are studied for all five magnitudes and the resulting amplitude ratios and phase differences are discussed. The differences in the Fourier decomposition parameters between the five magnitudes are shown to be relatively small. Comparisons of the Fourier decomposition parameters for the two halo RR Lyrae stars with recently published data for field RR Lyrae stars show no systematic differences.

INTRODUCTION

With the purpose of studying bump progression sequences and other properties of Cepheid type variables we obtained UBVRI photometry of the Population II Cepheid KZ Centauri with period 1.52 d and used Fourier decomposition technique for analysis of these observations (Petersen & Hansen, 1984) and of mean light curves for RR Lyrae stars in the globular cluster ω Centauri (Petersen, 1984).

In the present study we investigate two halo RR Lyrae stars for which no accurate photometry has been available until now. The main purpose is to derive the Fourier decomposition parameters and compare amplitude ratios and phase differences calculated for U, B, V, R, and I.

THE OBSERVATIONS OF U AND V CAELI

The UBVRI photometry was obtained during six nights from November 29th to December 5th, 1982, by means of the Danish 1.5 m telescope at La Silla. Both the observations and the data reductions were performed by the methods described in Petersen & Hansen (1984).

LIGHT AND COLOUR CURVES

Fig. 1 presents the light and colour curves of V Caeli. It is seen that the light curves in all bands are remarkably similar. To a good approximation they are simply scaled versions of one another, just as we found for KZ Centauri. The U and B curves are virtually identical, the only systematic variation in U-B being a small decrease close to light maximum followed by a similar increase. In these respects U Cae is very similar to V Cae.

In V Cae a gradual change in light curve form from U to I is indicated in the phase interval $f = 0.70 - 0.90$: While the U curve may have a small positive slope, the B curve seems practically horizontal and the V, R, and I curves show negative, gradually decreasing slopes.

Figure 1. Light and colour curves of V Caeli. Eighth order Fourier fits are shown in all cases and individual observations are given except for R, U, and R-I.

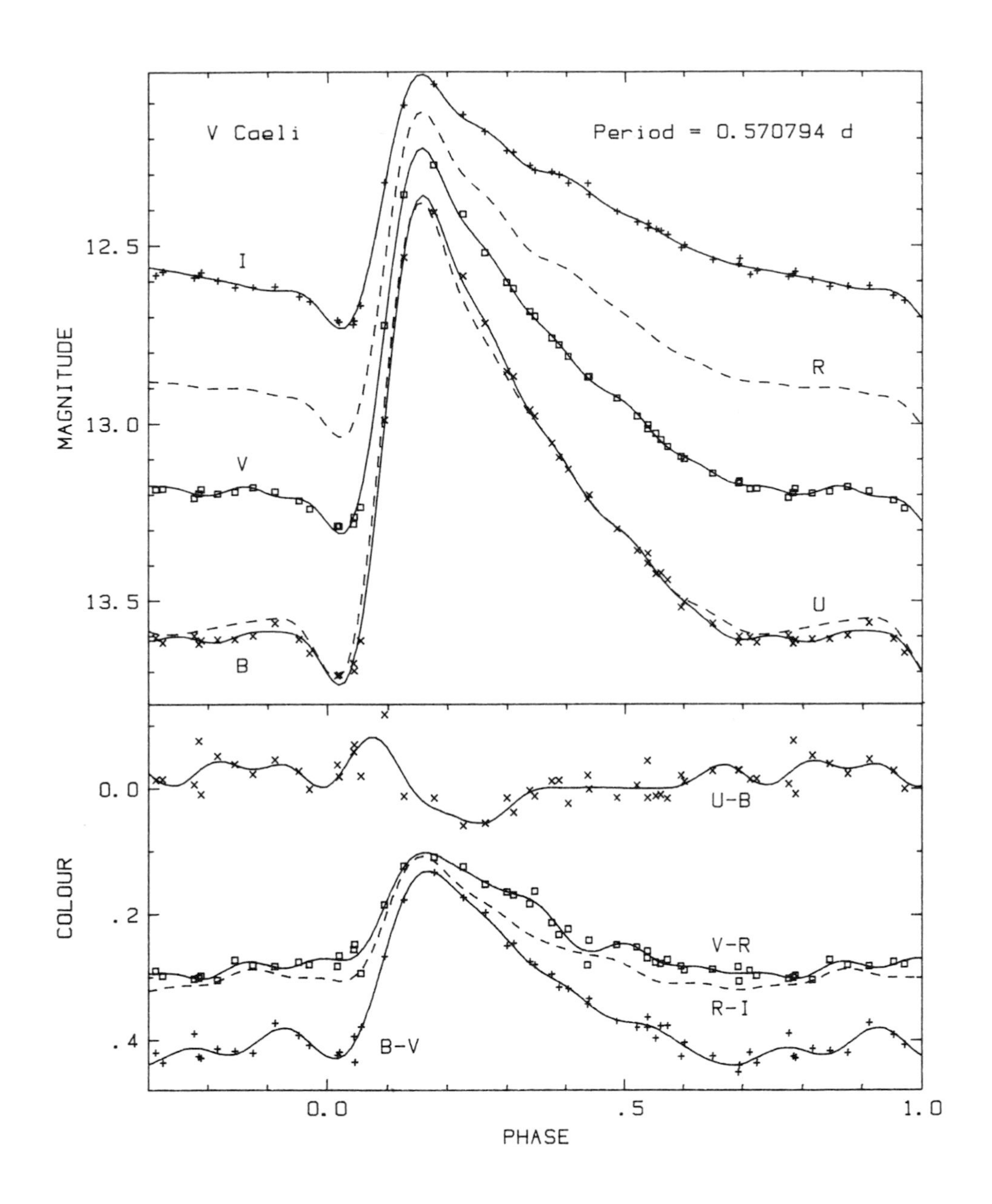

In the following we also find small systematic changes in the Fourier decomposition parameters.

Photographic light curves for U and V Cae have been published by Hoffmeister (1956). For U Cae Hoffmeister gives the light variation $m_{ph} = 11.6 - 13.1$ with a scatter of about ±0.2. This agrees well with our result $B = 11.4 - 13.0$. Hoffmeister remarks that the large scatter he found present around light minimum is probably real. Our observations (six nights) show a scatter somewhat larger than expected from the photometric accuracy, but smaller than about ±0.04. We cannot decide whether this scatter is due to a non-repeating light curve, a small period change or somewhat unfavourable observing conditions. In order to study this problem we plan to observe U and V Cae again at a later date. For V Cae Hoffmeister gives $m_{ph} = 11.9 - 13.1$ while we find $B = 12.3 - 13.7$. Due to the large scatter in the photographic light curves no detailed comparison of light curve features can be performed.

Comparing our observations of U and V Cae with the very detailed light curves for field RR Lyrae stars, published by Lub (1977), in the period interval 0.42 - 0.57 d, we note a nice qualitative agreement in light curve features. In particular the small dip occuring in U and V Cae just before the steep rise to maximum also occurs in most of Lub's curves.

FOURIER DECOMPOSITIONS

Table 1 gives the results of 8th order Fourier decompositions of the light curves of U and V Caeli calculated by the methods described in Petersen & Hansen (1984). The values given for the V magnitude should correspond to the data given by Simon & Teays (1982) for RR Lyrae field stars. A comparison of the phase differences φ_{21} and φ_{31} and the

Table 1. Amplitude ratios, R_{kl}, and phase differences φ_{kl}, for U and V Caeli calculated for decompositions of order N = 8. For V magnitudes these quantities are given directly. In other cases we give Δ_{kl} defined as the value for the magnitude in question minus the corresponding value for the V magnitude.

STAR	MAG	R_{kl} or Δ_{kl}			φ_{kl} or Δ_{kl}		
		21	31	41	21	31	41
U Cae	V	.50	.36	.22	3.78	7.95	5.88
	U	-.01	.00	.00	-.08	-.14	-.27
	B	-.02	-.02	-.01	-.06	-.12	-.16
	R	.01	.01	.01	.11	.18	.21
	I	.01	.02	.01	.26	.43	.63
V Cae	V	.47	.34	.22	3.87	8.06	6.02
	U	-.02	.00	.01	-.13	-.18	-.19
	B	-.01	.00	-.01	-.11	-.18	-.22
	R	.01	.03	.03	.11	.22	.22
	I	.01	.02	.03	.33	.66	.70

amplitude ratio R_{21} by means of Simon & Teays' figures show perfectly normal values for U and V Caeli.

Table 1 shows that the differences in Fourier decomposition parameters calculated for different photometric systems are relatively small in particular for the amplitude ratios. Comparison of the data for U Cae with those for V Cae show a good agreement. In all cases the differences have same sign, but the U Cae data have on the average somewhat smaller numerical values.

Similar data have been obtained for the Population II Cepheid KZ Centauri with period 1.52 d by Petersen & Hansen (1984). Due to the conspicuous differences in light curve characteristics between KZ Cen and the RR Lyrae stars, it is not surprising that we find quite different values of R_{k1} and φ_{k1}. But it is interesting that the differences in R_{k1} and φ_{k1} between the values determined for the V magnitude and the other bands are very similar. For R_{k1} these differences are very small, and for φ_{k1} we find in all cases same sign and numerical values that are only slightly larger on the average for KZ Cen than for the RR Lyrae stars.

Presupposing that photographic magnitudes correspond to B, we can for RR Lyrae stars estimate the differences in Fourier decomposition parameters expected between m_{ph} and V curves by the values given for B in Table 1. It is seen that these differences are only 0.00 to -0.02 for R_{k1} and -0.10 to -0.15 for φ_{21} and φ_{31}. As remarked in Petersen (1984) uncertainties of this order of magnitude are not important in discussions of bump progression sequences.

REFERENCES

Hoffmeister, C. (1956). Veröffentlichungen Sonneberg, *Band 3*, 5.
Lub, J. (1977). Astron. Astrophys. Suppl., *29*, 345.
Petersen, J.O. (1984). Astron Astrophys., submitted.
Petersen, J.O. & Hansen, L. (1984). Astron. Astrophys., in press.
Simon, N.R. & Teays, T.J. (1982). Astrophys. J., *261*, 586.

COMPARISON OF THE PULSATION PROPERTIES OF
THE RR LYRAE STARS IN ω CENTAURI
WITH THOSE OF CLASSICAL CEPHEIDS

J.O. Petersen
Copenhagen University Observatory, Øster Voldgade 3
DK-1350 Copenhagen K, Denmark

INTRODUCTION

In the last few years several studies have shown that Fourier decomposition technique is a powerful method for quantitative description of light curves of pulsation variables. This technique was introduced by Simon & Lee (1981), who showed that amplitude ratios and phase differences provide a very useful description of the Hertzsprung progression for classical Cepheids. Recently, Simon & Teays (1982) discussed 70 RR Lyrae field stars.

In the present study I analyse 130 photographic mean light curves of RR Lyrae variables in ω Centauri taken from Martin (1938). I wish (i) to compare the Fourier decomposition parameters of the ω Cen RR Lyrae stars with those of the field variables as studied by Simon & Teays, (ii) to discuss the evidence for progression sequences among the ω Cen variables and (iii) to compare the basic pulsation properties of the RRab variables in ω Cen with those of classical Cepheids.

THE FUNDAMENTAL MODE RR LYRAE VARIABLES

I calculate the Fourier decomposition parameters defined by Simon & Lee (1981) by the method described in Petersen & Hansen (1984). Here I only present the analyses of the lowest order phase difference, φ_{21}, and amplitude ratio, R_{21}. The higher order Fourier parameters give results that are rather similar; they will be discussed in more detail elsewhere (Petersen, 1984).

Fig. 1 shows φ_{21} and R_{21} as functions of the pulsation period. These distributions can be compared with the corresponding distributions for field RRab stars as given by Simon & Teays (1982). About half the field stars have periods below 0.50 d compared to only two of the ω Cen variables. But for periods 0.50 - 0.75 d a close agreement is found both for φ_{21} and R_{21}. Within about ±0.10 for the phase difference and ±0.03 for the amplitude ratio no systematic differences are seen.

In the period interval 0.75 - 0.90 d I find smooth variations in φ_{21} and R_{21} rather than a drastic change at a period of about 0.80 d as indicated by XZ Ceti. This result also holds for higher order Fourier parameters (Petersen, 1984).

For the field stars Simon & Teays noted some evidence for a Cepheid-like progression for periods 0.55 - 0.75 - 0.82 d. In particular they pointed out that XZ Ceti might be associated with the fundamental mode - second overtone resonance in the bump progression sequence.

However, I can now safely follow the sequence to 0.90 d and find no indication of a close resonance. Clearly, the long period RRab stars in ω Cen do not resemble XZ Ceti.

To me it is reassuring that the resonance, after all, does not occur in RRab stars at a period of about 0.80 d. Firstly, the BL Herculis variables probably form a simple continuation of the RRab stars toward higher periods. And several investigations (e.g. Petersen & Hansen, 1984; and references therein) agree that this resonance occur at a period 1.5 - 1.6 d. Secondly, comparing with Simon & Lee's (1981) Fourier description of the well established bump progression for the classical Cepheids, one should expect a rise in φ_{21} to about 5.4 (and a decrease in R_{21} to about 0.1) before the resonance appears. And the RRab stars in ω Cen show a rise to only about 4.2 at a period of 0.80 d.

I will now consider the classical Cepheid sequences starting at a period of 3 d and ending at the resonance period 9 d. The φ_{21} sequence is well represented by a straight line from $\varphi_{21} = 4.0$ to 5.4.

Figure 1. φ_{21} and R_{21} versus period for 75 RRab stars and 3 BL Herculis stars in ω Centauri and the field variable XZ Ceti. Full lines and the area shown on the lower panel represent mean sequences defined by the classical Cepheids transformed to the ω Cen variables (see text).

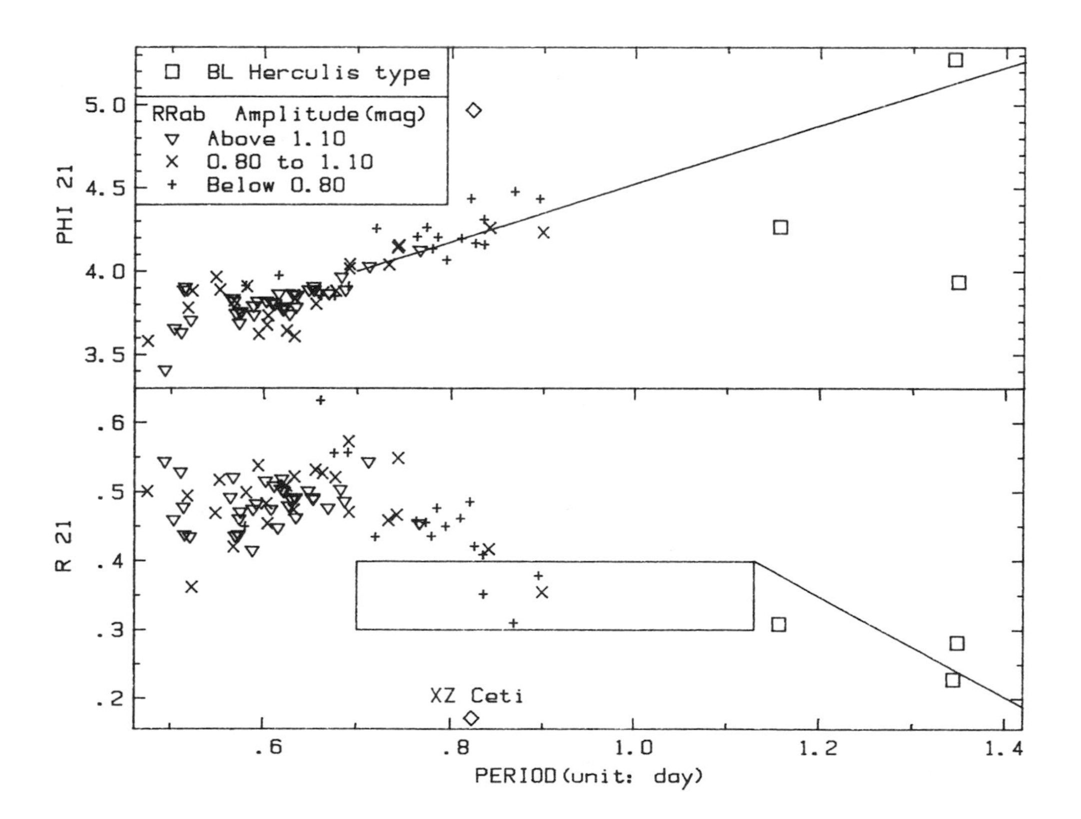

This sequence can be transformed into an equivalent *expected* sequence for the ω Cen variables using the plausible assumptions: (a) φ_{21} varies in the same way along the progression sequences and R/M is a good pulsation parameter determining φ_{21} (Cogan, 1970); (b) for the ω Cen variables the mass is about 0.5 solar masses, and (c) the relevant resonance period is 1.5 d corresponding to 9 d for the classical Cepheids; (d) well established properties of classical Cepheids. The resulting sequence is shown on Fig. 1 together with similar predictions for R_{21} also based upon (a) - (d) (see Petersen (1984) for details).

ω Cen contains three BL Herculis type variables with period 0.9 - 1.5 d. As they ought to follow the bump progression they are also plotted in Fig. 1.

Fig. 1 shows a nice agreement between the observational data for ω Cen variables and the transformed schematical data for classical Cepheids. If we simply assume that the RRab sequences continue somewhat farther from the resonance than the corresponding Cepheid sequences, I assess the agreement to be very satisfactory. And also the BL Her stars follow the sequences reasonably well. From Fig. 1 I conclude that there is strong evidence for a bump progression sequence from a period of about 0.5 d through the RRab region and the BL Her variables to the resonance at about 1.5 d and that this sequence is essentially identical with the well established sequence for classical Cepheids from 3 d to 9 d.

This result seems to be confirmed by a study of theoretical RRab pulsation models by Stothers (1981). His standard model with period 0.529 d gives velocity curves for the various zones which very convincingly show the formation of a secondary bump by a Christy wave in the way that is well known from models of short period classical Cepheids.

THE FIRST OVERTONE RR LYRAE VARIABLES

The large number (55) of RRc variables in ω Cen combined with their relatively large period interval (0.25 - 0.53 d) makes it possible now to demonstrate systematic variations in φ_{21} and R_{21} also for them (Fig. 2; details in Petersen, 1984). It seems interesting that the systematics in many respects is the same as in the well studied case of the classical Cepheids, if we assume a resonance at about 0.45 d. However, since the RRc variables are first overtone pulsators, effects from the fundamental mode - second overtone resonance seems impossible. Therefore, it is tempting to propose the presence here of another resonance. I must emphasize that both the data base used and and the resonance explanation need confirmation from independent investigations. I think that further studies of the RRc variables could be important for the understanding of the bump mechanism.

Figure 2. φ_{21} and R_{21} vs. period for 55 RRc variables in ω Centauri.

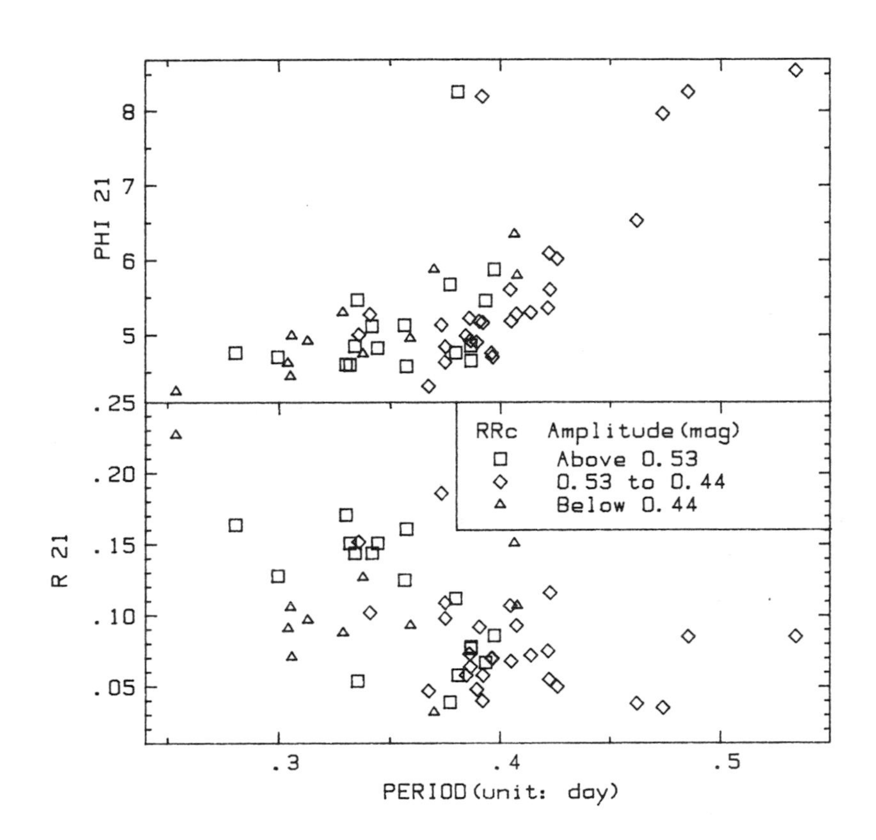

REFERENCES

Cogan, B.C. (1970). Astrophys. J., *162*, 139.
Martin, W.C. (1938). Leiden Annalen, *Vol. XVII*, Part 2.
Petersen, J.O. (1984). Astron. Astrophys., submitted.
Petersen, J.O. & Hansen, L. (1984). Astron. Astrophys., in press.
Simon, N.R. & Lee, A.S. (1981). Astrophys. J., *248*, 291.
Simon, N.R. & Teays, T.J. (1982). Astrophys. J., *261*, 586.
Stothers, R. (1981). Astrophys. J., *247*, 941.

THE EFFECTS OF CONVECTION IN RR LYRAE STARS

R. F. Stellingwerf
Mission Research Corporation
Albuquerque, New Mexico

Abstract: The effect of convection in RR Lyrae stars has been investigated using nonlinear models that include the effects of time dependence, turbulent pressure, convective overshooting, and convection-pulsation interaction. The convective structure reduces to that of mixing-length theory in the limit of no time dependence and thick convective zones. Initial tests suggested that convection has an important effect on the stability and pulsation mode of these stars. We now have investigated the nonlinear nature of convection at limiting amplitude. Convection serves as a limiting amplitude mechanism for these stars by causing turbulence near the phase of minimum radius that damps the driving. A comparison with Geneva observations of RR Lyrae shows good agreement in the phase dependence and amplitude of the turbulent motions.

STABILITY

The effect of convection on the stability of RR Lyrae models is shown in Figure 1. Here the dashed lines are the linear growth rates for purely radiative models, while the solid lines show the growth rates with convection included, for modes 0 and 1. Several effects are evident. The red edge is present in the convective models, and is strongly mode-dependent. The blue edges have shifted a bit, an effect that becomes more important as helium abundance is lowered. These results are described in detail in Stellingwerf 1982a,b, 1984a,b.

LIMITING AMPLITUDE

A convective fundamental mode model at effective temperature 6500K (type b) has been integrated for 100 periods to limiting amplitude. The variation of the velocity and luminosity are shown in Figures 2 and 3 for the limit cycle motion. These are typical of an RR Lyrae "b" type star. Of most interest is the time dependent behavior of the convection, shown in Figure 4. Here the convective velocity and luminosity are plotted versus exterior mass at twelve phases of the motion.

Figure 1

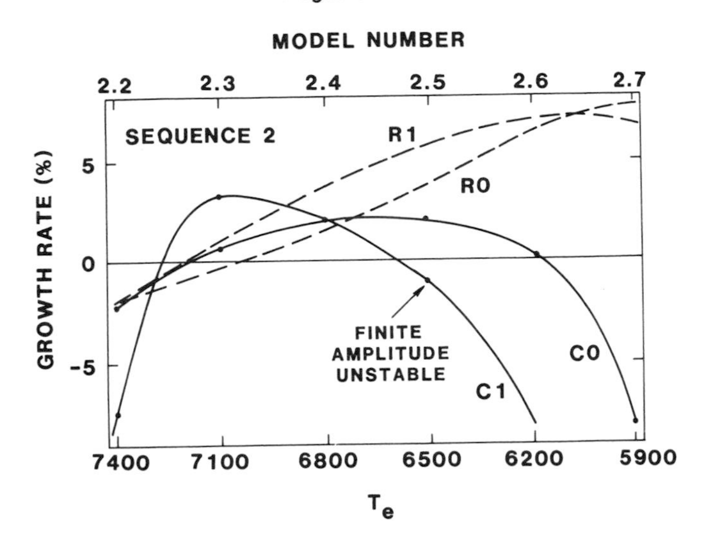

Figure 2

Figure 3

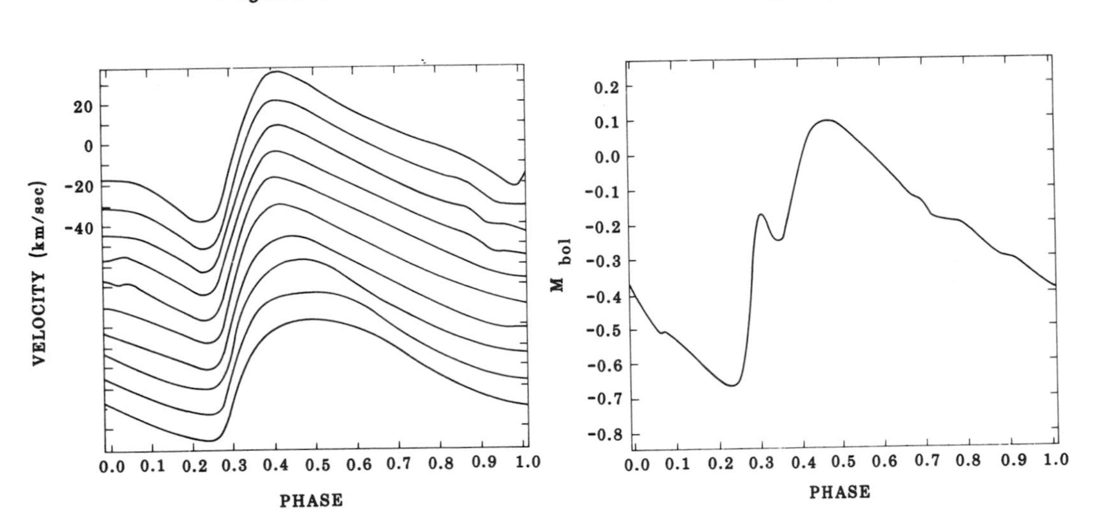

The structure at phase 0.8 (maximum radius) is very close to that of the static model. Following this phase, the convection strengthens, then remains essentially constant through phase 0.225. The arrows show the location of the ionization zones. At phase 0.275, just before minimum radius, the convection suddenly turns on very strongly - the two unstable zones merge, and convection carries all of the flux over a substantial region of the star. Then, at phase 0.325, just following minimum radius, the convection quite rapidly decreases nearly to zero. This behavior is apparently caused by the very strong compression and expansion occurring near the phase of minimum radius.

Figure 4

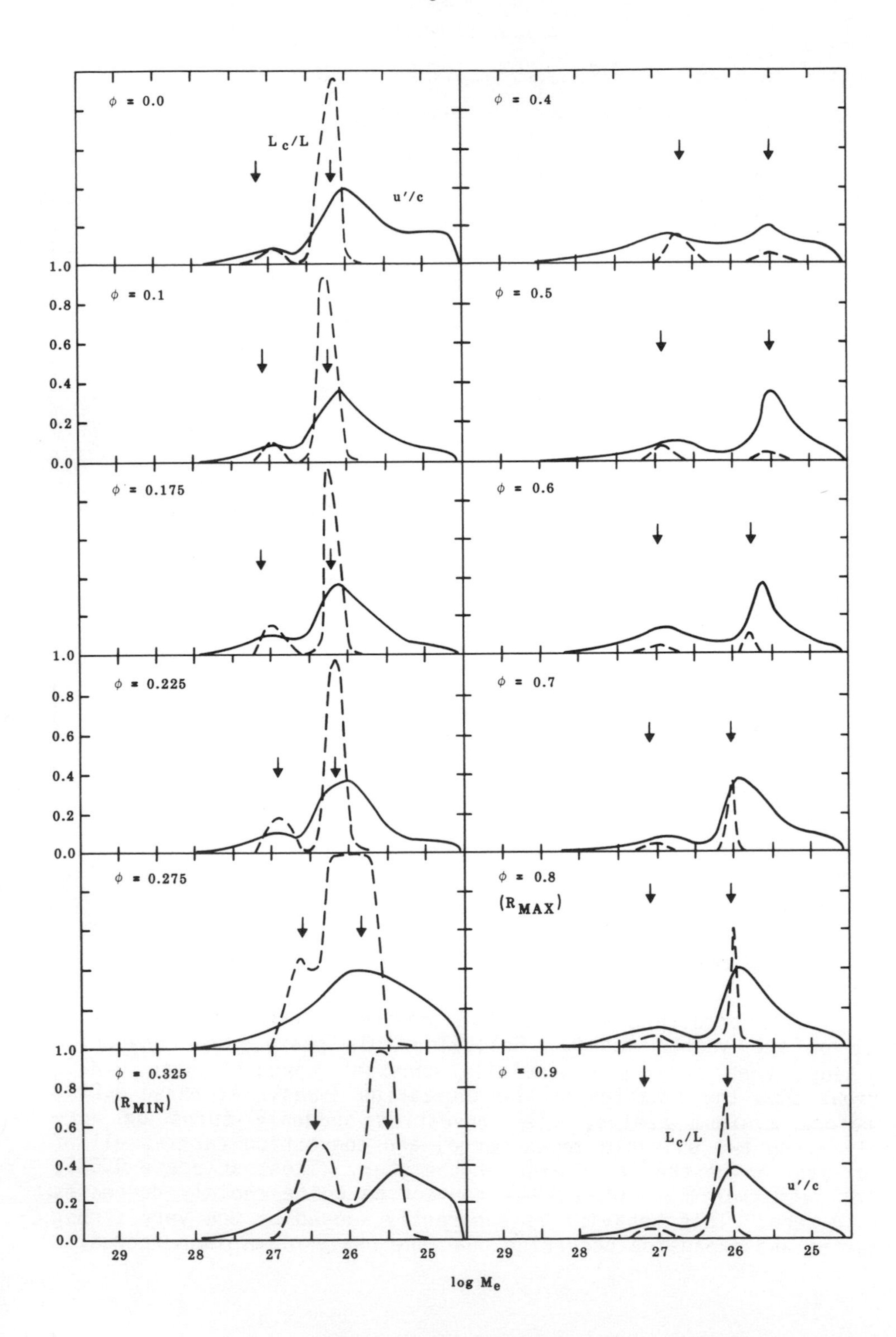
φ = 0.0
L_c/L
u′/c
φ = 0.4
φ = 0.1
φ = 0.5
φ = 0.175
φ = 0.6
φ = 0.225
φ = 0.7
φ = 0.275
φ = 0.8
(R_{MAX})
φ = 0.325
(R_{MIN})
φ = 0.9
L_c/L
u′/c
1.0
0.8
0.6
0.4
0.2
0.0
29
28
27
26
25
log M_e

The variation of the rms convective velocity in the atmosphere of the model versus phase of oscillation is shown in Figure 5. The range for this model is 5-10 km/sec. Also shown are preliminary data from Geneva depicting the measured variation of line widths corrected for projection. The peak in turbulence near minimum radius, as well as the subsequent decline are visible. These features are more prominent in several Cepheid observations reported by Benz and Mayor, 1982.

Figure 5

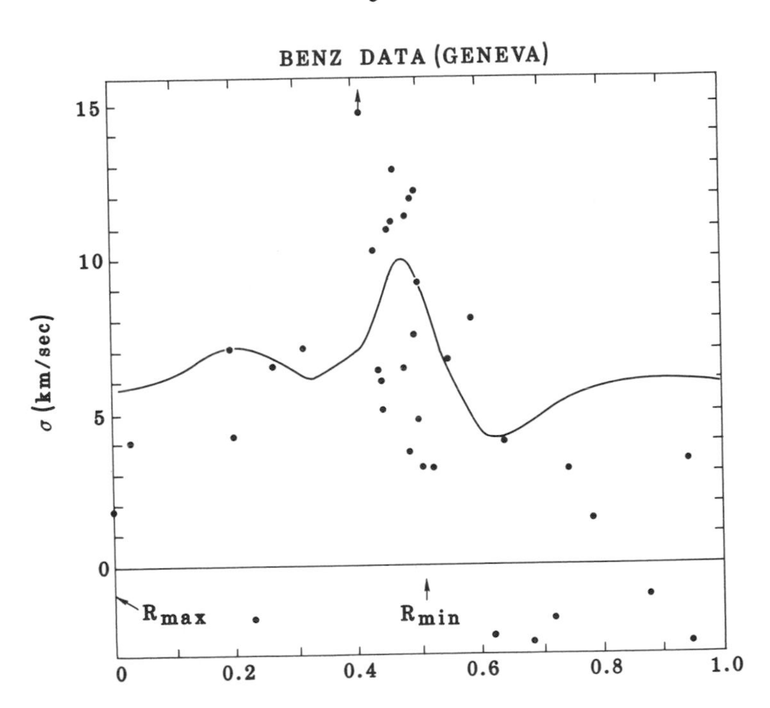

COMPARISON WITH RR LYRAE OBSERVATIONS

REFERENCES

Stellingwerf, R. F. (1982a). Ap.J. 262, 330.
Stellingwerf, R. F. (1982b). Ap.J. 262, 339.
Stellingwerf, R. F. (1984a). Ap.J. 277, 322.
Stellingwerf, R. F. (1984b). Ap.J. 277, 327.
Benz, W. and Mayor, M. (1982). Astr. Ap., 111, 224.

THE LONG TERM BEHAVIOUR OF TWO VARIABLES IN THE GLOBULAR CLUSTER M56.

Amelia Wehlau, Philip Rice, Marcia Wehlau,
Department of Astronomy,
The University of Western Ontario.

Helen Sawyer Hogg,
David Dunlap Observatory,
University of Toronto.

Of twelve variables known in Messier 56 in Lyra, two, V1 and V6 are the subject of this paper. V1 was discovered by Shapley (1920) and first determined by Sawyer Hogg (1942) to be a Cepheid with a period of 1.5 days. V6 was discovered by Sawyer Hogg (1940) from her early plates at the David Dunlap Observatory. Later Sawyer Hogg (1949) showed it to be an RV Tauri type with a period of 90.02 days, one of the first such to be identified in a globular cluster. A.H.Joy (1949) determined spectral type and radial velocity for both of these variables.

At the present time twenty Population II Cepheids with periods shorter than five days are known to exist in galactic globular clusters. These variables, commonly called BL Her stars after the more metal-rich field star of that name, are believed to be evolving away from the horizontal branch toward the asymptotic branch. A recent study by Wehlau and Bohlender (1982) of period changes for twelve of these stars found that all of them showed either increasing periods or no detectable period change. This lack of decreasing periods agrees well with the evolutionary theory.

A number of these stars have only recently been discovered as discussed in the paper by Clement et al. (1984b) presented at this Colloquium. Period changes for these stars as well as for a few others chiefly at southern declinations cannot be determined at the present time because of a lack of older observations. However older observational data do exist for V3 in M10 and V1 in M56. The variable in M10 is discussed in the paper by Clement et al. (1984a) also given at this meeting. This star shows a more complicated period change, the period first increasing and then decreasing. However this variable has a longer period of 7.90 days and thus falls between the two groups of Population II Cepheids found in globular clusters and it may not be a true BL Her star.

In the case of M56 two long series of observations by Sawyer Hogg and Rosino existed and therefore plates of the cluster were recently taken by A. Wehlau with the 1.2 m telescope of the Observatoire de Haute Provence and the 1.2 m telescope of The University of Western Ontario, principally to investigate the period change, if any, of V1.

Fig.1 shows three light curves for V1 using data from plates taken by Sawyer Hogg, Rosino and Wehlau, respectively. The light curve shown for 1947-49 is based on published data by Rosino (1949) and the other two light curves are based on measures made by P. Rice with the iris diaphragm photometer of The University of Western Ontario using an unpublished photoelectric sequence of Harris, Olszewski and Schommer (1981). In order to compare the light curves all the magnitudes given by Rosino have been increased by $0^{m}.2$. In agreement with the other globular cluster BL Her stars investigated, V1 shows an increasing period as can be seen by the shifts in the phase of maximum.

The light curve for V1 displays a bump on the descending branch as well as a dip shortly before maximum. The bump is similar to those seen on the light and velocity curves of other BL Her stars with periods close to 1.5 days (Carson & Stothers, 1982) which have been used to determine "bump" masses of between 0.55 and 0.60 solar masses. The dip, sometimes referred to as the "artificial viscosity dip," appears on the model light curves of Davis (1982) and Carson & Stothers who suggest it is due to higher opacity of the atmospheric Hydrogen caused by a transiting shock. It has been seen in the light curves of some other BL Her stars and of the much longer period (16.4 days) Cepheid X Cygni (Davis et al., 1981).

The long series of plates of M56 taken by Sawyer Hogg with the 1.88, 0.48 and 0.40 m telescopes of the University of Toronto and extending from 1935 through 1975 makes it possible to study the long term

Fig.1. Light curves of Variable 1 from three different epochs showing shifts in phase of maximum due to an increasing period.

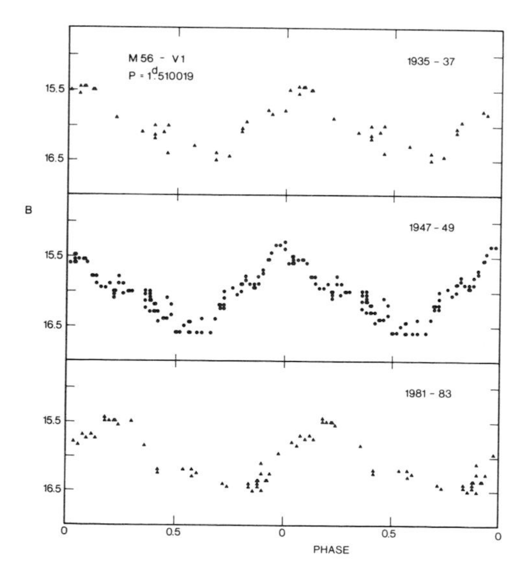

behaviour of V6, the RV Tauri star. Fig.2 shows eight light curves for this variable with various symbols representing the measures made by the different observers. The published magnitudes of Rosino (1944, 1949) are shown as filled dots and the eye estimates made by Sawyer Hogg from her plates taken from 1935 through 1949 are shown as open circles. Triangles represent eye estimates based on the new sequence as made by M. Wehlau from Sawyer Hogg's plates taken from 1950 through 1975. It can be seen that there have been several reversals of primary and secondary minima in the light curve but that the period has remained stable at the value of 90.02 days.

The data used for this paper will be published later along with a more complete discussion of the results.

The authors are grateful for the support of the National Science and Engineering Research Council of Canada. A. Wehlau wishes to thank the director and staff of the Observatoire de Haute Provence for their hospitality and assistance.

Fig.2. Light curves of the RV Tauri star, Variable 6, showing reversals of primary and secondary minima.

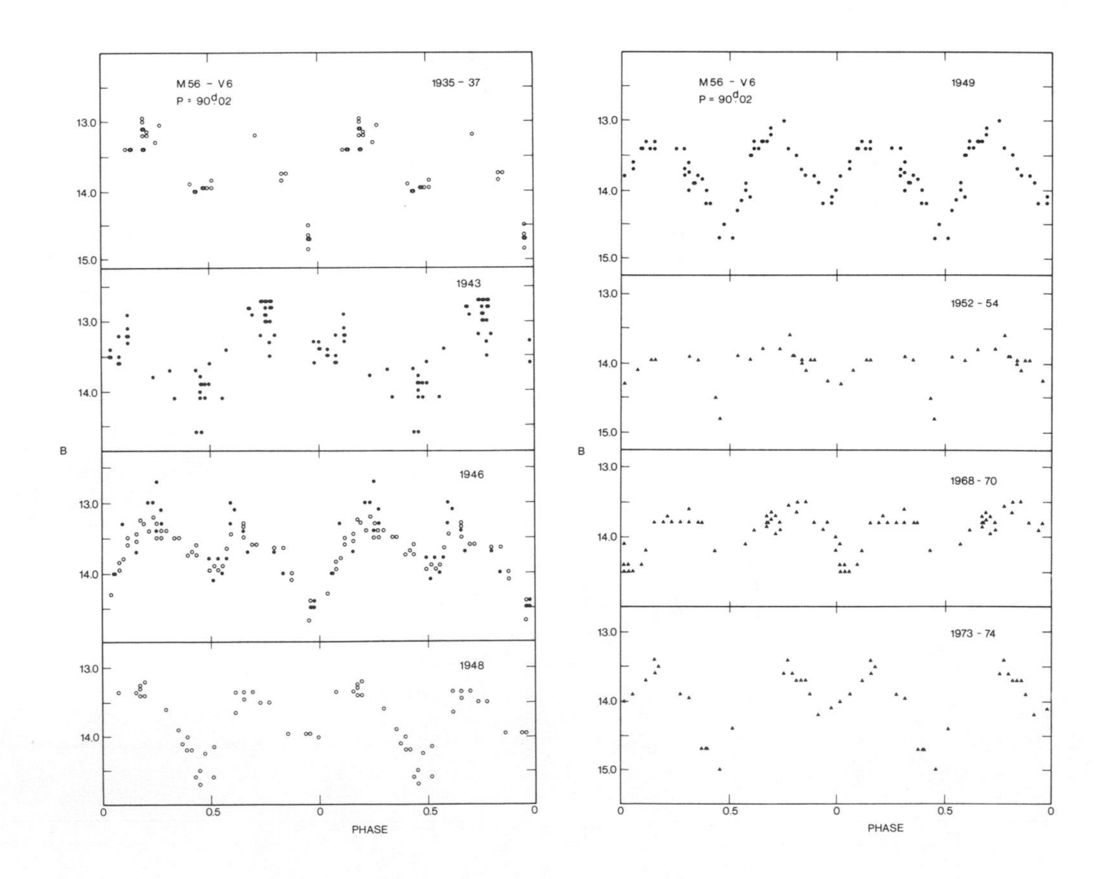

References

Carson, T.R., and Stothers, R. 1976, Astrophys.J., 204, 461.
Clement, C., Sawyer Hogg, H. and Lake, K. 1984a, in IAU Colloquium 82, Cepheids: Observation and Theory, edited by Barry F. Madore (University of Cambridge, Cambridge), p. 260.
Clement, C., Sawyer Hogg, H. and Wells, T. 1984b, in IAU Colloquium 82, Cepheids: Observation and Theory, edited by Barry F. Madore (University of Cambridge, Cambridge), p. 262.
Davis, C.G. 1982, in Pulsations in Classical and Cataclysmic Variable Stars, edited by J.P. Cox and C.J. Hansen (University of Colorado, Boulder), p.232.
Davis, C.G., Moffett, T.J. and Barnes, T.G. 1981, Astrophys.J., 246, 914.
Harris, H.C., Olszewski, E.W. and Schommer, R.A. 1981, private communication.
Joy, A.H. 1949, Astrophys.J., 110, 105.
Rosino, L. 1944, Pub.Oss.Astr.Bologna IV, No.7.
Rosino, L. 1949, Pub.Oss.Astr.Bologna V, No.12.
Sawyer, H.B. 1940, D.D.O. Pub. 1, No.5.
Sawyer, H.B. 1942, Pub.A.A.S., 10, 233.
Sawyer, H.B. 1949, J.Roy.Astr.Soc.Can., 43, 38.
Shapley, H. 1920, Astrophys.J., 52, 73.
Wehlau, A. and Bohlender, D. 1982, Astron.J., 87, 780.

Table 1. Known BL Her stars in globular clusters.

NGC	Var.	Period Days	Period Change $d/10^6$yr.	NGC	Var.	Period Days	Period Change $d/10^6$yr.
2419	18	1.58		6333 (M9)	12	1.34	
5139	43	1.15	+ 0.54±0.11				
(ωCen)	48	4.47	+15.8 ±1.6	6402	2	2.79	+ 0.34±1.09
	60	1.34	+ 0.62±0.09	(M14)	76	1.89	+ 7.4 ±0.4
	61	2.27	+ 0.52±0.26				
	92	1.34	+11.3 ±0.6	6656 (M22)	11	1.69	+ 0.01±0.19
6205	1	1.45	+ 0.05±0.19				
(M13)	2	5.11	+18.0 ±2.0	6715	1	1.34	
	6	2.11	+ 0.36±0.34	(M54)			
6273 (M19)	4	2.43		6752	1	1.38	
				6779	1	1.51	+ 3.5 ±0.2
6284	1	4.48		(M56)			
	4	2.82					
				7078	1	1.43	+ 4.7 ±0.2
				(M15)	72	1.13:	

Index of Authors

Index of Objects

Index of Topics